Advances in

ORGANOMETALLIC CHEMISTRY

VOLUME 22

CONTRIBUTORS TO THIS VOLUME

John S. Bradley

Michael I. Bruce

Donald J. Darensbourg

Rebecca A. Kudaroski

Hester E. Oosthuizen

Eric Singleton

A. Geoffrey Swincer

H. Vahrenkamp

Advances in Organometallic Chemistry

EDITED BY

F. G. A. STONE

DEPARTMENT OF INORGANIC CHEMISTRY

THE UNIVERSITY

BRISTOL, ENGLAND

ROBERT WEST

DEPARTMENT OF CHEMISTRY

UNIVERSITY OF WISCONSIN

MADISON, WISCONSIN

VOLUME 22

1983

ACADEMIC PRESS

A Subsidiary of Harcourt Brace Jovanovich, Publishers

New York   London

Paris   San Diego   San Francisco   São Paulo   Sydney   Tokyo   Toronto

ACADEMIC PRESS, INC.
111 Fifth Avenue, New York, New York 10003

United Kingdom Edition published by
ACADEMIC PRESS, INC. (LONDON) LTD.
24/28 Oval Road, London NW1 7DX

LIBRARY OF CONGRESS CATALOG CARD NUMBER: 64–16030

ISBN 0–12–031122–4

PRINTED IN THE UNITED STATES OF AMERICA

83 84 85 86   9 8 7 6 5 4 3 2 1

Contents

The Chemistry of Carbidocarbonyl Clusters

JOHN S. BRADLEY

Vinylidene and Propadienylidene (Allenylidene) Metal Complexes

MICHAEL I. BRUCE and A. GEOFFREY SWINCER

The Activation of Carbon Dioxide by Metal Complexes

DONALD J. DARENSBOURG and REBECCA A. KUDAROSKI

Basic Metal Cluster Reactions

H. VAHRENKAMP

Metal Isocyanide Complexes

ERIC SINGLETON and HESTER E. OOSTHUIZEN

Contributors

Numbers in parentheses indicate the pages on which the authors' contributions begin.

JOHN S. BRADLEY (1), *Exxon Research and Engineering Company, Linden, New Jersey 07036*

MICHAEL I. BRUCE (59), *Jordan Laboratories, Department of Physical and Inorganic Chemistry, University of Adelaide, Adelaide, Australia 5001*

DONALD J. DARENSBOURG (129), *Department of Chemistry, Texas A&M University, College Station, Texas 77843*

REBECCA A. KUDAROSKI (129), *Department of Chemistry, Texas A&M University, College Station, Texas 77843*

HESTER E. OOSTHUIZEN* (209), *National Chemical Research Laboratory, Council for Scientific and Industrial Research, Pretoria 0001, Republic of South Africa*

ERIC SINGLETON (209), *National Chemical Research Laboratory, Council for Scientific and Industrial Research, Pretoria 0001, Republic of South Africa*

A. GEOFFREY SWINCER (59), *Jordan Laboratories, Department of Physical and Inorganic Chemistry, University of Adelaide, Adelaide, Australia 5001*

H. VAHRENKAMP (169), *Institut für Anorganische Chemie, Albert-Ludwigs-Universität, D-7800 Freiburg, Federal Republic of Germany*

*Previously Hester E. Swanepoel.

ADVANCES IN ORGANOMETALLIC CHEMISTRY, VOL. 22

The Chemistry of Carbidocarbonyl Clusters

JOHN S. BRADLEY

Exxon Research and Engineering Company
Linden, New Jersey

I

INTRODUCTION

In 1962 Dahl and co-workers reported the synthesis and structure of an iron cluster with a unique structural feature. $Fe_5C(CO)_{15}$, the new molecule, contained, in addition to its carbonyl ligands, a single carbon atom bound only to the five iron atoms in the molecule. This, the first carbidocarbonyl cluster, remained the only example of its type for several years, until the isolation, by Lewis and co-workers, of $Ru_6C(CO)_{17}$ and its derivatives. Then in the mid-1970s, the late Paolo Chini and co-workers began reporting their successes in the synthesis and structural characterization of a number of cobalt and rhodium carbidocarbonyl clusters, and a similar acceleration occurred in the rate of growth of the analogous chemistry of iron, ruthenium, and osmium. As the array of carbidocarbonyl clusters grew, so their significance as a class became more obvious. What were once regarded as structural oddities began to assume a more central role in cluster chemistry and at a time when the ostensible similarities between metal cluster compounds and metallic crystallites were being explored in terms of both structure, reactivity, and catalytic potential. This "cluster–surface analogy"

1

has proved to be a driving force in cluster chemistry, and although the goal of homogeneous cluster catalysis analogous to highly active and versatile heterogeneous catalysis by metal surfaces has proved to be an elusive one, the structural similarities between cluster molecules and metallic particles continue to be developed as new and larger clusters are synthesized. This instructive relationship is also manifested in the structural chemistry of carbidocarbonyl clusters when compared with transition metal carbides.

An additional significance for carbidocarbonyl clusters has appeared more recently with the discovery of the fascinating reactivity of carbon atoms in clusters when they are exposed to reactive molecules in low nuclearity carbidocarbonyl clusters. These observations followed on the heels of the recognition of the crucial role played by surface bound carbon atoms in metal-catalyzed carbon monoxide hydrogenation, and so a new area of overlap between cluster chemistry and surface chemistry has arisen. Moreover, in this case the comparisons between organometallic and surface chemistry may lie in reactivity and not merely structural similarities.

The rapid expansion in carbidocarbonyl cluster chemistry has enabled the chemistry of these compounds to be addressed in its own right rather than as a subgroup of cluster chemistry as a whole. A recent survey of this area by Muetterties has appeared (1), covering the literature up to the end of 1979, and in it some of the aspects of carbide clusters referred to in this introduction are described. However, the rate of growth of the field in the past three years in both the synthesis and reactivity of these molecules makes a second review appropriate. I have incorporated results reported up to mid-1982, and although the earlier work is included with the hope of providing a comprehensive survey, emphasis is given to the most recent observations.

The metals that so far have been reported to form carbidocarbonyl clusters are iron, ruthenium, osmium, cobalt, rhodium, and most recently rhenium. The iron clusters will be dealt with separately since they provide some unique chemistry involving the reactivity of the carbido carbon atom. Ruthenium and osmium are covered together in a second section, cobalt and rhodium in a third, and rhenium in a fourth. Where X-ray structural data are available a perspective drawing of the molecule is given. For some of the smaller clusters, this may seem redundant, but the structural units established for these compounds are observed in more complicated frameworks in the larger molecules. I have not attempted to describe the minutiae of carbonyl coordination geometry in these clusters unless such considerations affect some other aspect of the molecular structure. Spectroscopic data for carbide clusters are being reported more frequently than when this area was previously reviewed, and a section is devoted to these observations. Finally, some general observations are made on the chemistry of these molecules, based on the previous sections.

II

CARBIDOCARBONYL CLUSTERS OF IRON

The first carbidocarbonyl transition metal cluster to be recognized was $Fe_5C(CO)_{15}$ (**1**), which was isolated in very low yield from the reaction of triiron dodecacarbonyl with methylphenylacetylene and characterized by X-ray diffraction by Dahl and co-workers (*2*). The molecule (Fig. 1) comprises a square pyramidal Fe_5 core with the carbide situated .08 Å below the center of a square face. Each iron atom bears three terminal carbonyls. Improved syntheses of **1** by protonation (*3*) or oxidation of $[Fe_6C(CO)_{16}]^{2-}$ (*4*) (see below) have been reported.

The reactions of **1** with phosphines and phosphites proceed smoothly at room temperature yielding products with varying degrees of substitution, $Fe_5C(CO)_{15-n}(L)_n$, $n = 1-3$, depending on the phosphorus donor ligand L (*5*). No structural data are available for the substituted clusters.

All attempts to protonate the partially exposed carbon atom in **1** have been unsuccessful (*1, 5*). Reaction with base yields $[Fe_5C(CO)_{14}]^{2-}$ (**2**), which was isolated as its tetraethylammonium salt (*5*). The dianion may also be synthesized by the reaction of $Fe_5C(CO)_{15}$ and $[Fe(CO)_4]^{2-}$, and although its structure is as yet undetermined, it presumably retains a square pyramidal core (*4*).

Muetterties has reported the reaction of **2** with a number of mono- or di-

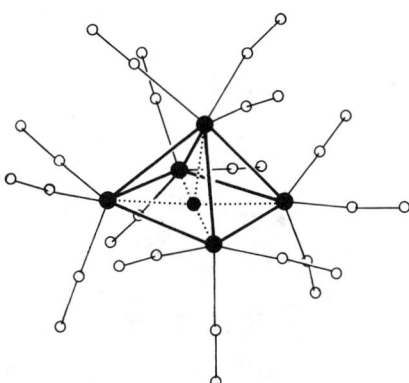

FIG. 1. $Fe_5C(CO)_{15}$, **1** (*2*). As will be the case for all perspective drawings in this survey, metal atoms are represented by larger filled circles, the carbide carbon atom as a small filled circle, and carbonyls as small open circles, with metal–metal bonds as heavy lines to emphasize the core geometry. Mean Fe–Fe distances: apical–basal = 2.63 Å, intrabasal = 2.66 Å. Fe_{apical}–$C_{carbide}$ = 1.96 Å, mean Fe_{basal}–$C_{carbide}$ = 1.89 Å. The carbide carbon lies 0.08 Å below the basal plane.

nuclear transition metal complexes to yield octahedral $[MFe_5C(CO)_xL_y]^{z-}$ clusters [Eq. (1)] (4).

$$
\begin{array}{lll}
\text{Fe}_2(\text{CO})_9 & \rightarrow & [\text{Fe}_6\text{C(CO})_{16}]^{2-} \\
\text{Cr(CO})_3(\text{pyridine})_3 & \rightarrow & [\text{CrFe}_5\text{C(CO})_{17}]^{2-} \\
\text{Mo(CO})_3(\text{THF})_3 & \rightarrow & [\text{MoFe}_5\text{C(CO})_{17}]^{2-} \\
\text{W(CO})_3(\text{MeCN})_3 & \rightarrow & [\text{WFe}_5\text{C(CO})_{17}]^{2-} \\
\text{Rh}_2(\text{CO})_4\text{Cl}_2 & \rightarrow & [\text{RhFe}_5\text{C(CO})_{16}]^{-} \\
[\text{Fe}_5\text{C(CO})_{14}]^{2-} + \text{Rh}_2(\text{COD})_2\text{Cl}_2 & \rightarrow & [\text{RhFe}_5\text{C(CO})_{14}(\text{COD})]^{-} \quad (1) \\
\text{Ir}_2(\text{COD})_2\text{Cl}_2 & \rightarrow & [\text{IrFe}_5\text{C(CO})_{14}(\text{COD})]^{-} \\
\text{Ni(COD})_2 & \rightarrow & [\text{NiFe}_5\text{C(CO})_{13}(\text{COD})]^{2-} \\
\text{Ni(COD})_2, \text{CO} & \rightarrow & [\text{NiFe}_5\text{C(CO})_{15}]^{2-} \\
\text{Pd}_2(\text{C}_3\text{H}_5)_2\text{Cl}_2 & \rightarrow & [\text{PdFe}_5\text{C(CO})_{14}(\text{C}_3\text{H}_5)]^{-} \\
\text{Cu(MeCN})_4\text{BF}_4 & \rightarrow & [\text{CuFe}_5\text{C(CO})_{14}(\text{MeCN})]^{-}
\end{array}
$$

With $Fe_2(CO)_9$ the reaction yields the previously known dianion $[Fe_6C(CO)_{16}]^{2-}$, 3, which was originally synthesized by the reaction of $Fe(CO)_5$ with a variety of metal carbonyl anions [e.g., $Mn(CO)_5^-$] in high boiling solvents (6). The structure of the tetramethylammonium salt of 3 is shown in Fig. 2 (7). The dianion contains an essentially octahedral core of iron atoms encapsulating the carbido carbon atom, and of the sixteen carbonyl ligands twelve are terminal and four asymmetrically bridging (one being highly disordered). The central carbon is not equidistant from the six iron atoms, and the six $Fe-C_{carbide}$ distances range from 1.82 to 1.97 Å. $[MoFe_5C(CO)_{17}]^{2-}$, one of the bimetallic carbidocarbonyl anions derived from 2, has been structurally characterized as its tetraethylammonium salt (4). The cluster is essentially octahedral, but the carbide carbon atom is

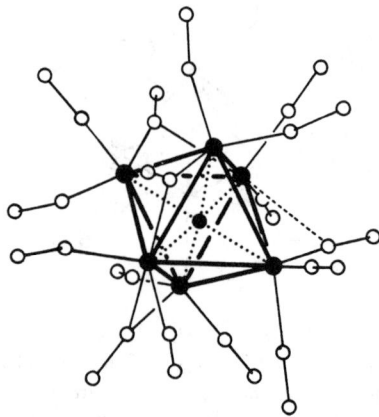

FIG. 2. $Fe_6C(CO)_{16}^{2-}$, 3, as in its $(CH_3)_4N^+$ salt (7). Unbridged Fe–Fe bonds average 2.69 Å; bridged bonds average 2.61 Å. $Fe-C_{carbide}$ varies from 1.82 to 1.97 Å. Broken line indicates long contact to disordered terminal carbonyl which tends toward bridging position.

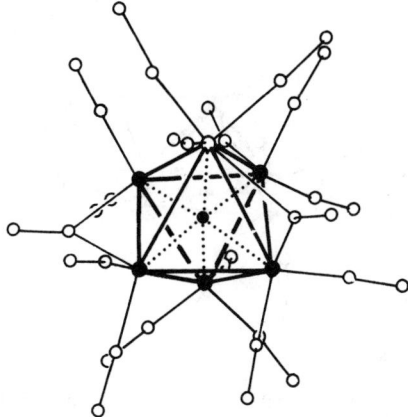

FIG. 3. $MoFe_5C(CO)_{17}^{2-}$, as in its $(C_2H_5)_4N^+$ salt (4). The molybdenum atom is denoted by open circle in the octahedral M_6C core. The carbide carbon is displaced by 0.10 Å from the four coplanar equatorial iron atoms toward the apical molybdenum atom. Mean Fe_{eq}–Fe_{eq} = 2.67 Å; mean Fe_{ax}–Fe_{eq} = 2.64 Å; mean Mo–Fe_{eq} = 2.92 Å. Mean Fe_{eq}–$C_{carbide}$ = 1.89 Å; Fe_{ax}–$C_{carbide}$ = 1.95 Å; Mo–$C_{carbide}$ = 2.12 Å.

displaced by 0.10 Å from the center of the equational Fe_4 plane toward the molybdenum atom (Fig. 3). Of the seventeen carbonyl ligands two are asymmetrically bridging (one to molybdenum) and fifteen terminal.

Three of these octahedral bimetallic MFe_5C anions undergo partial fragmentation on oxidation with ferric ion, as does $[Fe_6C(CO)_{16}]^{2-}$, losing one iron vertex and yielding $MFe_4C(CO)_n$ products [Eq. (2)] (4).

$$\begin{array}{ccc} [Fe_6C(CO)_{16}]^{2-} & & Fe_5C(CO)_{15} \\ [CrFe_5C(CO)_{17}]^{2-} & \xrightarrow{\ Fe^{3+}\ } & CrFe_4C(CO)_{16} \\ [MoFe_5C(CO)_{17}]^{2-} & & MoFe_4C(CO)_{16} \\ [RhFe_5C(CO)_{16}]^{-} & & [RhFe_4C(CO)_{14}]^{-} \end{array} \qquad (2)$$

All three of the $MFe_4C(CO)_n$ clusters have ^{13}C NMR properties characteristics of a square pyramidal geometry (as found for $Fe_5C(CO)_{15}$) with facile CO exchange both localized on each $M(CO)_n$ vertex and between the basal metal atoms, but not between apical and basal sets. The crystal structure of $[RhFe_4C(CO)_{14}]^-$ was determined, and confirmed the square pyramidal geometry of the cluster (Fig. 4), the rhodium atom being located in the basal plane. It is noteworthy that the carbido carbon protrudes 0.19 Å from the $RhFe_3$ plane, substantially further than the analogous distance of 0.08 Å in $Fe_5C(CO)_{15}$.

Iron is so far unique in that it alone forms carbidocarbonyl clusters containing only four metal atoms, and the discovery of this class of compounds and the reactivity of its members has provided one of the most

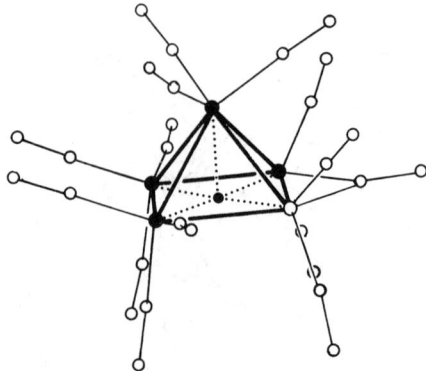

FIG. 4. $RhFe_4C(CO)_{14}^-$, as in its $(C_2H_5)_4N^+$ salt (4). Rhodium atom denoted as an open circle in the base of square pyramidal core. The salt contains two enantiomeric forms of the anion, producing a disordered structure in which iron and rhodium equally occupy each end of the carbonyl bridged edge of the basal plane (Rh–Fe = 2.779 Å). Fe–M bonds to disordered metal atoms (Rh/Fe) average 2.63 Å, and remaining Fe–Fe bonds average 2.62 Å. Metal–carbide distances are Fe_{apical}–C = 1.98 Å, Fe_{basal}–C = 1.87 Å, Rh/Fe–C = 1.94 Å. The carbide lies 0.19 Å below the $RhFe_3$ plane.

intriguing developments in cluster chemistry over the last few years. It was reported in 1979 that the cluster bound carbon atom, which was apparently inert in Fe_5C and Fe_6C clusters, could display novel chemistry if exposed by removal of some of its surrounding metal atoms, reducing the coordination number of the carbon atom to four in an open butterfly geometry. The first observation (8) of the reactivity of a carbide carbon atom in this environment resulted from the partial fragmentation of the octahedral carbide cluster $[Fe_6C(CO)_{16}]^{2-}$, 3. When treated with tropylium bromide, a mild one electron oxidizing agent, in methanol, 3 loses two iron vertices (as ferrous ion), exposing the central carbon atom, which reacts with CO and methanol yielding the μ^4-carbomethoxymethylidyne cluster, 4 [Eq. (3)].

$$[Fe_6C(CO)_{16}]^{2-} \xrightarrow[\text{CH}_3\text{OH, 25°C}]{\text{C}_7\text{H}_7^+\text{Br}^-} [Fe_4(CO)_{12}C \cdot CO_2CH_3]^- + 2Fe^{2+} + 3CO \qquad (3)$$

(3)   (4)

The structure of 4 (Fig. 5) comprises an open butterfly arrangement of four iron atoms, each bearing three terminal carbonyls (8). The hitherto encapsulated carbon atom is equidistant (1.99 ± 0.03 Å) from the metal atoms and is now bonded to the carbomethoxy group. The reaction may be visualized as in Scheme 1. The exposed carbon atom undergoes carbon–carbon bond formation with carbon monoxide to give a ketenylidene group,

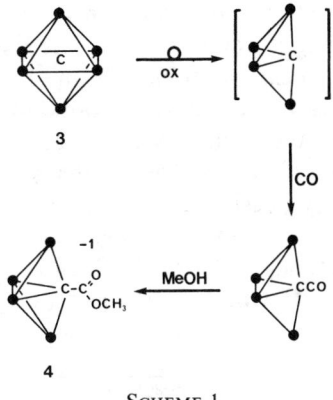

SCHEME 1

which reacts with methanol to form the carbomethoxy derivative. Although no direct evidence was presented for the intermediacy of the ketenylidene, it is supported by the similar reaction of $[Co_3(CO)_9CCO]^+$ with alcohols yielding the carbomethoxy derivative $Co_3(CO)_9CCO_2R$ (9) (see also Section VII,C).

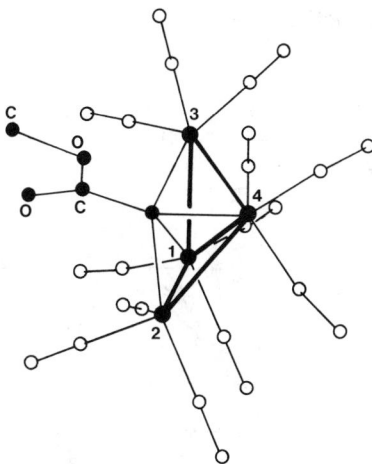

FIG. 5. $Fe_4(CO)_{12}C(CO_2CH_3)^-$, 4, as in its $(C_2H_5)_4N^+$ salt (8). A butterfly core of four iron atoms bears twelve terminal carbonyls, three on each iron. Fe(1)–Fe(4) [2.590(1) Å] is significantly longer than the other Fe–Fe bonds [2.503(9) Å mean]. The methylidyne carbon is 2.027(9) Å from Fe(2), 2.012(9) Å from Fe(3), 1.957(9) Å from Fe(1), and 1.952(9) Å from Fe(4). Fe(1), Fe(4), and the carbomethoxy group are coplanar. The dihedral angle between Fe(1)(3)(4) and Fe(1)(2)(4) is 130°; Fe(2)CFe(3) = 148°.

The carbomethoxymethylidyne group in **4** may be hydrogenated to methyl acetate, providing a unique cluster based synthesis of an organic molecule essentially from carbon monoxide and hydrogen.

The reactivity of the carbon atom when bound to an open Fe_4 cluster has an obvious relevance to the chemistry of surface bound carbon atoms in heterogeneous catalysis (10), and is of great interest. Several members of the Fe_4C family have been synthesized and structurally characterized, and the chemistry of these clusters is under investigation in several laboratories.

The parent neutral carbido-cluster of this class, $Fe_4C(CO)_{13}$, **5,** is most readily synthesized from **4** by protonation [Eq. (4)] (11).

$$[Fe_4(CO)_{12}C \cdot CO_2CH_3]^- \xrightarrow[\text{methylcyclohexane}]{CF_3SO_3H} Fe_4C(CO)_{13} \qquad (4)$$

$$(4) \qquad\qquad\qquad (5)$$

This reaction is reversible by treating **5** with methanol to regenerate **4.** The molecule (Fig. 6) comprises an open butterfly shaped Fe_4C core, with three terminal carbonyls bound to each iron atom. The thirteenth carbonyl bridges the two iron atoms, which make up the "backbone" of the butterfly. Details of the structure are described below in comparison with other members of this family.

Several phosphine substituted derivatives of **5** have been synthesized by reaction of trimethylphosphine with **5** in *n*-pentane (12).

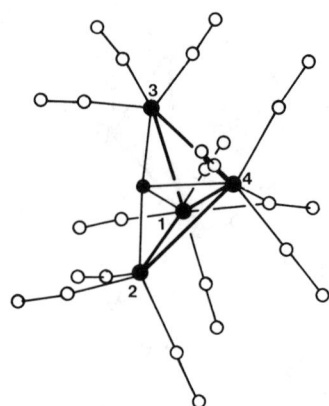

FIG. 6. $Fe_4C(CO)_{13}$, **5** (11). The butterfly Fe_4C core bears twelve terminal carbonyls and a thirteenth bridging Fe(1)–Fe(4). Unbridged Fe–Fe distances average 2.642(4) Å; Fe(1)–Fe(4) = 2.545(1) Å, Fe(1)–C = 1.998(4) Å, Fe(4)–C = 1.987(4) Å, Fe(2)–C = 1.797(4) Å; Fe(3)–C = 1.800(4) Å. Dihedral angle between Fe(1)(4)(3) and Fe(1)(4)(2) = 101°. Fe(2)CFe(3) = 175°.

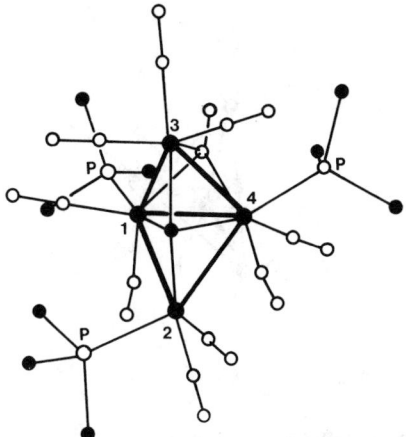

FIG. 7. $Fe_4C(CO)_{10}(P(CH_3)_3)_3$ (13). The three trimethylphosphine ligands are coordinated to Fe(1), Fe(2), and Fe(4). Of the ten carbonyl ligands nine are terminal, and one is triply bridging on the Fe(1)(3)(4) face. Fe(1)–Fe(4) = 2.528(1) Å, the shortest of the metal–metal bonds. The other Fe–Fe distances average 2.644(19) Å. Fe(1)–C = 1.979(4) Å, Fe(4)–C = 2.014(4) Å, Fe(2)–C = 1.785(4) Å, Fe(3)–C = 1.838(4) Å. The dihedral angle between Fe(1)(2)(4) and Fe(1)(3)(4) = 102.4°. Fe(2)CFe(3) = 174°.

$Fe_4C(CO)_{13-n}(PMe_3)_n$ ($n = 3,4$) have been fully characterized crystallographically. The structure of $Fe_4C(CO)_{10}(PMe_3)_3$ has been determined and is shown in Fig. 7 (13, 14). The carbido carbon is now bonded to two chemically inequivalent wingtip iron atoms Fe(2) and Fe(3), since Fe(3) remains unsubstituted by trimethylphosphine, and the values of Fe(2)–$C_{carbide}$ and Fe(3)–$C_{carbide}$ reflect this fact [1.785(5) and 1.838(5) Å, respectively]. These remain, however, significantly shorter than the bonds to Fe(1) and Fe(4) [1.979(5) and 2.014(5) Å], a difference characteristic of the Fe_4C clusters in which the carbon atom is bound only to iron atoms (see below).

The tetrakis(trimethylphosphine) derivative $Fe_4C(CO)_9(PMe_3)_4$ has been characterized by ^1H, ^{31}P, and ^{13}C NMR spectroscopy, and on the basis of the presence of two equally populated phosphine sites is assigned the symmetric structure, each iron atom bearing one phosphine ligand and two carbonyls (12). This was confirmed by X-ray diffraction (13).

Reduction of $Fe_4C(CO)_{13}$ by sodium amalgam yields $[Fe_4C(CO)_{12}]^{2-}$, 6 (14), which was previously synthesized by Tachikawa and Muetterties indirectly from $[Fe_5C(CO)_{14}]^{2-}$ (see below) (15). The structure of 6 (Fig. 8) is very similar to that of 5, but without the bridging carbonyl (16).

Reaction of 6 with methyl iodide yields the anionic acetylmethylidyne cluster, $[Fe_4(CO)_{12}C \cdot C(O)CH_3]^-$ (16), in a reaction which might proceed via an iron bound acetyl which migrates to the carbido carbon under

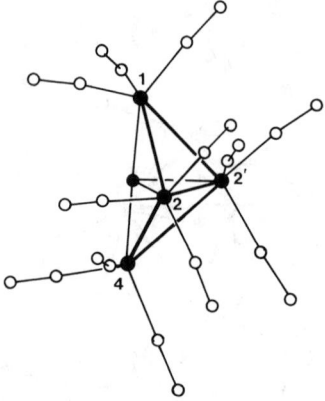

Fɪɢ. 8. $Fe_4C(CO)_{12}^{2-}$, **6**, as in its $Zn(NH_3)_4^{2+}$ salt (*16*). The dianion contains twelve terminal carbonyls. Fe(2) and Fe(2′) are related by a crystallographic mirror plane which contains Fe(1), Fe(4), and C. Fe(1)–Fe(2) = 2.637(1) Å; Fe(2)–Fe(4) = 2.653(1) Å; Fe(2)–Fe(2′) = 2.534 Å. Fe(1)–C = 1.810(7) Å; Fe(2)–C = 1.969(5) Å; Fe(4)–C = 1.786(7) Å. Dihedral angle between Fe(1)(2)(2′) and Fe(4)(2)(2′) = 101.5°. Fe(1)CFe(4) = 178°.

the influence of carbon monoxide (the source of which is unclear) as in Scheme 2.

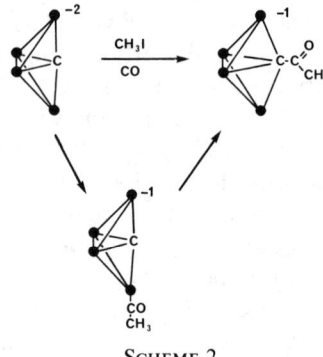

Sᴄʜᴇᴍᴇ 2

Oxidation of **6** under carbon monoxide gives **5** (*16*), and under hydrogen yields the hydridomethylidyne hydride cluster **7** [Eq. (5)] (*15*).

$$[Fe_4C(CO)_{12}]^{2-} \xrightarrow[CH_2Cl_2]{Ag^+/H_2} HFe_4(CO)_{12}CH \qquad (5)$$

(6)   (7)

This reaction is certainly among the most significant observations in

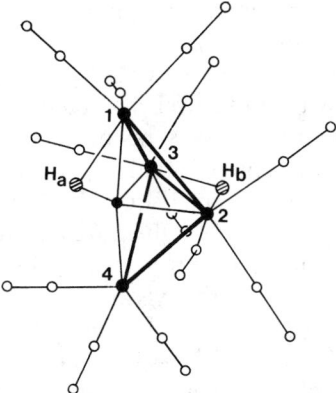

Fig. 9. $HFe_4(CO)_{12}CH$, **7**, as determined by neutron diffraction (*19*). The cluster contains twelve terminal carbonyls. Fe(1)–Fe(2) = 2.611(2) Å; Fe(1)–Fe(3) = 2.614(2) Å; Fe(2)–Fe(4) = 2.631(2) Å; Fe(3)–Fe(4) = 2.623(2) Å; Fe(2)–Fe(3) = 2.603(2) Å. Fe(1)–C = 1.927(2) Å; Fe(2)–C = 1.941(2) Å; Fe(3)–C = 1.949(2) Å; Fe(4)–C = 1.827(2) Å. Fe(1)–H_a = 1.753(4) Å; C–H_a = 1.191(4) Å. Fe(3)–H_b = 1.668(4) Å; Fe(2)–H_b = 1.670(4) Å. Dihedral angle between Fe(1)(2)(3) and Fe(4)(2)(3) = 110.5°. Fe(1)$\widehat{C}$Fe(4) = 170.5°.

recent cluster chemistry, providing as it does a molecular analog for the reaction of surface bound carbon atoms with hydrogen in heterogeneous methanation and Fischer–Tropsch catalysis of carbon monoxide hydrogenation. Alternative syntheses of **7** are provided by protonation of **6** (*15, 17*) or by refluxing **5** in toluene under hydrogen (*14*).

The structure of **7** is shown in Fig. 9. The hydrogen atoms were located in both X-ray (*18*) and neutron diffraction studies (*19*), one bridging Fe(2)Fe(3), the other in a three center FeHC unit, in which the methylidyne C–H vector tilts toward one wingtip iron atom in the Fe_4 butterfly.

The molecule is fluxional, exhibiting CO site exchange, CH exchange from one wingtip to the other, and methylidyne-bridging hydride exchange. This compound has also been synthesized from $[Fe_5C(CO)_{14}]^{2-}$, **2**, by protonation (*15*), or from the noncarbide cluster $[Fe_4(CO)_{13}]^{2-}$, again by protonation (see below) (*17*). The methylidyne hydrogen in **7** is quite acidic, and in methanol ionization is quantitative, yielding $HFe_4C(CO)_{12}^-$, **8** (*17*), [which is also formed when the carbomethoxymethylidyne cluster **4** is reduced with $BH_3 \cdot THF$ (*14*)] which may in turn be deprotonated to **6** [Eq. (6)] (*15*).

$$[Fe_5C(CO)_{14}]^{2-} \xrightarrow{HCl} HFe_4(CO)_{12}CH \xrightarrow{MeOH} [HFe_4C(CO)_{12}]^- \xrightarrow{Et_3N} [Fe_4C(CO)_{12}]^{2-} \quad (6)$$

$$(2) \qquad\qquad (7) \qquad\qquad (8) \qquad\qquad (6)$$

Again, the structure of **8** resembles that of **5**, but with bridging hydride in place of the bridging carbonyl (17).

The three isoelectronic clusters $Fe_4C(CO)_{13}$, **5**, $[Fe_4C(CO)_{12}]^{2-}$, **6**, and $[HFe_4C(CO)_{12}]^-$, **8**, provide the central characters in the Fe_4C family. Their structures have been determined by X-ray diffraction, and in all three cases there is a basic $Fe_4C(CO)_{12}$ core of relatively invariant geometry, with the open butterfly arrangement characteristic of Fe_4C clusters. Consistent with the predictions of Wade's rules (20) these 62 electron clusters, possessing seven skeletal electron pairs and four vertices, adopt an arachno structure based on the octahedron. The carbonyls are all terminal in groups of three on each iron atom, and the carbido carbon atom is situated centrally above the midpoint of the "backbone" Fe–Fe bond, midway between the two "wingtips." The dihedral angle between the two triangular wings of the butterfly averages $102° \pm 2°$ over the three clusters; the angle subtended at the carbide carbon atom by the "wingtip" iron atoms averages $176° \pm 2°$, and the Fe–$C_{carbide}$ distances a and b are almost constant (see Table I).

The dihedral angle δ is rather less than the 109° provided by a regular octahedral cavity (from which the butterfly is derived), and this compression is reflected in the disparity between the two sets of Fe–$C_{carbide}$ distances, a and b (see Table I). Note, however, that the *shorter* distances a are close to the Fe–$C_{carbide}$ distances found in the octahedral $[Fe_6C(CO)_{16}]^{2-}$ (7), and that the smaller dihedral angle results in effect from the lengthening of b in going to the Fe_4C cluster.

It will be apparent at this point that the synthesis of some of the Fe_4C clusters has enjoyed its share of serendipity (in the tradition of cluster chemistry). In two instances, however, rational mechanisms of complex reactions have been elucidated by the isolation and characterization of

TABLE I

CORE GEOMETRY OF Fe_4C CLUSTERS

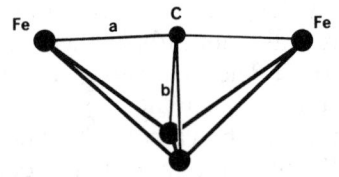

Cluster	Dihedral δ (deg)	FeCFe (α) (deg)	a (Å)	b (Å)	Ref.
$Fe_4C(CO)_{13}$, **5**	101	175	1.80	1.94	11
$[Fe_4C(CO)_{12}H]^-$, **8**	104	174	1.79	1.88	17
$[Fe_4C(CO)_{12}]^{2-}$, **6**	101	178.3	1.79	1.97	16

intermediates. The reactions are

1. The synthesis of $HFe_4(CH)(CO)_{12}$ by protonation of $[Fe_4(CO)_{13}]^{2-}$ (21).
2. The synthesis of $[Fe_4(CO)_{12}CCO_2CH_3]^-$ by tropylium bromide oxidation of $[Fe_6C(CO)_{16}]^{2-}$ in methanol (8).

In the first example, Shriver and co-workers have identified each step in the proton induced reduction of coordinated carbon monoxide in $[Fe_4(CO)_{13}]^{2-}$ to methane [Eq. (7)] (17).

$$[Fe_4(CO)_{13}]^{2-} + nH^+ \rightarrow CH_4 + H_2O + Fe^{2+} + CO + H_2 \cdots \qquad (7)$$

It was shown that the reaction occurs via formation of $HFe_4(CO)_{12}CH$ and its eventual protolysis yielding methane. The process is summarized in Scheme 3. Protonation of $[Fe_4(CO)_{13}]^{2-}$ results in the opening of the Fe_4 tetrahedron to a butterfly configuration in $[HFe_4(CO)_{13}]^-$. This anion, which had been structurally characterized by Manassero et al. (22), has the familiar $Fe_4C(CO)_{12}$ core, with the thirteenth carbonyl (containing the putative carbide carbon) adopting a four electron bridging geometry, spanning the wingtips of the Fe_4 butterfly, and a hydride bridging the remaining two iron atoms. Further protonation occurs at the oxygen atom of the unique carbonyl, forming $HFe_4(CO)_{12}(COH)$ (identified by ^{13}C and ^1H NMR at low temperature) (23). Reaction with an additional proton breaks the C–O bond with the liberation of water, forming the bona fide carbidocarbonyl cluster $[HFe_4C(CO)_{12}]^-$, 8, which again protonates to the hydridomethylidyne cluster, 7. Thus the formation of the carbide from coordinated CO was elegantly demonstrated.

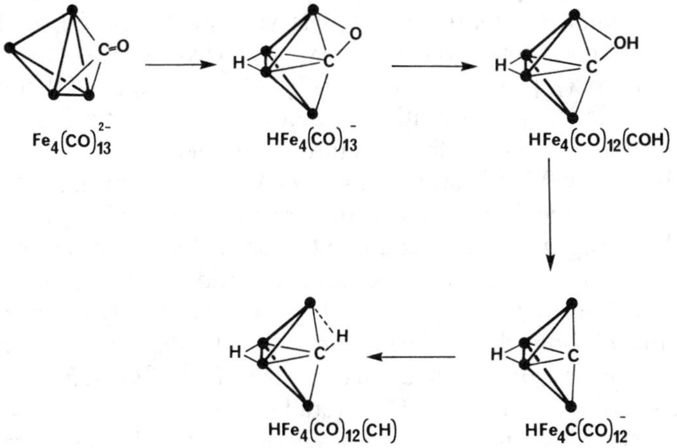

SCHEME 3

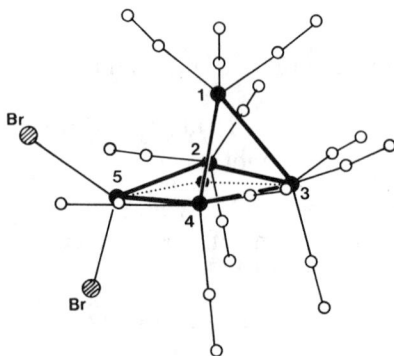

FIG. 10. $Fe_5C(CO)_{12}Br_2^{2-}$, as in its $(C_2H_5)_4N^+$ salt (25). The Fe_5 framework comprises an Fe_4C butterfly [Fe(1),(2),(3),(4)] capped by Fe(5), which is bound to two bromine atoms. Each of the butterfly iron atoms bears three terminal carbonyls. The Fe–Fe distances within the Fe_4C unit resemble those in $Fe_4C(CO)_{12}^{2-}$, 6. The Fe(2)–Fe(3) vector, common to the two triangular "wings," is 2.531(1) Å [cf. 2.534(1) Å in 6]. The other Fe–Fe bonds average 2.661 Å (2.645 Å in 6). The bonds to the $FeBr_2$ vertex are Fe(2)–Fe(5) = 2.680(1) Å, Fe(4)–Fe(5) = 2.707 Å. The Fe(1)–Fe(5) distance is 3.156(1) Å. The dihedral between Fe(1)(2)(3) and Fe(1)(4)(3) is 103.6°, (101° in 6) and Fe(2)$\widehat{C}$Fe(4) = 177.4° (cf. 178° in 6). Fe(5) lies slightly below the Fe(2)(3)(4) plane.

In the second example, the synthesis of $[Fe_4(CO)_{12}CCO_2CH_3]^-$, 4, from $[Fe_6C(CO)_{16}]^{2-}$, 3, the mechanism of the reaction has been elucidated by the isolation and interconversion of several carbide containing intermediates (24). Initial oxidation of $[Fe_6C(CO)_{16}]^{2-}$ by tropylium ion removes one iron vertex as Fe^{2+}, yielding $Fe_5C(CO)_{15}$ [as observed for the same oxidation using ferric ion (4)]. Nucleophilic attack by bromide results in the substitution of two halide ions for three carbonyls, yielding $[Fe_5C(CO)_{13}Br_2]^{2-}$. This unusual cluster was obtained in minute yield as a by-product of the synthesis of 4 from 3 [Eq. (2)], and its structure (Fig. 10) (25) suggested that it was an intermediate in the fragmentation of $Fe_5C(CO)_{15}$ to an Fe_4C cluster. The most significant feature of the structure is the long Fe(1)–Fe(5) distance, 3.16 Å, longer by 0.5 Å than the other Fe–Fe bonds. This distortion from the square pyramidal geometry of the precursor $Fe_5C(CO)_{15}$ presages the loss of the $FeBr_2$ group, leaving an Fe_4C cluster. Indeed, reaction between $Fe_5C(CO)_{15}$ and tetraethylammonium bromide in acetone proceeds smoothly with formation of $[Fe_4C(CO)_{12}]^{2-}$, 6, presumably via the intermediacy of the dibromo cluster. Oxidation of 6 under the conditions of Eq. (2) (i.e., in the presence of carbon monoxide) yields $Fe_4C(CO)_{13}$, 5, which reacts with methanol to yield 4 (11). The overall sequence is shown in Scheme 4.

Each intermediate in this process has been structurally characterized and the interconversions demonstrated. The implied necessity of bromide to the

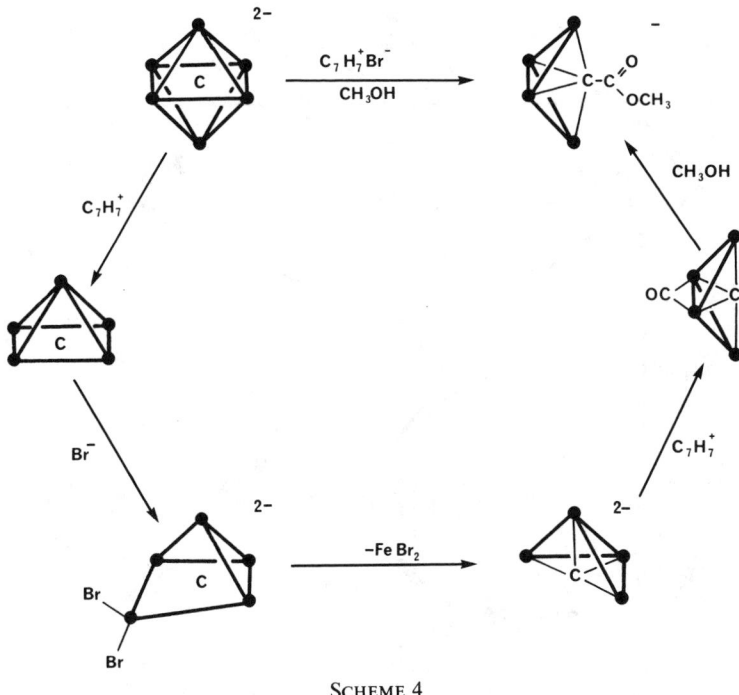

SCHEME 4

fragmentation was also demonstrated by the failure to observe the formation of **4** from **3** if tropylium fluorophosphate is used as oxidizing agent. Addition of bromide ion to the reaction mixture produces a clean transformation from **3** to **4**, and in fact this is the preferred synthesis of $[Fe_4(CO)_{12}CCO_2CH_3]^-$ in our laboratory (26).

The reactions of the Fe_4C clusters are dominated by the reactivity of the exposed carbon atom (Scheme 5). A number of organometallic derivatives have been synthesized, and it is clear that the scope of this chemistry will continue to broaden.

There are so far no examples of trigonally coordinated carbon atoms in Fe_3C clusters, which is not surprising. However, the intermediacy of such a species is implied in the reported isolation of the ketenylidene cluster $Fe_3(CO)_{10}(CCO)$, from the oxidation of $[Fe_4C(CO)_{12}]^{2-}$ in the absence of effective reagents. The identification of this novel cluster rests as yet on mass spectroscopic evidence (16).

The smallest member of the iron carbide family is neither a carbonyl nor a cluster, but is included here since its structure is an example of the lowest coordination number for a carbon atom in a transition metal complex. The

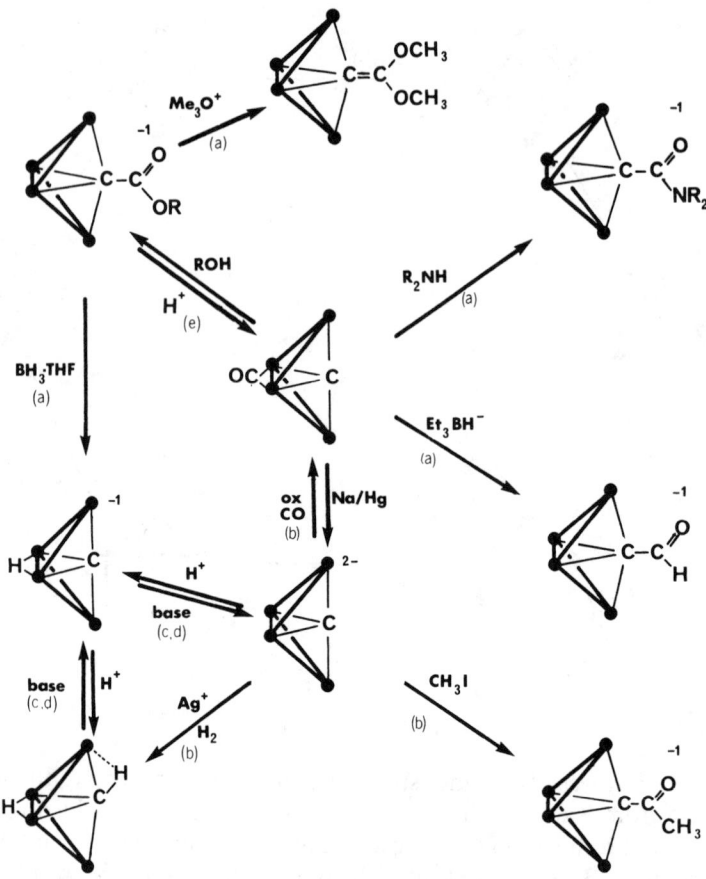

SCHEME 5.   (a) Ref. *14*; (b) Ref. *16*; (c) Ref. *17*; (d) Ref. *15*; (e) Ref. *8*.

reaction of tetraphenylporphyrin iron(II) with carbon tetraiodide in the presence of an excess of iron powder yields **9** in 80% yield [Eq. (8)] (*27*).

$$(TPP)Fe + CI_4 \xrightarrow{\text{Fe powder}} (TPP)Fe{=}C{=}Fe(TPP) \qquad (8)$$

$$(9)$$

The identity of **9** as a carbon bridged metalloporphyrin was originally suggested on the basis of both spectroscopic and chemical data, and this was subsequently confirmed by an X-ray diffraction study (*27a*). The existence of **9** as a stable species was predicted by Hoffman and co-workers (*28*) on the basis of extended Hückel calculations, predicting an Fe–C–Fe angle of 180°, which the structural results confirmed.

III

CARBIDOCARBONYL CLUSTERS OF RUTHENIUM AND OSMIUM

The lowest nuclearity carbidocarbonyl clusters of ruthenium and osmium are $Ru_5C(CO)_{15}$, **10**, and $Os_5C(CO)_{15}$, **11**. The former is synthesized in high yield by the carbonylation of $Ru_6C(CO)_{17}$ (see below) under precisely determined conditions of temperature and pressure [Eq. (9)] (29).

$$Ru_6C(CO)_{17} \xrightarrow[\text{70°C, 3 hr}]{\text{CO, 80 atm}} Ru_5C(CO)_{15} + Ru(CO)_5 \qquad (9)$$
$$(10)$$

The osmium analog was obtained in moderate yield by pyrolysis of $Os_6(CO)_{18}$ or $Os_3(CO)_{12}$ (30). Both the ruthenium (29) and osmium clusters (31) are isostructural with the original iron analog, **1** (2), the metal atoms describing a square pyramid, each vertex bearing three terminal carbonyls. The carbon atom lies fractionally below the center of the basal plane of the cluster, protruding 0.11 Å below the Ru_4 plane in **10** and 0.12 Å below the Os_4 plane in **11** [cf. a value of 0.08 Å for $Fe_5C(CO)_{15}$ (2)] (see Fig. 11).

Reactions of **10** with iodine, acetonitrile, triphenylphosphine, and hydrogen halides have been reported (Scheme 6) (32).

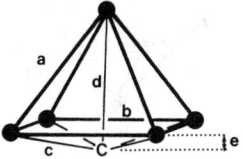

FIG. 11. Comparison of the structures of $M_5C(CO)_{15}$ (M = Fe, Ru, Os) (2, 29, 31):

M	a	b	c	d	e	Ref.
Fe	2.62(4)	2.66(3)	1.96(3)	1.89(2)	0.08	2
Ru	2.3	2.85	2.04		0.11	20
Os	2.85(3)	2.88(2)	2.06		0.12	31

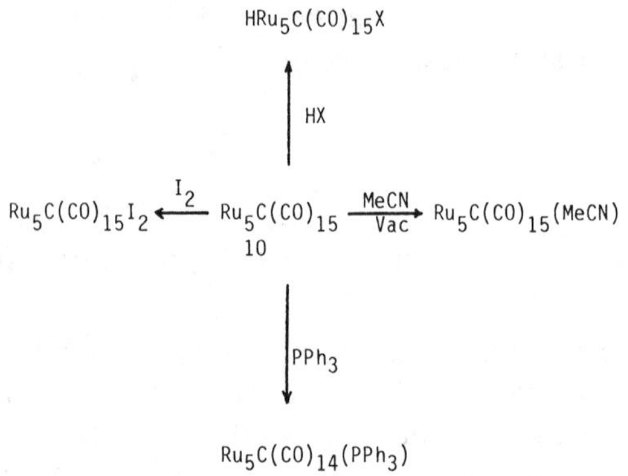

$$HRu_5C(CO)_{15}X$$

$$\uparrow HX$$

$$Ru_5C(CO)_{15}I_2 \xleftarrow{I_2} \underset{10}{Ru_5C(CO)_{15}} \xrightarrow{\frac{MeCN}{Vac}} Ru_5C(CO)_{15}(MeCN)$$

$$\downarrow PPh_3$$

$$Ru_5C(CO)_{14}(PPh_3)$$

SCHEME 6

The structure of the triphenylphosphine substitution product is known (*29*). The square pyramidal Ru_6C core is retained, and the phosphine replaces one carbon monoxide from a basal $Ru(CO)_3$ group in **10**. The carbide carbon atom is situated 0.19 Å below the square face of the pyramid (cf. 0.11 Å in **10**).

Heating **10** under argon regenerates $Ru_6C(CO)_{17}$, and reaction with hydrogen yields a hydride cluster tentatively identified as $H_2Ru_5C(CO)_{15}$ (*29*).

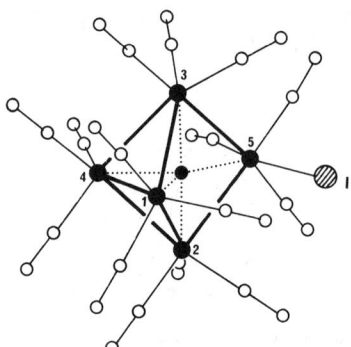

FIG. 12. $Os_5C(CO)_{15}I^-$, **12**, as in its $(PPh_3)_2N^+$ salt (*31*). The Os_5C core describes an open bipyramidal structure, with Os(1), Os(4), and Os(5) occupying three of the five equatorial sites of a pentagonal bipyramid, Os(2) and Os(3) being in the two apical sites. The fifteen carbonyls (three per metal atom) are all terminal. Os(1)–Os(2) = 2.903(1) Å; Os(1)–Os(3) = 2.921(1) Å; Os(1)–Os(4) = 2.748(1) Å; Os(2)–Os(4) = 2.899(1) Å; Os(2)–Os(5) = 2.933(1) Å; Os(3)–Os(4) = 2.896(1) Å; Os(3)–Os(5) = 2.934(1) Å. The two Os–C bonds from Os(2) and Os(3) are shorter [1.978(12) and 1.995(12) Å] than those in the equatorial plane [Os(1)–C = 2.108(12) Å; Os(4)–C = 2.108(12) Å; Os(5)–C = 2.174(12) Å].

Both **10** and **11** react with halide ions to give $[M_5C(CO)_{15}X]^-$ products. The iodopentaosmium cluster $[Os_5C(CO)_{15}I]^-$, **12**, has been fully characterized by X-ray diffraction (Fig. 12) and adopts an open arachno structure based on a pentagonal bipyramid but lacking two equatorial vertices (*31*).

The structure of the ruthenium analog to **12** has not been determined crystallographically, but infrared spectroscopic evidence (see Section VI,A) and analogy with the osmium cluster led to the conclusion that $[Ru_5C(CO)_{15}X]^-$ has the same open bipyramidal structure (*29*). The geometries adopted by these clusters are consistent with Wade's rules (*20*), since the halide ion adds two electrons to the 74 electron square pyramidal clusters, resulting in eight skeletal electron pairs to be accomodated by the five vertex cluster. This results in the breaking of the metal–metal bonds giving the observed open structures (with increased access to the carbon atom, an important factor in any attempt to observe reactions at the carbon site) (*1, 8*). The $HRu_5C(CO)_{15}X$ (X = Cl, Br) clusters have been assigned open pyramidal structures similar to $[Ru_5C(CO)_{15}X]^-$ (see below) on the basis of infrared spectroscopy (*32*) (see Section VI,A).

Further examples of Os_5C clusters which adopt this more open framework are provided by three products of the pyrolysis of $Os_3(CO)_{11}P(OMe)_3$ [Eq. (10)].

$$Os_3(CO)_{11}P(OMe)_3 \xrightarrow{\Delta} HOs_5C(CO)_{14}(OP(OMe)_2)$$

(13)

$$+ HOs_5C(CO)_{13}(OP(OMe)_2)(P(OMe)_3)$$

(14)

$$+ HOs_5C(CO)_{13}(OP(OMe)OP(OMe)_2)$$

(15)

$$+ \text{other products} \qquad\qquad (10)$$

The structures of **13** (*33*), **14** (*34*), and **15** (*35*) are shown in Figs. 13, 14, and 15. Each of the products is a 76-electron five-vertex cluster; in **13** the phosphorus ligand donates three electrons to the cluster, **14** is the trimethylphosphite substitution product of **13**, and in **15** the phosphorus ligand contributes five electrons to the cluster. As with the isoelectronic $[Os_5C(CO)_{15}I]^-$, the Os_5C cores take open structures based on a pentagonal bipyramid in which only three of the equatorial vertices are present. The structures of the Os_5C cores are very similar for **13**, **14**, and **15**. Mean Os–Os distances are 2.90 ± 0.06 Å for all three, and the carbido carbon atom is

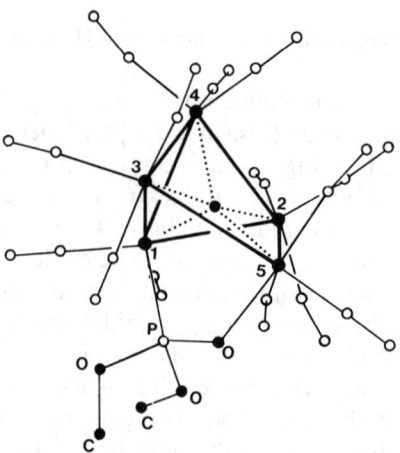

FIG. 13. HOs$_5$C(CO)$_{14}$(OP(OCH$_3$)$_2$), **13** (*33*). The Os$_5$C core is similar to that in **12** (Fig. 12). Os(2) and Os(3) are the apices and Os(1), Os(4), and Os(5) are three of the equatorial vertices of a pentagonal bipyramid [alternately Os(5) caps the Os$_4$C butterfly described by Os(1), Os(2), Os(3), and Os(4)]. The longest Os–Os bonds are to Os(5) [Os(2)–Os(5) = 2.928(2) Å, Os(3)–Os(5) = 2.937(2) Å], which bears three carbonyls as well as the phosphonate oxygen atom. Of the five remaining metal–metal bonds Os(1)–Os(4) is significantly longer [2.914(2) Å] than the other four [which average 2.878(8) Å], which probably reveals the location of the hydrogen atom. The mean Os–C$_{carbide}$ distance is 2.06(6) Å.

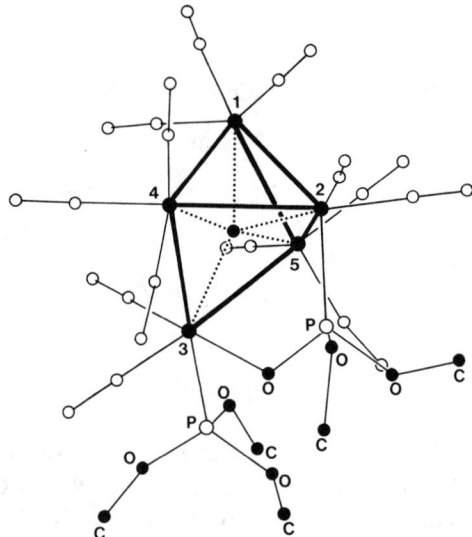

FIG. 14. HOs$_5$C(CO)$_{13}$(OP(OCH$_3$)$_2$)(P(OCH$_3$)$_3$), **14** (*34*). The Os$_5$C core is similar to that in Fig. 13. In this case Os(4) and Os(5) occupy apical sites and Os(1), Os(2), and Os(3) are three of the equatorial sites of the incomplete pentagonal bipyramid. The trimethylphosphite ligand is coordinated to Os(3), which also bears two carbonyls and the phosphonate oxygen atom. The hydride is assigned a bridging position on Os(1)–Os(2). The two metal–metal bonds to Os(3) are significantly longer [2.954(20) Å] than the remaining five [mean 2.877(9) Å]. The mean Os–C$_{carbide}$ distance is 2.07(7) Å.

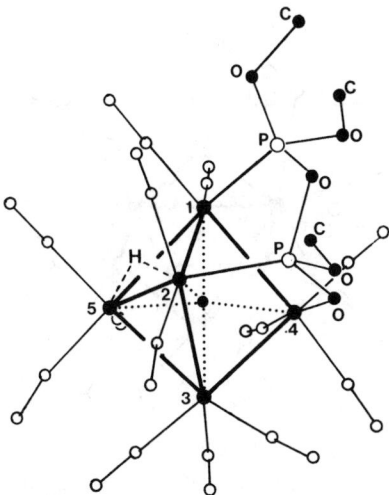

FIG. 15. HOs$_5$C(CO)$_{13}$(OP(OCH$_3$)OP(OCH$_3$)$_2$), **15** (*35*). The Os$_5$C core is similar to both **13** and **14**. Os(1) and Os(3) are in apical sites, and Os(2), Os(4), and Os(5) are in equatorial sites. The tridentate phosphorus ligand coordinates to Os(1) via one phosphorus atom, to Os(2) via the second phosphorus atom, and to Os(4) via the phosphonate oxygen. The Os(1)–Os(4) bond is the longest in the core [2.951(3) Å], and the second bond to Os(4) [Os(3)–Os(4)] is also significantly longer [2.911(3) Å] than the other Os–Os bonds [mean 2.852(18) Å]. The Os(2)–Os(5) bond [2.927(3) Å] is assigned to the bridging hydride. The mean Os–C$_{carbide}$ distance is 2.06(6) Å.

approximately equidistant from the five osmium atoms in each cluster, with mean Os–C$_{carbide}$ distances of 2.06 ± 0.07 Å. The central carbon atom lies in the equatorial plane of each bipyramidal molecule, midway between the two apical osmium atoms.

The structures of **13**, **14**, and **15** provide an increase in spatial access to the carbon atom when compared with the 74 electron square pyramidal analog Os$_5$C(CO)$_{15}$, as was found for [Os$_5$C(CO)$_{15}$I]$^-$.

The structural modifications caused by the addition of electrons to these clusters raise the prospect of the synthesis of Ru$_4$C and Os$_4$C clusters by the fragmentation of larger clusters, as was found for the iron case, but no such compounds have yet been reported.

The first reported examples of M$_6$C carbidocarbonyl clusters were originally found in ruthenium chemistry. Ru$_6$C(CO)$_{17}$, **16**, and its arene derivities Ru$_6$C(CO)$_{14}$(arene) (arene = toluene, xylene, and mesitylene) (*36, 37*) were synthesized in modest yields by refluxing Ru$_3$(CO)$_{12}$ in the requisite arene, **16** becoming the major product when a saturated hydrocarbon such as decane was the solvent.

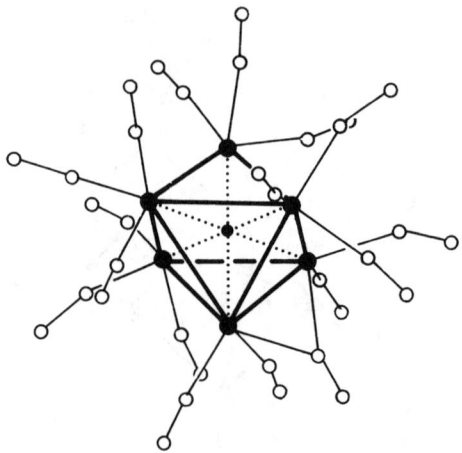

FIG. 16. $Ru_6C(CO)_{17}$, **16** (38). The molecule contains one bridging and sixteen terminal carbonyls coordinated to an octahedral Ru_6C core. The Ru–Ru distances range from 2.827(5) to 3.034(5) Å, averaging 2.90(6) Å. The mean Ru–$C_{carbide}$ distance is 2.05(7) Å.

The structures of both **16** and its mesitylene substituted analog **17** are known (Figs. 16 and 17). The six ruthenium atoms in **16** form an octahedron with the carbon atom at the center (38), whereas in **17** the carbon atom is situated somewhat closer to the ruthenium vertex bearing the arene (39).

An improved synthesis of **16**, by the reaction of $Ru_3(CO)_{12}$ with ethylene, has been reported, with yields of 70% (40).

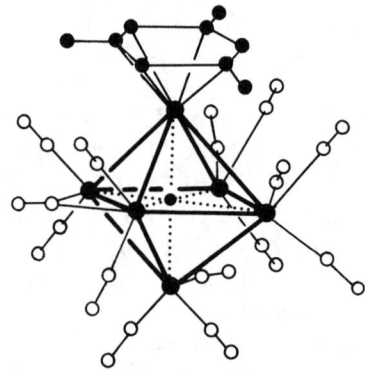

FIG. 17. $Ru_6C(CO)_{14}$(mesitylene), **17** (39). As in the case of **16**, one of the carbonyl ligands bridges an edge of the octahedral Ru_6C core. The arene ligand is bound to a ruthenium atom which forms a triangular face with the carbonyl-bridged Ru–Ru edge. The Ru–Ru distances average 2.88(4) Å and the mean Ru–$C_{carbide}$ distance is 2.04 Å. The carbon atom is displaced from the center of the octahedron by 0.12 Å, toward the arene-bearing vertex.

The dianion $[Ru_6C(CO)_{16}]^{2-}$, **18**, isoelectronic with **16**, has been prepared by treatment of **17** with sodium methoxide in methanol to yield $[Ru_6C(CO)_{16}CO_2Me]^-$ followed by reaction with aqueous base [Eq. (11)] (*40*).

$$Ru_6C(CO)_{17} \xrightarrow[\text{MeOH}]{\text{OMe}^-} [Ru_6C(CO)_{16}CO_2Me]^- \xrightarrow{\text{OH}^-} [Ru_6C(CO)_{16}]^{2-} \qquad (11)$$

(17)   (18)

Cluster **18** may also be synthesized, more conveniently, from the reaction of $Ru_3(CO)_{12}$ with $NaMn(CO)_5$ in refluxing diglyme (*41*).

Two structures of the dianion have been reported. The tetramethylammonium salt, **18a,** apparently differs from the tetraphenylarsonium salt **18b** in the precise disposition of the bridging carbonyl ligands. Cluster **18b** is reported to contain four bridging carbonyls (*40*), whereas **18a** has only three (*41*). However, a fourth carbonyl in the latter is highly disordered [as was the case for the isostructural iron analog (*7*)] tending toward an asymmetrically bridging position, and it seems likely that the two salts are in fact essentially identical, the apparent differences being due more to crystallographic artifacts than to distinguishable chemistry. The structure of **18b** is shown in Fig. 18.

The dihydride $H_2Ru_6C(CO)_{16}$ has been synthesized by protonation of **18** (*40, 41*) but no structure determination has been reported, characterization relying on analytic and spectroscopic evidence.

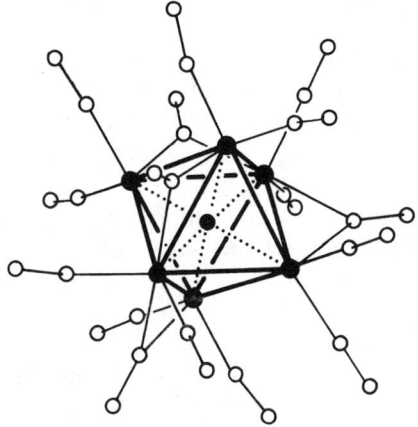

Fig. 18. $Ru_6C(CO)_{16}^{2-}$, **18**, as in its Ph_4As^+ salt (*40*). The anion has approximate C_{2v} symmetry, with four asymmetrically bridging and twelve terminal carbonyls bound to an octahedral Ru_6C core. Mean Ru–Ru distance is 2.89 Å, the mean Ru–$C_{carbide}$ distance is 2.05 Å.

Several derivatives of $Ru_6C(CO)_{17}$ have been synthesized. Phosphine and phosphite derivatives $Ru_6C(CO)_{17-n}(P)_n$ have been prepared by direct reaction of **16** with a wide range of phosphorus ligands, and the degree of substitution is dependent on the phosphorus donor (42–44). With PPh_3 in refluxing hexane only the monosubstitution product is obtained, whereas a disubstitution product is formed with PPh_2Et under the same conditions. $P(OMe)_3$ displaces up to four moles of CO from **16**, and all four $Ru_6C(CO)_{17-n}[P(OMe)_3]_n$ clusters have been isolated. The crystal structure of one phosphine derivative is known. $Ru_6C(CO)_{17}(PPh_2Et)$, **19** (44), retains the octahedral structure of **16**, and like **16** has one bridging carbonyl ligand, the remainder being terminal. The phosphine is coordinated to a ruthenium atom which makes a triangular face with the bridged Ru–Ru bond. The carbide carbon atom in **19** is not significantly displaced from its position in **16** and occupies a cavity of similar dimensions [mean Ru–C = 2.060(26) Å for **19**, 2.05 Å for **16**].

A widely applicable method for synthesizing $Ru_6C(CO)_{17-n}L$ complexes is available [Eq. (12)]. $[Ru_6C(CO)_{16}]^{2-}$, **18**, is easily oxidized by mild oxidants such as ferrocenium cation. Even in the absence of added CO, **18** is transformed to **16** in high yield, demonstrating the highly reactive nature of the coordinatively unsaturated $[Ru_6C(CO)_{16}]$, the probable intermediate (41). Under one atmosphere of CO the reaction is quantitative [providing a convenient high yield synthesis of $Ru_6C(CO)_{17}$ from $Ru_3(CO)_{12}$].

$$[Ru_6C(CO)_{16}]^{2-} + Cp_2Fe^+ \xrightarrow{\ L\ } Ru_6C(CO)_nL + Cp_2Fe$$
$$L = CO, \text{ tertiary phosphine, } n = 16$$
$$L = C_6H_5C{\equiv}C_6H_5, C_6H_5C{\equiv}CH, CH_3C{\equiv}CC_2H_5, n = 15 \qquad (12)$$

The acetylene adducts $Ru_6C(CO)_{15}(RC_2R')$ are interesting in that the organic ligand contributes four electrons to the cluster by bonding to an Ru_3 face via two σ-bonds and one π-bond, as exemplified by $Ru_6C(CO)_{15}$ $(C_6H_5C{\equiv}CH)$, the structure of which was determined by X-ray diffraction (Fig. 19) (45, 46).

A further example of the synthesis of an organometallic derivative of **16** by oxidation of **18** is provided by the reaction of **18** with tropylium bromide [Eq. (13)] (41).

$$[Ru_6C(CO)_{16}]^{2-} + 2C_7H_7^+ \rightarrow Ru_6C(CO)_{14}(C_{14}H_{14}) + 2CO \qquad (13)$$

In this reaction the incoming ligand is the organic redox product, bitropyl $(C_{14}H_{14})$, which coordinates to one Ru_3 face of the octahedral Ru_6C core, each double bond of one C_7H_7 ring bonding to a separate ruthenium atom (47) (Fig. 20). The carbido carbon is significantly closer to the substituted face of the cluster (mean Ru–C = 2.04 Å) than to the unsubstituted face (mean Ru–C = 2.08 Å).

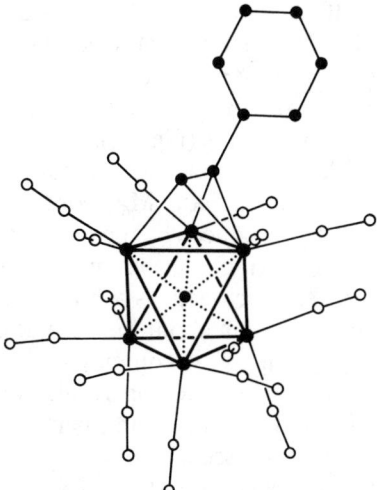

FIG. 19. $Ru_6C(CO)_{15}(PhC \equiv CH)$ (45, 46). The six ruthenium atoms define a slightly distorted octahedron, with the carbide carbon at the center (mean $Ru-C = 2.04(2)$ Å). The phenylacetylene ligand is bound to one triangular face of the cluster. $Ru-Ru$ bond lengths fall into two groups: those on the face bearing the acetylene average 2.80(4) Å, and the remaining metal–metal bonds average 2.92(5) Å.

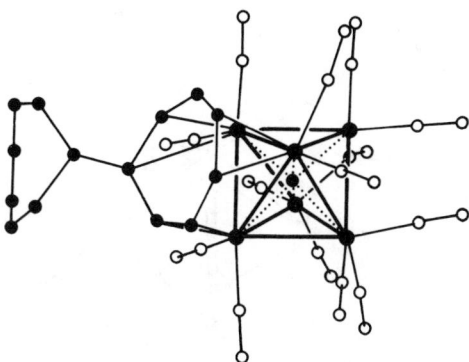

FIG. 20. $Ru_6C(CO)_{14}(C_{14}H_{14})$ (47). One C_7H_7 ring of the bitropyl molecule is bound to a triangular face of the octahedral Ru_6C core. Thirteen of the carbonyls are terminal, one bridging. The eight unbridged $Ru-Ru$ bonds average 2.93 Å, the carbonyl bridged bond is 2.778 Å, and the three bonds between the metal atoms bound to the organic ligand average 2.90 Å. The carbide carbon is closer to these ruthenium atoms (2.04 Å) than to the unsubstituted ruthenium atoms (2.08 Å).

Another organometallic derivative of **19** is formed as a minor by-product in the synthesis of **16** by treatment of $Ru_3(CO)_{12}$ with ethylene (*48*). $Ru_6C(CO)_{15}(CH_3CH = CH—CH = CHCH_3)$ produced in this reaction is an octahedral cluster, with the *trans,trans*-2,4-hexadiene molecule coordinated to adjacent ruthenium atoms (Fig. 21).

An interesting example of pronounced distortion which can result when a multielectron donor coordinates to the M_6C core is provided by the structure of $HRu_6C(CO)_{15}(SEt)_3$, produced by the reaction between $Ru_6C(CO)_{17}$ and EtSH (*49*) (Fig. 22). The Ru_6C cage has opened by the motion of Ru(3) about the Ru(1)–(2) edge, which acts as a hinge, resulting in a distorted trigonal bipyramidal coordination about the carbide carbon, with the sixth ruthenium atom bridging an axial–equational bond. This is a further example of a distortion of a closed M_nC core resulting from the coordination of multielectron donors, and producing an open cage cluster with a more accessible carbide carbon atom (see above).

Two bimetallic clusters derived from $[Ru_6C(CO)_{16}]^{2-}$ have been prepared, in which eight and thirteen metal atoms are present. The reaction of $Cu(MeCN)_4^+BF_4^-$ with **18** results in the addition of two copper vertices to the Ru_6C core [Eq. (14)] (*50*).

$$2Cu(MeCN)_4^+ + [Ru_6C(CO)_{16}]^{2-} \longrightarrow (MeCN)_2Cu_2Ru_6C(CO)_{16} \qquad (14)$$

$$(18) \qquad\qquad\qquad (20)$$

The structure of **20** is shown in Fig. 23. The Ru_6C octahedron remains

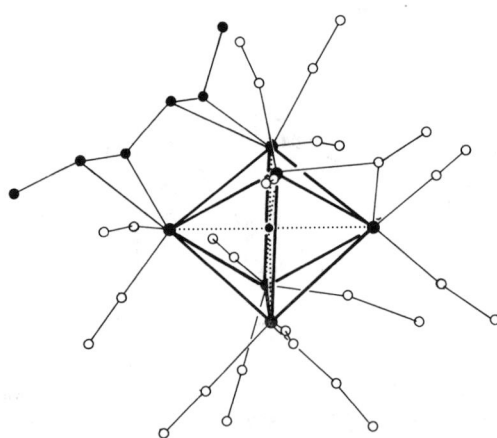

FIG. 21. $Ru_6C(CO)_{15}(CH_3CH{=}CH—CH{=}CHCH_3)$ (*48*). This structure is related to that of **16** by the substitution of two carbonyls on adjacent ruthenium atoms by the hexadiene molecule. The Ru–Ru distances average 2.91(5) Å; the Ru–$C_{carbide}$ distances average 2.06 Å.

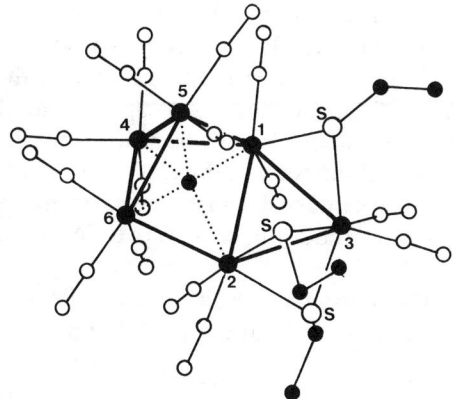

FIG. 22. $HRu_6C(CO)_{15}(SEt)_3$ (*49*). The carbide carbon is coordinated by five of the ruthenium atoms in a distorted trigonal bipyramidal array (mean Ru–C = 2.07 ± 0.08 Å). Ru(2)(4)(5) define the equatorial plane; Ru(1) and Ru(6), the axial sites. Ru(3) bridges the axial–equatorial bond Ru(1)–Ru(2) [3.049(1) Å]. Ru(1)–Ru(3) [3.052(1) Å] is bridged by one thiolate ligand, Ru(2)–Ru(3) [3.012(1) Å] by two. Ru(2)–Ru(6) = 2.908(1) Å. The remaining Ru–Ru bonds average 2.86 ± 0.04 Å. The hydride ligand bridges Ru(4)–Ru(5) [Ru(4)–H = 1.82 Å, Ru(5)–H = 1.75 Å]. The carbonyl ligands are all terminal.

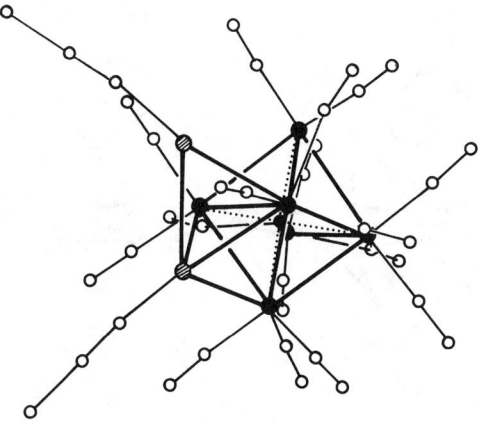

FIG. 23. $(CH_3CN)_2Cu_2Ru_6C(CO)_{16}$, **20** (*50*). The distorted octahedral Ru_6C core is capped by two directly bonded copper atoms [Cu–Cu = 2.693(1) Å], one on an Ru_3 face, the second on the $CuRu_2$ face so formed. The Ru–Ru distances range from 2.798(1) to 3.072(1) Å (mean 2.89 Å). Ru–$C_{carbide}$ distances range from 2.031(4) to 2.073(4) Å (mean 2.05 Å). There are thirteen terminal carbonyls, and three asymmetrically bridging Ru–Ru edges. No Cu–CO contacts are short enough to imply bonding interactions.

intact, with the coordination geometry of the sixteen CO ligands staying remarkably similar to that in **18**. The two copper atoms are directly bonded (Cu–Cu = 2.692 Å), one capping a triangular face of the Ru_6C core and providing a new $CuRu_2$ face, and the second capping a newly formed $CuRu_2$ face of the $CuRu_6$ monocapped octahedron. The ruthenium–ruthenium bond lengths reflect the asymmetry of the cluster. The proximity of the two unbridged copper atoms is unusual, and in contrast with their disposition in an ostensibly analogous rhodium cluster $(MeCN)_2Cu_2Rh_6C(CO)_{15}$ (see below).

A second bimetallic derivitive of $[Ru_6C(CO)_{16}]^{2-}$ is produced by the reaction of **18** with thallium(III) nitrate in methanol [Eq. (15)] (*51*).

$$Tl^{3+} + 2[Ru_6C(CO)_{16}]^{2-} \xrightarrow[25°C]{MeOH} [Tl(Ru_6C(CO)_{16})_2]^- \qquad (15)$$

The bimetallic anion was characterized crystallographically as its $AsPh_4^+$ salt (*52*). The cluster (Fig. 24) comprises two $Ru_6C(CO)_{16}$ octahedra, linked via one edge to a central thallium atom. The geometry round the thallium atom is determined by nonbonding contacts between the carbonyl oxygen atoms on the adjacent $Ru_6C(CO)_{16}$ units. The octahedra are disposed to allow an effectively staggered conformation of the oxygens, resulting in a torsional angle of 35.9° between the two Ru–Ru edges directly bonded to the thallium atom.

No osmium analogs to the octahedral Ru_6C clusters have been reported.

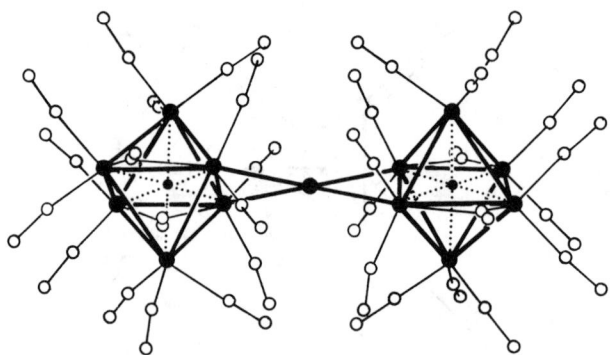

FIG. 24. $TlRu_{12}(C)_2(CO)_{32}^-$, as in its Ph_4As^+ salt (*52*). Two Ru_6C octahedra are connected via one edge of each to the central thallium atom. The Ru–Ru distances within the octahedron range from 2.813(1) to 2.938(1) Å, with the exception of the two Ru–Ru edges bound to thallium (3.098 Å mean). The torsional angle at thallium is 35.9°, and is determined by the nonbonded interactions between the carbonyls on the adjacent $Ru_6C(CO)_{16}$ units. There are two bridging and fourteen terminal carbonyls on each Ru_6 unit. The carbide carbons lie near the centers of their respective Ru_6 cages; mean $Ru–C_{carbide}$ = 2.05 Å.

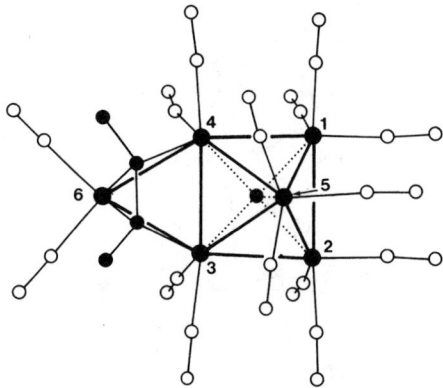

FIG. 25. $Os_6C(CO)_{16}(CH_3C\equiv CCH_3)$ (53). Five osmium atoms (1–5) define a square pyramid with the carbide carbon ca. 0.20 Å below the basal plane equidistant from the four basal osmium atoms [mean Os–C = 2.04(2) Å], and 2.20(2) Å from the apical osmium. The sixth osmium atom bridges one basal edge [Os(3)–Os(4)] lying below the basal plane. The acetylene ligand is bound to the Os(3)(4)(6) triangle, which makes an angle of 25.3° with the four basal metal atoms. Unbridged Os–Os distances average 2.87(5) Å; Os–Os distances in the Os(3)(4)(6) triangle average 2.77(3) Å.

An alternative structure is observed for the only Os_6C cluster so far known. $Os_6C(CO)_{16}(CH_3C\colon CCH_3)$ (53) has been isolated in 5% yield from the reaction of $Os_6(CO)_{18}$ with ethylene at 175°C for 60 hr. The structure of the cluster (Fig. 25) comprises an edge bridged square pyramid of osmium atoms with the bridging atom lying below one edge of the basal plane, the new triangle of osmium atoms making an angle of 25.3° with that plane. That is the first and so far the only example of such a geometry [although an edge bridging metal atom is also found in $Rh_8C(CO)_{19}$ (see p. 40)]. However, whereas the Ru_6C group represents the most common state of aggregation for ruthenium carbidocarbonyl clusters, osmium provides a well-established and growing class of larger carbide clusters, emanating (as have many advances in ruthenium carbidocarbonyl chemistry) from the Cambridge group of Lewis, Johnson, and co-workers.

$H_2Os_7C(CO)_{19}$ (54), the first example known of an M_7C carbide cluster, was identified by mass spectroscopy and infrared spectroscopy, among the many products obtained by heating $Os_3(CO)_{12}$ to 230°C with small amounts of water in a sealed tube. No structural information is yet available. $Os_8C(CO)_{21}$ was similarly characterized, being present among the products of $Os_3(CO)_{12}$ pyrolysis (55).

Four $Os_{10}C$ clusters have been isolated. $[Os_{10}C(CO)_{24}]^{2-}$, 21, is prepared in 80% yield by the vacuum pyrolysis of $Os_3(CO)_{11}$(pyridine) at 250°C for

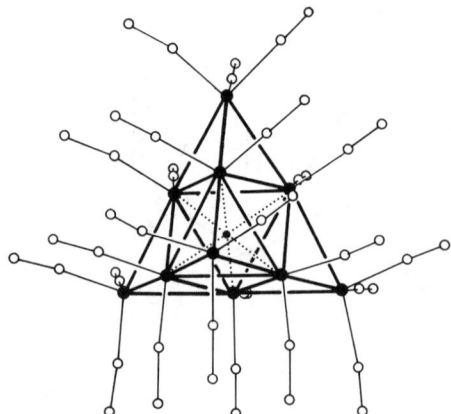

FIG. 26. $Os_{10}C(CO)_{24}^{2-}$, 21, as in its $(PPh_3)_2N^+$ salt (56). The cluster comprises an octahedral Os_6C core, with four additional osmium atoms capping tetrahedrally disposed faces of the octahedron. The overall symmetry is close to T_d. Within the Os_6C octahedron the Os–Os bondlengths average 2.88(1) Å; those from the capping metal atoms average 2.79(1) Å. The longer metal–metal bonds allow the accommodation of the carbon atom [mean $Os–C_{carbide}$ = 2.04(3) Å]. All the carbonyls are terminal.

24 hr (56). The structure of the bis(triphenylphosphine)iminium salt has been determined and is shown in Fig. 26.

The carbido carbon resides at the center of an Os_6 octahedron, with the remaining four osmium atoms capping tetrahedral faces of the Os_6C core. The overall symmetry is approximately tetrahedral, comprising three layers of one, three, and six osmium atoms, respectively. The carbonyl ligands are all terminal, with each of the octahedrally disposed metal atoms bearing two and each tetrahedrally disposed capping metal atom bearing three.

Protonation of 21 yields $H_2Os_{10}C(CO)_{24}$ (57). The crystal structure of the dihydride has not been determined, but analysis of the vibrational spectrum of the cluster in the region associated with motion of the interstitial carbon atom has led to the conclusion that the symmetry of 21 is reduced on protonation, probably by protonation of the central Os_6 octahedron (see Section VI,A) (57). Cluster 21 also reacts with iodine to yield sequentially $[Os_{10}C(CO)_{24}I]^-$, 22, and $Os_{10}C(CO)_{24}I_2$, 23, the result of electrophilic attack by I^+ on the dianion [Eq. (16)] (58).

$$[Os_{10}C(CO)_{24}]^{2-} + 2I_2 \rightarrow [Os_{10}C(CO)_{24}I]^- + I_3^-$$

$$(21) \qquad\qquad\qquad (22)$$

$$(22) + 2I_2 \qquad \rightarrow \qquad Os_{10}C(CO)_{24}I_2 + I_3^- \qquad\qquad (16)$$

$$(23)$$

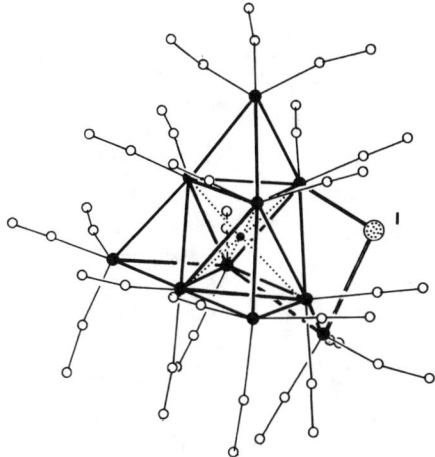

FIG. 27. $Os_{10}C(CO)_{24}I^-$, **22**, as in its $(Ph_3P)_2N^+$ salt (*58*). The cluster is related to **21** by the addition of an iodine atom across one of the Os–Os bonds from a capping metal atom to an Os_6C core metal atom. The octahedral Os–Os bonds average 2.90(5) Å; the bonds from the capping atoms to the core metal atoms average 2.80(3) Å.

The two iodo clusters, whose structures were determined by X-ray diffraction (Figs. 27 and 28), are two further examples of the opening of a closed cluster framework by the bond breaking effect of adding electron donating ligands. The iodine atoms in these clusters are regarded as three electron

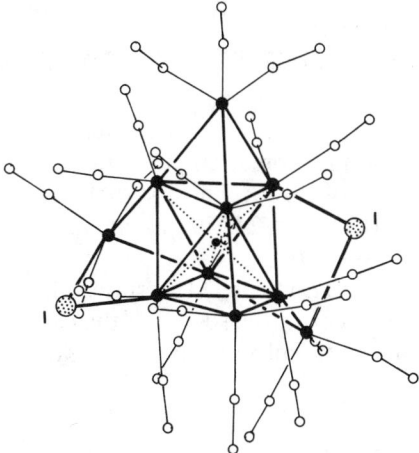

FIG. 28. $Os_{10}C(CO)_{24}I_2$, **23** (*58*). The structure is derived from that of **22** by the addition of an iodine atom to a second Os–Os bond between a capping metal atom and the octahedral Os_6C core. Os–Os bonds within the octahedron average 2.92(5) Å; the $Os-C_{carbide}$ distances average 2.07 Å.

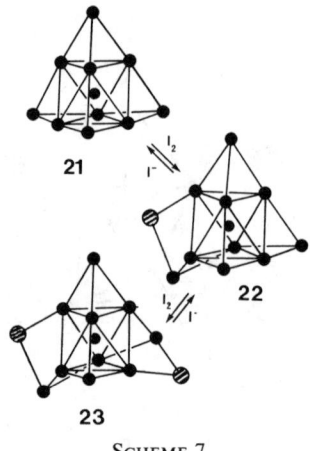

21

22

23

SCHEME 7

donors, and so **22** and **23** possess two and four electrons, respectively, more than the closed dianion **21**. The iodine atom in **22** has added to one of the capping Os_4 tetrahedra breaking one $Os-Os$ bond and bridging the wingtips of the resulting Os_4 butterfly. The second iodine atom repeats this process at a second Os_4 site in **23**. The opening of the capping tetrahedra is reversible by addition of iodide (see Scheme 7) (*58*).

IV

CARBIDOCARBONYL CLUSTERS OF COBALT AND RHODIUM

The chemistry of the carbidocarbonyl clusters of cobalt and rhodium (none is known for iridium) is predominantly the work of Italian school of the late Paolo Chini and colleagues. The first rhodium cluster of this type to be reported was $[Rh_6C(CO)_{15}]^{2-}$, **24** (*59*), isolated as a minor by-product in the synthesis of $[Rh_7(CO)_{16}]^{3-}$ (and originally misformulated as $[Rh_3(CO)_{10}]^-$). The formation of **24** resulted from the reaction of $[Rh_7(CO)_{16}]^{3-}$ with chloroform (the source of the carbon atom), present as an impurity in the reaction solvent, and **24** is now synthesized by this reaction [Eq. (17)].

$$Rh_4(CO)_{12} \xrightarrow[CO]{NaOH/MeOH} [Rh_7(CO)_{16}]^{3-} \xrightarrow{CHCl_3} [Rh_6C(CO)_{15}]^{2-} \qquad (17)$$

(24)

The structure of the benzyltrimethylammonium salt of **24** has been determined, and it provided the first example of a trigonal prismatic metal atom cluster (Fig. 29). The Rh–Rh distances within the trigonal bases of the prismatic Rh_6C core are slightly shorter [2.776(3) Å] than the interbasal distances [2.817(2) Å]. The carbide atom resides at the center of the prismatic Rh_6 cage [mean $Rh-C_{carbide} = 2.134(4)$ Å]. All nine of the Rh–Rh bonds are bridged by carbonyl ligands, the remaining six carbonyls being terminal, one on each rhodium atom.

The cobalt analog to **24**, $[Co_6C(CO)_{15}]^{2-}$, is synthesized from $(CO)_9Co_3CCl$ and $Co(CO)_4^-$ in THF at 70°C, and is isomorphous with **24**, as is the mixed anion $[Co_3Rh_3C(CO)_{15}]^-$, and so all three of these $[M_6C(CO)_{15}]^{2-}$ clusters are presumed to be isostructural (*60*).

The reaction of $Cu(MeCN)_4^+BF_4^-$ with **24** results in the capping of the two trigonal faces of the prismatic Rh_6C core by MeCN · Cu vertices [Eq. (18)] (*61*).

$$[Rh_6C(CO)_{15}]^{2-} + 2Cu(MeCN)_4^+ \rightarrow (MeCN)_2Cu_2Rh_6C(CO)_{15} + 6MeCN \qquad (18)$$

$$(24)$$

Neither the Rh–Rh bond lengths nor the configuration of the carbonyl ligands is significantly affected by the addition of the two copper capping atoms to **24**. The structure of $(MeCN)_2Cu_2Rh_6C(CO)_{15}$ is shown in Fig. 30.

On refluxing in isopropanol **24** loses carbon monoxide producing

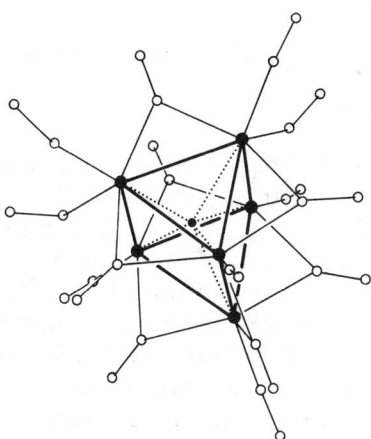

FIG. 29. $Rh_6C(CO)_{15}^{2-}$, **24**, as in its $(PhCH_2N(CH_3)_3)^+$ salt (*59*). The rhodium atoms describe a trigonal prism. The interbasal Rh–Rh bonds average 2.776(2) Å; those within the trigonal bases average 2.817(2) Å. The mean $Rh-C_{carbide}$ distance is 2.13 Å. Six carbonyls are terminal (one on each rhodium atom), and the remaining nine bridge the metal–metal bonds.

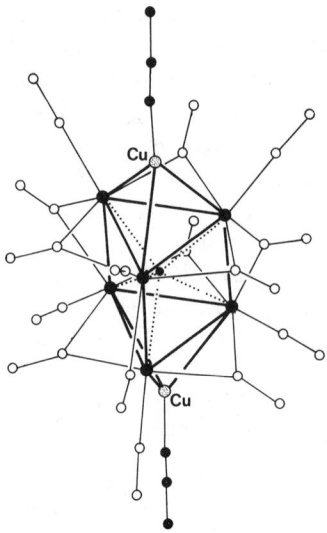

FIG. 30. $(CH_3CN)_2Cu_2Rh_6C(CO)_{15}$ (*61*). The trigonal prismatic Rh_6C core of **24** (Fig. 29) is essentially unperturbed by the additional copper atoms that cap the two trigonal faces of the prism. Rh–Rh distances average 2.765(1) Å (basal) and 2.810(1) Å (interbasal). The Cu–Rh bonds average 2.660(1) Å. The mean $Rh–C_{carbide}$ distance is 2.13 Å.

$[Rh_6C(CO)_{13}]^{2-}$, **25**, and this reaction is readily reversible at room temperature [Eq. (19)] (*62*).

$$[Rh_6C(CO)_{15}]^{2-} \underset{25°C}{\overset{^iPrOH,\ reflux}{\rightleftharpoons}} [Rh_6C(CO)_{13}]^{2-} + 2CO \tag{19}$$

(24)

With decarbonylation the trigonal prismatic **24** loses four electrons and adopts the octahedral geometry expected for an 86 electron cluster, but with a marked distortion (Fig. 31). The Rh–Rh distances vary from ~2.75 Å for those with bridging carbonyls, to ~3.10 Å for the four longest (unbridged) bonds. This distortion might be explained by the steric requirements of the interstitial carbon atom, since the cavity provided by an octahedron is a small one (giving an effective radius of 0.60 Å for the carbide atom, compared with 0.74 Å in the prismatic **24**), but other carbidocarbonyl clusters with similarly small but more regular octahedral cavities are known (see Section VII,B).

The oxidation of $[Co_6C(CO)_{15}]^{2-}$ with ferric ion results in a similar decarbonylation, and rearrangement of the trigonal prism to an octahedron,

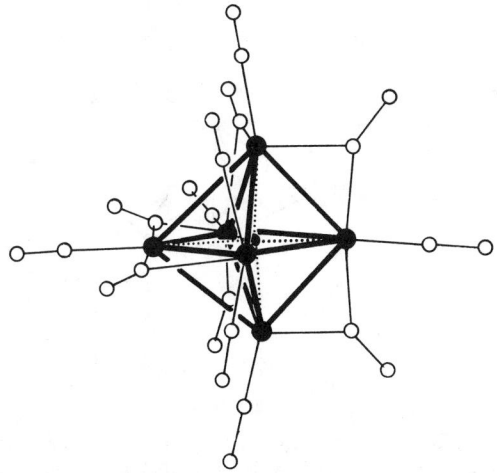

FIG. 31. $Rh_6C(CO)_{13}^{2-}$, 25, as in its Ph_4P^+ salt (62). The Rh_6C core is a distorted octahedron, with Rh–Rh bonds varying from 2.733(2) to 3.188(2) Å (mean 2.91 Å). $Rh–C_{carbide}$ distances range from 1.99(1) to 2.15(1) Å (mean 2.05 Å). There are six edge bridging and seven terminal carbonyls.

giving the paramagnetic 87 electron cluster $[Co_6C(CO)_{14}]^-$, **26**, [Eq. (20)] (63), which is also obtained from the reaction of $Co_3(CO)_9CCl$ with $[Co(CO)_4]^-$ in diethyl ether (64).

$$[Co_6C(CO)_{15}]^{2-} + FeCl_3 \xrightarrow{\text{THF, 25°C}} [Co_6C(CO)_{14}]^- + FeCl_2 + CO + Cl^- \quad (20)$$

(26)

The structure of **26** is shown in Fig. 32. The Co_6C core of **26** displays an interesting distortion from a regular octahedron, with one cobalt–cobalt bond $[Co(1)–Co(2) = 2.916$ Å] much longer than the others, which average ~2.63 Å. The central carbon atom lies at a mean distance of 1.85 Å from four of the cobalt atoms, but 1.93 Å from the metal atoms which make up the long edge. On the basis of these data the unpaired electron has been assigned to an antibonding orbital associated with the $Co(1)–Co(2)$ bond (63).

Further oxidation of **26** with tropylium bromide in methanol results in partial fragmentation of the Co_6C core (65), similarly to the analogous reaction of $[Fe_6C(CO)_{16}]^{2-}$ (see above) (8). In this case, however, three cobalt vertices are lost leaving a tricobalt fragment bearing a trigonally bound carbon atom. The isolated product, $Co_3(CO)_9C \cdot CO_2CH_3$ presumably arises

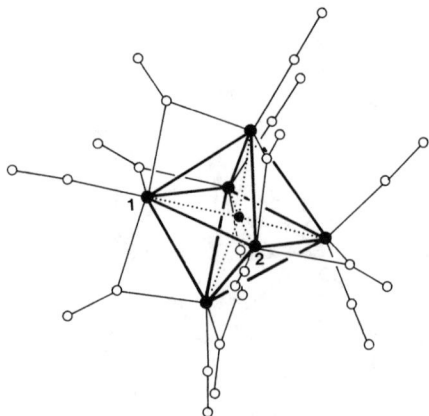

FIG. 32. $Co_6C(CO)_{14}^-$, **26**, as in its $(CH_3)_4N^+$ salt (63). The Co_6C core is distorted octahedron. One Co–Co bond [Co(1)–Co(2) = 2.916 Å] is longer than the eleven others (mean 2.63 Å). The bonds from Co(1) and Co(2) to the carbide carbon are longer (1.93 Å average) than those from the other cobalt atoms (1.85 Å). There are six edge-bridging and eight terminal carbonyls.

from the formation of the acylium ion $[Co_3(CO)_9CCO]^+$ which is known to react with methanol to yield the carbomethoxymethylidyne cluster (9) [Eq. (21)].

$$[Co_6C(CO)_{14}]^- \xrightarrow{C_7H_7^+Br^-} [Co_3(CO)_9]^+ \rightarrow [Co_3(CO)_9CCO]^+$$
$$Co_3(CO)_9CCO_2CH_3 \longleftarrow \qquad\qquad (21)$$

With both cobalt and rhodium, clusters of higher nuclearity are accessible from the hexanuclear anions. Condensation of $[Co_6C(CO)_{15}]^{2-}$ with $Co_4(CO)_{12}$ in isopropanol at 60°C results in the isolation of $[Co_8C(CO)_{18}]^{2-}$ [Eq. (22)] (64).

$$2[Co_6C(CO)_{15}]^{2-} + Co_4(CO)_{12} \xrightarrow{^iPrOH,\ 60°C} 2[Co_8C(CO)_{18}]^{2-} + 6CO \qquad (22)$$

The anion (Fig. 33) comprises two parallel rhombs of metal atoms at an interplanar distance of 2.12 Å. The two layers are staggered so as to produce an antiprismatic geometry, which may be derived from the trigonal prismatic starting material by the capping of two tetragonal faces and the breaking of the Co–Co bond common to these two faces (shown as a dotted line in Fig. 33).

In a regular antiprism the interstitial cavity provided by an edge of 2.52 Å (mean Co–Co distance) would be larger than optimal, and so the antiprism is distorted from a regular geometry to allow the formation of favorable

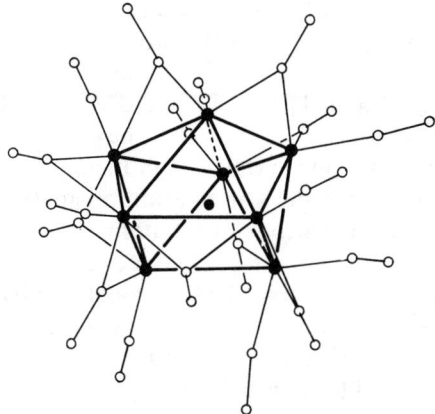

FIG. 33. $Co_8C(CO)_{18}^{2-}$, as in its $(PhCH_2N(CH_3)_4)^+$ salt (66). The eight cobalt atoms describe a tetragonal antiprism with Co–Co distances ranging from 2.464(4) to 2.598(4) Å (mean 2.52 Å). The central carbon atom forms four shorter (1.99 Å) and four longer (2.15 Å) bonds to the cobalt atoms.

Co–C carbide contacts with four of the cobalt atoms (mean 1.99 Å) at the expense of the other four (mean 2.15 Å) (66).

The largest cobalt carbide clusters yet synthesized are $[Co_{11}C_2(CO)_{22}]^{3-}$, **27** (67), and $[Co_{13}C_2(CO)_{24}]^{4-}$, **28** (68), prepared by refluxing $[Co_6C(CO)_{15}]^{2-}$ in diglyme for 8 hr. The structure of **27**, determined for its benzyltrimethylammonium salt, is shown in Fig. 34 (67). The $Co_{11}C_2$ unit is

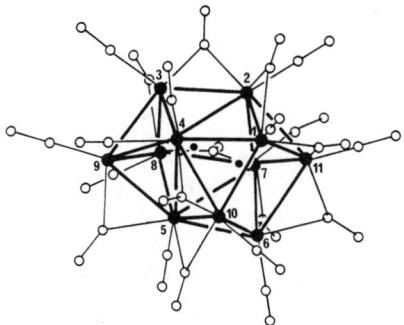

FIG. 34. $Co_{11}C_2(CO)_{22}^{3-}$, **27**, as in its $(PhCH_2N(CH_3)_3)^+$ salt (67). The $Co_{11}C$ core comprises a fused tetragonal antiprism and trigonal prism, sharing one tetragonal face (see text), of which two opposite edges are elongated [Co(2)–Co(4) and Co(5)–Co(7), average 3.09 Å]. The remaining cobalt–cobalt bonds average 2.48 Å with bridging carbonyls (eleven) and 2.60 Å unbridged. The two encapsulated carbon atoms are 1.62 Å apart, and the Co–C distances vary from 1.86 to 2.26 Å.

most instructively seen as a tetragonal antiprism and trigonal prism sharing one tetragonal face [Co(2)(4)(5)(7)] (two edges of which are elongated) and capped by one cobalt atom [Co(9)] on a second tetragonal face [Co(3)(4)(5)(8)] of the trigonal prism. [This description allows a comparison to be made with $Rh_{12}C_2(CO)_{25}$ (see below).] The carbide carbon atoms are at a distance of 1.62 Å from one another, slightly longer than a normal C–C single bond length. One resides in the antiprismatic cavity at average distances of 2.10 Å from the four metal atoms in the unshared face and 2.26 Å from those in the shared face. The second carbon atom lies in the prismatic cavity, 2.13 Å (av) from the four cobalt atoms in the shared face and 1.86 Å from the two prism edge metal atoms. The carbonyl ligands are divided between eleven bridging and eleven terminal, all the cobalt atoms accordingly being bound to three carbonyls.

The same synthetic procedure used to produce **27** also yields paramagnetic $[Co_{13}C_2(CO)_{24}]^{4-}$, **28**, which was also characterized crystallographically as its benzyltrimethylammonium salt (*68*). The anion (Fig. 35) contains a $Co_{13}C_2$ core which is best described as two trigonal prisms sharing one vertex, Co(1), mutually rotated so that the metal atoms bonded to Co(1) by an interbase edge in each prism [Co(2) and Co(3)] attains a capping position over a square face of the other. Two further cobalt atoms, Co(4) and Co(5), cap second square faces on each prism. The two carbide carbon atoms occupy the two prismatic cavities, well out of bonding range of each other, in contrast to the more recognizable C_2 unit found in **27**. The carbide atoms are 1.90–2.06(2) Å from the prismatic cobalt atoms, and 2.32 Å and 2.61 Å

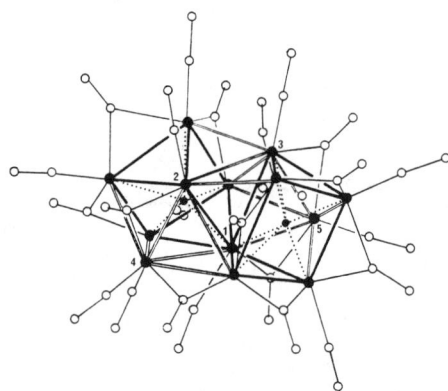

FIG. 35. $Co_{13}C_2(CO)_{24}^{4-}$, **28**, as in its $(PhCH_2N(CH_3)_3)^+$ salt (*68*). The two encapsulated carbon atoms occupy trigonal prismatic cavities, which share a common vertex [Co(1)]. The two prisms, which are outlined with solid lines for clarity, are rotated with respect to one another (see text), and two further cobalt atoms [Co(4) and Co(5)] assume capping positions. Average Co–Co distance = 2.57 Å; average Co–C distance = 1.98 Å.

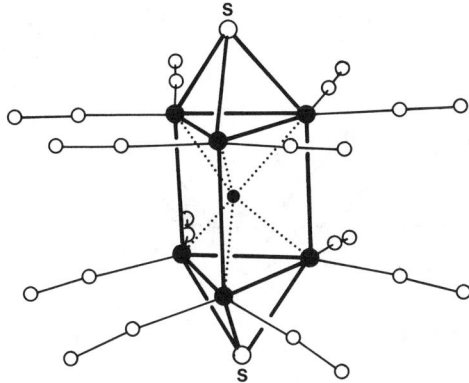

FIG. 36. $Co_6C(CO)_{12}S_2$, **29** (71). The trigonal prism of cobalt atoms is capped on its triangular faces by sulfur atoms. Basal Co–Co distances average 2.432 Å; interbasal Co–Co distances average 2.669 Å. Mean Co–$C_{carbide}$ = 1.94 Å; mean Co–S = 2.192 Å.

from the capping cobalt atoms [Co(4) and Co(5)], and the effective radius of the carbon atom is 0.69 Å.

Cobalt is unique among the metals forming carbidocarbonyl clusters in that clusters containing both carbon and sulfur atoms are also known. Bor and Stanghellini have reported the isolation of $S_2Co_6C(CO)_{12}$, **29** (69–71), and $SCo_6C_2(CO)_{14}$, **30** (72), from the reaction of $Co_2(CO)_8$ with carbon disulfide. The structure of **29** is based on the trigonal prismatic core geometry suggested for $[Co_6C(CO)_{15}]^{2-}$ (see above), with a sulfur atom capping each trigonal face of the Co_6C prism (Fig. 36). No structural data are available for $[Co_6C(CO)_{15}]^{2-}$, but the same order of metal–metal bond lengths is found for **29** as was the case for $[Rh_6C(CO)_{15}]^{2-}$, the isostructural rhodium analog to the cobalt precursor. The interbasal Co–Co bonds are longer [2.669(2) Å] than the basal sets [2.432(2) Å]. The carbon atom lies at the center of the prismatic cavity, with a mean Co–$C_{carbide}$ distance of 1.94 ± 0.01 Å, implying a covalent radius of 0.72 Å for the carbide carbon atom (compared with 0.74 Å in the prismatic cavity of $[Rh_6C(CO)_{15}]^{2-}$).

The second sulfidocarbido cobalt cluster, $SCo_6C_2(CO)_{14}$ (72), has a much more open structure (Fig. 37), and may be seen as two pyramidal Co_3C units joined by two metal–metal bonds and a carbon–carbon bond, with a capping sulfur atom over the Co_4 face. Detailed structural data have not yet been published.

Like cobalt, rhodium also forms clusters of higher nuclearity by the oxidation of the trigonal prismatic anion $[Rh_6C(CO)_{15}]^{2-}$, **24**. Treatment of aqueous $K_2Rh_6C(CO)_{15}$ with ferric ion yields an as yet uncharacterized solid, which on dissolution in methylene chloride yields a number of crystalline products, the nature of which depends on which reaction time

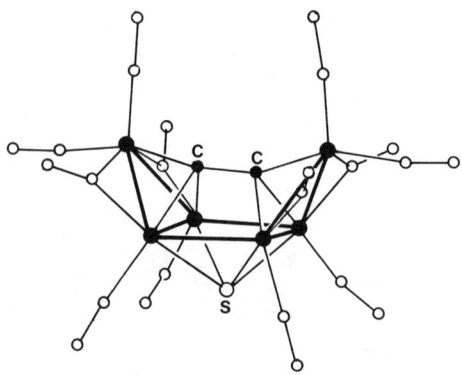

FIG. 37. $Co_6C_2(CO)_{14}S$, **30** (*72*).

and the atmosphere over the solution (*60*). Under carbon monoxide, $Rh_8C(CO)_{19}$ **31**, is obtained. The structure of the Rh_8C core is derived from that of its trigonal prismatic precursor **24** by the addition of one rhodium atom capping a square face of the prism, and a second bridging atom to the edge of a trigonal face (Fig. 38) (*73*). This geometry is a marked departure from the close packed structures which abound in cluster chemistry. Wades electron counting rules are not applicable in any obvious way to **31**, but the molecule does in fact conform to the 18-electron rule for the metal atoms as a whole (although not for each individual atom).

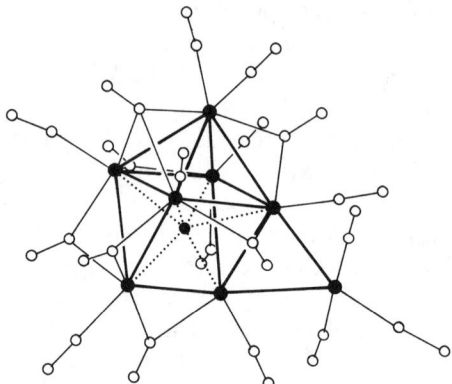

FIG. 38. $Rh_8C(CO)_{19}$, **31** (*73*). The carbide carbon occupies a trigonal prismatic cavity as in **24** (Fig. 29). The two additional rhodium atoms added to the Rh_6C core take up a tetragonal capping a trigonal edge bridging position. The geometry of the cluster is irregular, and Rh–Rh distances range from 2.699(3) to 2.913(3) Å (2.81 Å mean). There are two triply bridging carbonyls, six edge bridging, and eleven terminal. Rh–$C_{carbide}$ distances in the prismatic cavity average 2.13 Å.

The size of the prismatic cavity, which contains the carbon atom, is similar to that in **24**, giving a mean Rh–C distance of 2.127 Å (cf. 2.134 Å in **30**). Overall the geometry of the prism has changed slightly on adding the two rhodium atoms, the interbasal distances now being shorter (2.730 Å mean) than the basal distances (2.807 Å mean, excluding the Rh bridged edge), the opposite order to that found in **24**. On dissolving in acetonitrile **31** reverts to **24** with loss of $[Rh(CO)_2(MeCN)_2]^+$ (60).

When the oxidation of **24** is carried out under nitrogen instead of carbon monoxide, two larger rhodium clusters are produced. $Rh_{12}C_2(CO)_{25}$, **32** (74), forms slowly from solutions of the other product, $[Rh_{15}C_2(CO)_{28}]^-$, **33** (60, 75), which is unstable in solution.

The $Rh_{12}C_2$ core in **32** does not conform to any recognizable close packed array (Fig. 39) (74). The twelve rhodium atoms are arranged in three almost parallel layers of three, five, and four atoms, with an average separation between the layers of 2.2 Å. The apparently irregular structure of **32** bears a simple relationship with that of $[Co_{11}C_2(CO)_{22}]^{3-}$, **27** (see above), by the addition of a capping atom as shown in Fig. 40. It is noteworthy that these two complex but related structures are as yet the only examples known of carbidocarbonyl clusters containing a directly bonded interstitial C_2 unit. The irregular geometry of the cluster seems to be related to the presence of the two carbide atoms, which reside in the same irregular cavity as found for

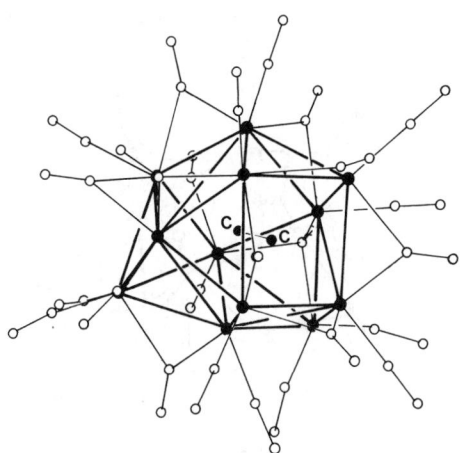

FIG. 39. $Rh_{12}C_2(CO)_{25}$, **32** (74). The irregular Rh_{12} array is related to that of $Co_{11}C_2(CO)_{22}^{3-}$, **27** (see Figs. 34 and 40) by the addition of one capping metal atom. The C–C distance between the encapsulated carbon atoms is 1.48(2) Å. The Rh–C distances (fourteen) fall into two categories: short (2.22 Å average) and long (2.58 Å average). Rh–Rh distances range from 2.67 to 2.97 Å (mean 2.79 Å). There are fourteen terminal carbonyls, ten edge bridging, and one face bridging.

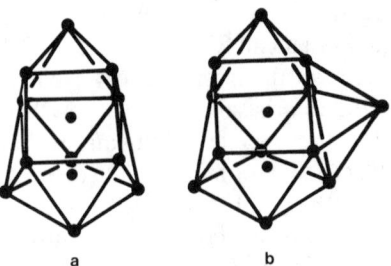

FIG. 40.   The relationship between (a) $Co_{11}C_2(CO)_{22}^{3-}$ and (b) $Rh_{12}C_2(CO)_{25}$.

27, at a C–C distance of 1.48(2) Å. The C_2 unit is, in this case, shorter than expected for a single C–C bond, but longer than that in the C_2^{2-} ion (1.19–1.35 Å).

Cluster 32 forms spontaneously from solutions of the larger cluster $[Rh_{15}C_2(CO)_{28}]^-$, 33, which is the largest carbidocarbonyl cluster yet isolated. It is prepared, as mentioned above, by the mild oxidation of $[Rh_6C(CO)_{15}]^{2-}$ under nitrogen, and was isolated as its hydroxonium salt (75). The structure of the anion is shown in Fig. 41. The geometry of the $Rh_{15}C_2$ core may be viewed as comprising two octahedra of rhodium atoms with a common vertex (the interstitial rhodium atom), plus four additional

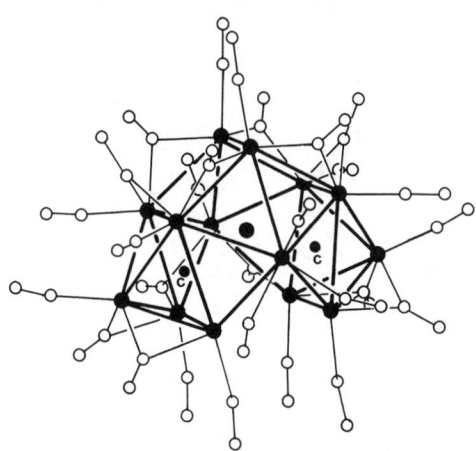

FIG. 41. $Rh_{15}C_2(CO)_{28}^-$, 33, as in its H_3O^+ salt (75). Bonds to the carbon atoms (shaded circles) and the interstitial rhodium atom are omitted. The cluster is most easily visualized as a tetracapped pentagonal bipyramid. Two octahedral cavities with a common vertex are provided by the capped square faces and the interstitial rhodium atom. The carbon atoms, which are remote from one another, occupy these octahedral cavities, with Rh–C distances ranging from 1.93 to 2.12 Å (mean 2.04 Å). Rh–Rh distances range from 2.738 to 3.332 Å (mean 2.87 Å).

metal atoms added in a plane including the common vertex but perpendicular to a vector joining the centroids of the octahedra. These four atoms form tetrahedra by capping trigonal faces of the two octahedra, and with the new trigonal faces so formed, resulting in an overall geometry closely approximating C_{2v}. The two carbon atoms reside in the octahedral cavities, with Rh–C distances averaging 2.04 Å. The two carbide carbon atoms are of course well out of bonding distance with each other. The twenty-eight carbonyl ligands are made up of fourteen terminal and fourteen bridging in a way that provides two bridges per metal atom. Cluster **33** thus provided the first example of a cluster containing both interstitial carbon and metal atoms.

V

CARBIDOCARBONYL CLUSTERS OF RHENIUM

Pyrolysis of $Et_4N[ReH_2(CO)_4]$ at 250°C in n-tetradecane yields a mixture of polynuclear products from which two carbidocarbonyl clusters of rhenium have been isolated. Extraction of the pyrolysis product mixture with tetrahydrofuran results in a residue of $(Et_4N)_3[Re_7C(CO)_{21}]$, **34** (76), and a solution from which may be isolated $(Et_4N)_2[Re_8C(CO)_{24}]$, **35**, (77), the structures of which are shown in Figs. 42 and 43. Both are based on an

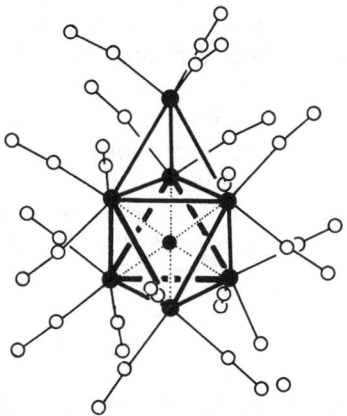

FIG. 42. $[Re_7C(CO)_{21}]^{3-}$, **34**, as in its Ph_4P^+ salt (76). The Re_7 core comprises a monocapped octahedron of rhenium atoms, with the carbide carbon in the octahedral cavity (mean Re–C = 2.13 ± 0.02 Å). The metal–metal bonds fall into several categories. Bonds from the capping atom to the capped face of the octahedron average 2.929 Å; those on the capped face, 2.955 Å; those between the capped face and the opposite uncapped face alternate longer (3.017 Å) and shorter (2.977 Å), and those in that uncapped face, average 3.080 Å. There are three terminal carbonyls on each metal atom.

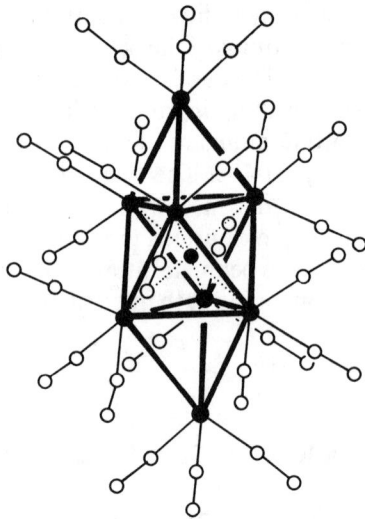

FIG. 43. [Re$_8$C(CO)$_{24}$]$^{2-}$, **35,** as in its Et$_4$N$^+$ salt (*77*). The Re$_8$ polyhedron comprises a trans-bicapped octahedron, with the carbide carbon at the center of the octahedral cavity (mean Re–C = 2.12 Å). Re–Re bond lengths average 2.993 Å within the octahedron and 2.970 Å for bonds to the capping atoms. There are three terminal carbonyls per metal atom, and the anion has overall D$_{3d}$ symmetry.

octahedral Re$_6$C core in which the carbide carbon resides at the center of the octahedral cavity. Assuming a radius of 1.50 Å for the rhenium atoms the apparent radii of the encapsulated carbon atoms in **34** and **35** are 0.62 Å and 0.63 Å, respectively, similar to the values observed for other octahedrally coordinated carbon atoms (see Table V). The additional rhenium atoms occupy trigonal faces of the core, giving mono- and bicapped octahedra. The relative disposition of the seventh and eighth rhenium atoms in **35** is attributed to the requirement for each metal atom to bear three carbonyls in accordance with electron counting and other considerations (*77*).

VI

SPECTROSCOPIC PROPERTIES OF CARBIDE CLUSTERS

The application of spectroscopic techniques familiar in organometallic chemistry to the unique features of carbidocarbonyl clusters has become more evident since the previous review of this field (*1*). Vibrational spectroscopy and ^{13}C NMR have yielded useful data relevant to the geometric and electronic environment of the carbon atom when bonded in the unusual manner characteristic of partial or total encapsulation in a transition metal cluster. In this section emphasis will be given to those data which relate to the properties of the lone carbon atom. For example, the application of vibrational spectroscopy in the C–O stretching region of the infrared spectrum of a carbidocarbonyl cluster is not considered appropriate to this discussion, whereas the similar application to the vibrational modes involving motion of the carbido carbon in its cavity will be discussed. Additional data can be obtained from the original literature (see Table V).

A. Vibrational Spectroscopy

In their original paper (2) on the structure of $Fe_5C(CO)_{15}$, Dahl and co-workers assigned two bands in the infrared spectrum of hydrocarbon solutions of the cluster, at 790 and 770 cm^{-1}, to ν_{FeC} modes. This assignment has been confirmed by a recent study of the infrared spectra of the series $M_5C(CO)_{15}$, (M = Fe, Ru, Os) (78). The room temperature spectra of the compounds (Table II) in the solid state are quite similar to each other, comprising three bands assigned as the a_1 and e modes (split in the solid state) expected for the C_{4v} symmetry of the isostructural clusters. At low temperature the ruthenium and osmium clusters exhibit five absorptions associated with M–C stretches, whereas the iron cluster retains its room temperature spectrum. This is ascribed to the presence of two types of cluster molecule in the crystal lattices of the ruthenium and osmium clusters which are isostructural with, but not isomorphous with, the iron analog in which all the molecules are identical.

This technique has also been applied to the products of hydrohalogenation of $Ru_5C(CO)_{15}$. The three ν_{RuC} absorptions of the parent cluster at 757, 738, and 730 cm^{-1} are reduced to two for $HRu_5C(CO)_{15}X$ at 823 and 685 cm^{-1} (X = Cl) and 823 and 687 cm^{-1} (X = Br). These observations were interpreted on the basis of a rearrangement of the square pyramidal $Ru_5C(CO)_{15}$ to a distorted bipyramidal geometry similar to that found for $[Os_5C(CO)_{15}I]^-$ (Fig. 12) (32).

The vibrational frequencies of the encapsulated carbon atom in $[Os_{10}C(CO)_{24}]^{2-}$, **21** (Fig. 26), and its protonated derivative $H_2Os_{10}C(CO)_{24}$ have been identified by Oxton et al. (57) and the assignments confirmed by isotopic enrichment. The carbon atom in the tetrahedral Os_{10} dianion resides in an octahedral cavity, and at room temperature a band at 753 cm^{-1} is observed and assigned to ν_{Os-C}. On protonation three absorptions of approximately equal intensity are observed in this region and ^{13}C enrichment of the central carbon atom identified these absorptions, at 772.8,

TABLE II

CARBON ATOM VIBRATIONAL FREQUENCIES IN CARBIDOCARBONYLS

Cluster	ν_{M_n-c} (cm^{-1})	Ref.
$Fe_5C(CO)_{15}$	770, 790 (cyclohexane)	2
	766, 775, 805 (CsBr disc)	78
$Ru_5C(CO)_{15}$	730, 738, 757 (CsBr disc)	78
$Os_5C(CO)_{15}$	757, 769, 793 (CsBr disc)	78
$HRu_5C(CO)_{15}Cl$	685, 823	32
$HRu_5C(CO)_{15}Br$	687, 823	32
$[Os_{10}C(CO)_{24}]^{2-}$	753 (CsI disc)	57
$H_2Os_{10}C(CO)_{24}$	735, 760, 773 (CsI disc)	57
$[Co_6C(CO)_{15}]^{2-}$	719, 772 (Nujol mull)	79
$Co_6C(CO)_{12}S_2$	548, 819 (CsI disc)	69, 71
$[Rh_6C(CO)_{15}]^{2-}$	653, 689 (Nujol mull)	79

760.3, and 735.4 cm^{-1}, as associated with the motion of the carbido carbon atom. The increase in the number of infrared active modes in $H_2Os_{10}C(CO)_{24}$ over the dianion is attributable to a lowering of the symmetry of the cluster on protonation, and since the carbide carbon is in the octahedral cavity of the cluster (56), the osmium atoms defining this cavity were suggested as possible sites of protonation.

The metal–carbide vibrational frequencies for the isostructural clusters $[M_6C(CO)_{15}]^{2-}$, (M = Co, Rh) have been assigned with the aid of ^{13}C labeling by Creighton et al. (79). The a_2'' and e' modes involving carbide motion in the trigonal prismatic M_6 cavity give rise to absorptions at 772 and 719 cm^{-1} (M = Co) and 689 and 653 cm^{-1} (M = Rh), respectively.

Bör and Stanghellini (69, 71) have reported the infrared spectrum of $Co_6{}^{13}C(CO)_{12}S_2$, identifying bands at 790 and 535.5 cm^{-1} (819 and 548 cm^{-1} for ^{12}C isotopomer) as the a_2'' and e' vibrations of the carbon atom in the trigonal prismatic cluster.

The values of $\nu_{MC_{carbide}}$ so far reported are contained in Table II.

B. ^{13}C NMR Spectroscopy

^{13}C NMR was first utilized in the location of an interstitial carbon atom in a carbidocarbonyl cluster by Heaton and co-workers, in their identification of the 264.7 ppm resonance of $[Rh_6C(CO)_{15}]^{2-}$, **24** (80), as due to the encapsulated carbon atom. The carbide signal, enhanced by selective enrichment using ^{13}CCl$_4$ in the synthesis of the dianion, showed the central five lines of the septet structure ($J_{Rh-C} = 13.7 \pm 2$ Hz) expected for the trigonal prismatic geometry of the cluster (Fig. 29). Reaction of **24** with acid at $-30°$C results in the protonation of one triangular face of the trigonal prismatic dianion as evidenced by the splitting of the carbide resonance (shifted downfield to 291.2 ppm) into a quartet of quartets (81).

For $[Co_6C(CO)_{15}]^{2-}$, the carbide resonance is shifted downfield from that found in its isostructural rhodium analog to 330.5 ppm (64).

The ^{13}C NMR spectrum of $[Rh_6C(CO)_{13}]^{2-}$ shows coupling of the carbide carbon to the six rhodium atoms and reflects the distortion of the Rh_6 core from octahedral symmetry (see Fig. 31). The carbide (90% ^{13}C) couples to two sets of rhodium atoms ($2Rh_A$, $4Rh_B$) producing an unresolved multiplet at 335.8 ppm ($-80°C$). At this temperature, coupling to the peripheral carbonyls (20% ^{13}C) is also observed (6.5 and 11 Hz), and reflects both the geometric and fluxional characteristics of the carbonyl groups (82).

The carbide resonances for several Ru_6C clusters have been observed. For $[Ru_6C(CO)_{16}]^{2-}$, 18, a shift of 288 ppm was reported (41). In light of the consistently greater shifts found for the iron carbide clusters (see below), 18 has recently been reexamined in the author's laboratory. The original value for the carbide resonance was found to be in error, and a correct value of 458.9 ppm observed (see Table III). This shift remains virtually unchanged

TABLE III

^{13}C CHEMICAL SHIFTS OF CARBIDE CARBON ATOMS

Cluster	$\delta_{carbide}$ (ppm)	Ref.
Iron		
$Fe_4C(CO)_{13}$ [a]	468.9	11
$[HFe_4C(CO)_{12}]^-$ [a]	464.2	14
$[Fe_4C(CO)_{12}]^{2-}$ [a]	478.0	14
$Fe_4C(CO)_{10}(PMe_3)_3$ [a]	471.5	14
$Fe_5C(CO)_{15}$ [a]	486.0	14
$[Fe_6C(CO)_{16}]^{2-}$ [b]	484.6	14
Ruthenium		
$[Ru_6C(CO)_{16}]^{2-}$ [b]	458.9	(d)
$(MeCN)_2Cu_2Ru_6C(CO)_{16}$ [c]	459.0	50
Cobalt		
$[Co_6C(CO)_{15}]^{2-}$ [c]	330.5	64
Rhodium		
$[Rh_6C(CO)_{15}]^{2-}$ [b]	266.7	78
$[HRh_6C(CO)_{15}]^-$ [a]	291.2	79
$[Rh_6C(CO)_{13}]^{2-}$ [e]	335.8	80
Rhenium		
$[Rh_8C(CO)_{24}]^{3-}$ [c]	431.3	77

[a] CD_2Cl_2 solution.
[b] $(CD_3)_2CO$ solution.
[c] THF-d_8 solution.
[d] Revised from 288 ppm reported previously in Ref. (41).
[e] $-80°C$.

on the addition of two capping vertices ($CH_3CN \cdot Cu$) to **18**, producing the Cu_2Ru_6C cluster **20** (Fig. 23) in which the encapsulated carbon resonance is found at 458 ppm (*50*). The position of the corresponding resonance for $(MeCN)_2Cu_2Rh_6C(CO)_{15}$, the rhodium analog of **20**, has not been reported.

Several iron carbidocarbonyl clusters have been examined by ^{13}C NMR (*14*), and all show large shifts to low field (see Table III). There seems to be no correlation between charge on the molecule and $\delta_{C_{carbide}}$. In the series $Fe_4C(CO)_{13}$, $[HFe_4C(CO)_{12}]^-$ and $[Fe_4C(CO)_{12}]^{2-}$, which are very similar in geometry (see Table I), the carbide resonances are found at 469.8, 464.2, and 478.0 ppm. The cause of these large shifts is not yet known, but some light will undoubtedly be shed on this phenomenon by molecular orbital calculations, magnetic suceptibility and temperature-dependent solid-state NMR studies.

VII

GENERAL OBSERVATIONS

On the basis of the information described above, some general observations may be made about carbidocarbonyl clusters. The source of the carbon atom, its structural requirements, and its chemical properties are all of interest, as is the effect that it has on the chemistry of the cluster as a whole.

A. *The Source of the Cluster Bound Carbon Atom*

Two means have been identified by which a carbon atom may become incorporated into a cluster. In the case of iron and of ruthenium carbidocarbonyls, the clusters are synthesized by reactions of the metal carbonyls in quite high boiling solvents (diglyme, di-*n*-butyl ether, arenes). The use of ^{13}CO-enriched $Fe(CO)_5$ in the synthesis of $[Fe_6C(CO)_{16}]^{2-}$ results in the incorporation of an encapsulated ^{13}C atom into the cluster (measured by ^{13}C NMR), demonstrating that carbon monoxide was the source of the interstitial atom in these clusters (*8*). It seems probable that a reaction of CO to carbon and CO_2 occurs (a homogeneous analog to the Boudart reaction), e.g.,

$$6Fe(CO)_5 \rightarrow [Fe_6C(CO)_{16}]^{2-} + 12CO + CO_2$$

A similar experiment using ^{13}C-labeled $Ru_3(CO)_{12}$ in the synthesis of $[Ru_6C(CO)_{16}]^{2-}$ gave identical results (*41*). In the synthesis of $Ru_6C(CO)_{17}$ by pyrolysis of $Ru_3(CO)_{12}$, CO_2 is detected during the reaction (49, 83), suggesting again that the disproportionation of carbon monoxide is the source of the carbon atom, and it seems probable that this is true for all carbidocarbonyl clusters synthesized under similar conditions.

In the case of rhodium, however, it was demonstrated early that in the synthesis of $[Rh_6C(CO)_{15}]^{2-}$ the encapsulated carbon atom originated as chloroform, which had reacted with the rhodium carbonyl anion $[Rh_7(CO)_{16}]^{3-}$ (59). In the cobalt analog, $[Co_6C(CO)_{15}]^{2-}$, the carbon atom is derived indirectly from carbon tetrachloride [via $Co_3(CO)_9CCl$] (60). Both these syntheses are performed under mild conditions, and there are apparently no examples of carbidocarbonyl clusters of cobalt or rhodium prepared directly from the metal carbonyls under pyrolysis conditions.

B. Structural Aspects of Cluster Bound Carbon Atoms

As a structural unit, the carbon atom in carbidocarbonyl clusters seems sterically undemanding. As with the interstitial metal carbides, the carbon atom is found predominantly in trigonal prismatic and antiprismatic (octahedral) cavities (84). A prismatic array of six metal atoms provides a larger interstitial cavity than does an octahedral cluster with the same number of atoms, and in the crystallographically characterized clusters with the carbon atom in a trigonal prismatic cavity the effective radius of the central carbon atom is greater than that found in octahedral clusters (see Table IV). It is significant that in $[Co_8C(CO)_{18}]^{2-}$ the idealized tetragonal antiprism of eight cobalt atoms (mean Co–Co 2.52 Å), which would provide a large interstitial cavity, is distorted to allow four shortened and presumably preferred Co–C distances of 1.99 Å (av), giving a radius of 0.73 Å for the carbide carbon (64). If we take this value as the optimum, then it is clear that in the trigonal prismatic clusters in Table IV, the carbon atom occupies a cavity close in size to its preferred value.

In clusters with octahedrally encapsulated carbon atoms, the observed values for the radius of the interstitial atom are much smaller. An illuminating comparison is provided by contrasting the optimum carbon radius in $[Co_8C(CO)_{18}]^{2-}$ (0.74 Å) with that for $Co_6C(CO)_{14}^{2-}$ (0.56 Å) (63), which has an octahedral core. Indeed all the octahedral carbido clusters have similarly compressed interstitial cavities (Table IV) and as has been pointed out previously, this fact militates against invoking minimum steric requirements of the carbon atom to explain distortions in octahedral carbido clusters.

It is also interesting in this respect to compare the geometry of the Fe_4C family with that of the hexanuclear and pentanuclear homologs

TABLE IV

EFFECTIVE RADII OF CARBIDE CARBON ATOMS IN CARBIDOCARBONYL CLUSTERS

Cluster	Mean M–M[a] (Å)	Mean M–C (Å)	$r_{carbide}$	Ref.
	1. Octahedrally Coordinated Carbides			
$[Fe_6C(CO)_{16}]^{2-}$	2.68	1.89	0.55	7
$[Ru_6C(CO)_{16}]^{2-}$	2.89	2.05	0.61	40, 41
$Ru_6C(CO)_{17}$	2.90	2.05	0.60	38
$Ru_6C(CO)_{14}$(mesitylene)	2.88	2.04	0.60	39
$Ru_6C(CO)_{14}$(bitropyl)	2.91	2.06	0.61	47
$Ru_6C(CO)_{15}$(hexadiene)	2.90	2.06	0.61	48
$Ru_6C(CO)_{15}(PhC{\equiv}CH)$	2.89	2.04	0.60	45, 46
$Ru_6C(CO)_{16}(PPh_2Et)$	2.91	2.05	0.61	44
$(CH_3CN)_2Cu_2Ru_6C(CO)_{16}$	2.90	2.05	0.60	50
$[TlRu_{12}C_2(CO)_{32}]^-$	2.90	2.05	0.60	52
$[Os_{10}C(CO)_{24}]^{2-}$	2.88	2.04	0.60	56
$[Os_{10}C(CO)_{24}I]^-$	2.90	2.05	0.61	58
$Os_{10}C(CO)_{24}I_2$	2.91	2.07	0.61	58
$[Co_6C(CO)_{14}]^-$	2.66	1.89	0.56	64
$[Rh_6C(CO)_{13}]^{2-}$	2.91	2.05	0.60	62
$[Rh_{15}C_2(CO)_{28}]^-$	2.87	2.04	0.61	75
	2. Four- and Five-Coordinate Carbides Based on Octahedra			
$Fe_5C(CO)_{15}$	2.65	1.90	0.62	2
$Fe_4C(CO)_{13}$	2.62	1.90	0.59	11
$[Fe_4C(CO)_{12}]^{2-}$	2.62	1.88	0.57	16
$[HFe_4C(CO)_{12}]^-$	2.62	1.89	0.58	17
$Ru_5C(CO)_{15}$	2.84	2.04	0.62	29
$Ru_5C(CO)_{15}(PPh_3)$	2.85	2.06	0.64	29
$Os_5C(CO)_{15}$	2.87	2.06	0.63	31
	3. Trigonal Prismatic Carbides			
$Co_6C(CO)_{12}S_2$	2.51	1.94	0.69	71
$[Co_{13}C_2(CO)_{24}]^{4-}$	2.58	1.98	0.69	68
$[Rh_6C(CO)_{15}]^{2-}$	2.79	2.13	0.74	59
$Rh_8C(CO)_{19}$	2.79	2.13	0.73	73
$(CH_3CN)_2Cu_2Rh_6C(CO)_{15}$	2.78	2.13	0.74	61

[a] Mean M–M bond distance between metal atoms within bonding distance of carbide.

$[Fe_6C(CO)_{16}]^{2-}$, and $Fe_5C(CO)_{15}$. If it were the case that the octahedral cavity in the dianion was too small for the adequate accommodation of the carbon atom, it would be expected that in the more open Fe_5C and Fe_4C clusters some relaxation would occur allowing the carbon atom to adopt a less crowded position. In fact, the effective radius of the carbon atom in $Fe_5C(CO)_{15}$ is 0.57 Å (2), an increase of only 0.02 Å from that in $[Fe_6C(CO)_{16}]^{2-}$ (7), and in the even more open structures of $Fe_4C(CO)_{13}$ (11), $[HFe_4C(CO)_{12}]^-$ (17) and $[Fe_4C(CO)_{12}]^{2-}$ (16), the average value is still only 0.58 Å. (These values are derived using a mean of the five Fe–Fe distances for Fe_4C molecules.)

So it would seem that the lone carbon atom in these molecules may adopt a number of geometries with a range of spatial requirements. It remains to be seen whether or not there is any correlation between the effective radius of the carbon atom and its chemical or spectroscopic properties.

There are as yet no thermochemical data to support the notion that the carbido carbon atom imparts additional stability to the cluster that surrounds it, but on the basis of observations made in the course of synthetic studies with these compounds, it is clear that the presence of the interstitial atom provides additional robustness to the molecule (86). Similar assertions have been made for clusters with other interstitial atoms (87).

The chemical consequences of the carbide's presence are more easily recognizable. The fact that the carbon atom donates four electrons to the cluster without taking up any peripheral space in the coordination sphere of the cluster allows the accommodation of fairly large ligands on the cluster. Examples of this effect are found in $Ru_6C(CO)_{15}$(hexa-2,4-diene) and $Ru_6C(CO)_{14}$(bitropyl), in which six and seven carbon organic ligands are coordinated to two and three adjacent metal atoms. Furthermore, the presence of the carbide in a sterically innocuous position allows the assembly of a coordinatively saturated 86-electron Fe_6 cluster, $[Fe_6C(CO)_{16}]^{2-}$, whereas the analogous binary carbonyl $[Fe_6(CO)_{18}]^{2-}$ is not known. This disparity is explained by the spatial requirements of the carbonyl coordination sphere on the relatively small octahedral Fe_6 core (88). A similar observation is provided by the difference between the structures of the two bimetallic clusters $(MeCN)_2Cu_2Ru_6C(CO)_{16}$ and $(C_6H_5CH_3)_2Cu_2Ru_6(CO)_{18}$ (50). In both cases, the ruthenium atoms form an octahedral core. In the former carbide containing cluster the copper atoms occupy adjacent positions on the Ru_6 core, and are directly bonded to one another, whereas in the noncarbide analog, the two copper atoms are situated on opposite faces of the Ru_6 octahedron. Again this may reflect the spatial constraints imposed by the presence of eighteen carbonyls in the latter case, in comparison to the relatively flexible carbonyl configuration allowed in the former, where the local perturbation caused by the presence of two adjacent heterometal atoms is more easily accommodated. Another example is provided by the geometry of the Re_8C core of $[Re_8C(CO)_{24}]^{3-}$ (77).

C. Chemical Reactivity of the Carbon Atom

This third aspect of the chemistry of carbidocarbonyl clusters is a relatively recently observed phenomenon. Carbon atom chemistry is found predominantly in the Fe_4C group of clusters, although increased activity in this area will undoubtedly uncover analogous examples with other metals.

The reactions at the carbon atom so far observed involved the formation of either C–H or C–C bonds. No mechanistic data, either kinetic or spectroscopic, have been established for either reaction, but for the latter case, some reasonable speculation is possible. In reactions of $Fe_4C(CO)_{13}$ with alcohols, the products are of the type $[Fe_4(CO)_{12}C \cdot CO_2R]^-$, and this suggests the intermediacy of a ketenylidene species $Fe_4(CO)_{12} (C{=}C{=}O)$. As mentioned in Section II there is no direct evidence for this, but analogies with the reactions of $H_2Os_3(CO)_{10}CCO$ and $[Co_3(CO)_9CCO]^+$ with alcohols, to give $H_3Os_3(CO)_{12}C \cdot CO_2R$ (85) and $Co_3(CO)_9CCO_2R$ (9), respectively, provide supporting precedents. These reactions may be interpreted in terms of a positively charged carbon atom, by analogy with the similar reactions of carbonium ions with carbon monoxide and alcohols to give esters. Consistent with this proposal, we have observed that increasing the electron density on the Fe_4C cluster by substituting trimethylphosphine for carbon monoxide prevents the formation of the carbomethoxymethylidyne cluster [Eq. (23)] (14).

$$Fe_4C(CO)_{13} \xrightarrow[25°C]{MeOH} [Fe_4(CO)_{12}C \cdot CO_2Me]^-$$

$$\Big\downarrow PMe_3 \tag{23}$$

$$Fe_4C(CO)_9(PMe_3)_4 \xrightarrow{MeOH} \text{no reaction}$$

The reversible formation of a C–H bond by protonation of $[HFe_4C(CO)_{12}]^-$ to $HFe_4(CH)(CO)_{12}$ (15, 17) is another instructive example of reactivity at the carbon atom. The exposed carbon atom in the anion proves to be a less basic site than is the methanol molecule, since the methylidyne cluster spontaneously deprotonates in methanol solution [Eq. (24)].

$$HFe_4(CO)_{12}CH + MeOH \rightarrow [HFe_4C(CO)_{12}]^- + MeOH_2^+ \tag{24}$$

The reactivity of the carbon atom raises questions about its chemical nature: should it be regarded as cationic, neutral (carbenoid perhaps) or anionic, as the name carbidocarbonyl has long implied. The available ^{13}C NMR data, in which the carbide resonances are significantly to low field, may be interpreted in terms of a deshielded positively charged carbon atom, and such a suggestion has been made (80). However, the cause of the shifts observed for cluster bound carbon atoms is not clear and may reflect the predominance of paramagnetic contributions to the chemical shift rather than simply deshielding at a cationic center.

More illuminating evidence on the nature of the carbon atom in this

TABLE V

SYNTHESIS STRUCTURE AND SPECTROSCOPIC PROPERTIES OF CARBIDOCARBONYL CLUSTERS

Cluster	Synthesis Ref.	Structure Ref.	IR	¹H NMR	¹³C NMR
Iron					
$Fe_4C(CO)_{13}$	*11, 16*	*11*	*11*		*11*
$[Fe_4C(CO)_{12}]^{2-}$	*14–16*	*16*	*15*		*15*
$[HFe_4C(CO)_{12}]^-$	*16, 17*	*17*	*15, 17*	*15*	
$Fe_5C(CO)_{15}$	*2–4*	*2*	*2, 3*		*4*
$[Fe_5C(CO)_{14}H]^-$	*15*				
$[Fe_5C(CO)_{14}]^{2-}$	*4, 5*		*5*		
$[Fe_5C(CO)_{12}Br_2]^{2-}$		*25*			
$Fe_5C(CO)_{14}(PPh_3)$	*5*		*5*	*5*	
$Fe_5C(CO)_{14}(PMe_2Ph)$	*5*		*5*	*5*	
$Fe_5C(CO)_{13}(PMe_2Ph)_2$	*5*		*5*	*5*	
$Fe_5C(CO)_{13}(P(OPh)_3)_2$	*5*		*5*	*5*	
$Fe_5C(CO)_{14}(P(OC_3H_7)_3)$	*5*		*5*	*5*	
$Fe_5C(CO)_{13}(P(OC_3H_7)_3)_2$	*5*		*5*	*5*	
$[Fe_6C(CO)_{16}]^{2-}$	*6*	*7*	*6*		
$[CrFe_5C(CO)_{17}]^{2-}$	*4*				
$[MoFe_5C(CO)_{17}]^{2-}$	*4*	*4*			
$[WFe_5C(CO)_{17}]^{2-}$	*4*				
$[RhFe_5C(CO)_{16}]^-$	*4*				*4*
$[RhFe_5C(CO)_{14}(COD)]^-$	*4*				
$[IrFe_5C(CO)_{14}(COD)]^-$	*4*				
$[NiFe_5C(CO)_{13}(COD)]^{2-}$	*4*				
$[NiFe_5C(CO)_{15}]^{2-}$	*4*				
$[PdFe_5C(CO)_{14}(C_3H_5)]^-$	*4*				
$[CuFe_5C(CO)_{14}(MeCN)]^-$	*4*				
$CrFe_4C(CO)_{16}$	*4*				
$MoFe_4C(CO)_{16}$	*4*				*4*
$[RhFe_4C(CO)_{14}]^-$	*4*	*4*			
Ruthenium					
$Ru_5C(CO)_{15}$	*29, 30*	*29*	*30*		
$Ru_5C(CO)_{14}(PPh_3)$	*29*	*29*			
$[Ru_5C(CO)_{15}I]^-$	*29*				
$Ru_5C(CO)_{15}I_2$	*29*				
$H_2Ru_5C(CO)_{15}$	*29*				
$HRu_5C(CO)_{15}Cl$	*29, 32*		*32*		
$HRu_5C(CO)_{15}Br$	*29, 32*		*32*		
$Ru_6C(CO)_{17}$	*36, 40, 42, 55*	*38*	*37*		
$Ru_6C(CO)_{14}(C_6H_6)$	*55*		*55*		
$Ru_6C(CO)_{14}(CH_3C_6H_5)$	*37*		*37*	*37*	
$Ru_6C(CO)_{14}((CH_3)_2C_6H_4)$	*37*		*37*	*37*	
$Ru_6C(CO)_{14}((CH_3)_3C_6H_3)$	*37*	*39*	*37*	*37*	
$Ru_6C(CO)_{16}(PPh_3)$	*44*		*44*		
$Ru_6C(CO)_{16}(PPh_2Et)$	*44*	*44*	*44*		
$Ru_6C(CO)_{16}(P(OCH_3)_3)$	*44*		*44*	*44*	

(Continued)

TABLE V (*Continued*)
SYNTHESIS STRUCTURE AND SPECTROSCOPIC PROPERTIES OF CARBIDOCARBONYL CLUSTERS

Cluster	Synthesis Ref.	Structure Ref.	IR	¹H NMR	¹³C NMR
$Ru_6C(CO)_{15}(PPh_3)_2$	44	44			
$Ru_6C(CO)_{15}(PPh_2Et)_2$	44	44			
$Ru_6C(CO)_{15}(P(OCH_3)_3)_2$	44	44		44	
$Ru_6C(CO)_{14}(P(OCH_3)_3)_3$	44	44		44	
$Ru_6C(CO)_{13}(P(OCH_3)_3)_4$	44	44		44	
$Ru_6C(CO)_{15}(PhC{\equiv}CH)$	45	45, 46	45		
$Ru_6C(CO)_{15}(hexa\text{-}2,4\text{-}diene)$	48	48	44	44	
$Ru_6C(CO)_{14}(bitropyl)$	41	47			
$HRu_6C(CO)_{15}(SEt)_3$	49	49			
$[Ru_6C(CO)_{16}]^{2-}$	40, 41	40, 41	40		40
$H_2Ru_6C(CO)_{16}$	40, 41			41	
$[Ru_6C(CO)_{16}CO_2CH_3]^-$	40				
$(CH_3CN)_2Ru_6C(CO)_{16}$	50	50	50	50	50
$[TlRu_{12}C_2(CO)_{32}]^-$	52	52	52		
Osmium					
$Os_5C(CO)_{15}$	30, 55	31	31		
$[Os_5C(CO)_{15}I]^-$	31	31	31		
$HOs_5C(CO)_{14}(OP(OCH_3)_2)$	34	33	34	34	
$HOs_5C(CO)_{13}(OP(OCH_3)_2)(P(OCH_3)_3)$	34	34	34	34	
$HOs_5C(CO)_{13}(OP(OCH_3)OP(P(OCH_3)_2)$	34	35	34	34	
$Os_6C(CO)_{16}(CH_3C{\equiv}CCH_3)$	53	53			
$H_2Os_7C(CO)_{19}$	54		54		
$Os_8C(CO)_{21}$	55		55		
$[Os_{10}C(CO)_{24}]^{2-}$	56	56	56		
$H_2Os_{10}C(CO)_{24}$	57		57		
$[Os_{10}C(CO)_{24}I]^-$	58	58			
$Os_{10}C(CO)_{24}I_2$	58	58			
Cobalt					
$[Co_6C(CO)_{15}]^{2-}$	60	60			60
$[Co_6C(CO)_{14}]^-$	63, 64	63, 64	64		
$[Co_8C(CO)_{18}]^{2-}$	64	64, 66	64		
$Co_{11}C_2(CO)_{22}$	67	67	67		
$[Co_{13}C_2(CO)_{24}]^{4-}$	68	68	68		
$Co_6C(CO)_{12}S_2$	71	71	71		
$Co_6C_2(CO)_{14}S$		72			
Rhodium					
$[Rh_6C(CO)_{15}]^{2-}$	59	59			76
$[Rh_6C(CO)_{13}]^{2-}$	62	62	62		78
$[HRh_6C(CO)_{15}]^-$	77		77	77	77
$(MeCN)_2Cu_2Rh_6C(CO)_{15}$	61	61	61		
$Rh_8C(CO)_{19}$	60	73	60		
$Rh_{12}C_2(CO)_{25}$	74	74	74		
$[Rh_{15}C_2(CO)_{28}]^-$	60, 75	75	74		

unique environment must await further spectroscopic and theoretical investigations which are lacking for these molecules. Table V is a guide to the literature on carbidocarbonyl clusters.

VIII

ADDENDUM

Since the completion of this review (mid-1982), the chemistry of carbidocarbonyl clusters has continued to expand rapidly. The task of the reviewer is made even more difficult as fascinating results continue to appear. In resisting the temptation to make a comprehensive update of the field, it would be remiss of me not to direct the reader's attention to the continued investigations of Lewis, Johnson, and co-workers in the chemistry of ruthenium and osmium carbidocarbonyls (89), the report by Longoni and co-workers (90) of the syntheses of the first nickel carbide clusters and some mixed nickel–cobalt carbides, the syntheses by Shapley of a new ruthenium dicarbide cluster $[Ru_{10}C_2(CO)_{24}]^-$ (91) and of $Os_6C(CO)_{17}$ (92), and the work of Shriver which implies the existence of a very reactive tri-iron carbide cluster (93).

ACKNOWLEDGMENTS

The author thanks Exxon Research and Engineering Company for continuing support for his research in this area. Thanks are also due to Professor V. G. Albano, Drs. B. T. Heaton and B. F. G. Johnson, Professors J. Lewis, G. Longoni, and E. L. Muetterties, Dr. J. N. Nicholls, and Professors D. F. Shriver and P. L. Stanghellini for communicating results prior to publication.

REFERENCES

1. M. Tachikawa and E. L. Muetterties, *Prog. Inorg. Chem.* **28**, 203 (1981).
2. E. H. Braye, L. F. Dahl, W. Hubel, and D. L. Wampler, *J. Am. Chem. Soc.* **84**, 4633 (1962).
3. R. P. Stewart, U. Anders, and W. A. G. Graham, *J. Organomet. Chem.* **32**, C49 (1971).
4. M. Tachikawa, A. C. Sievert, E. L. Muetterties, M. R. Thompson, C. S. Day, and V. W. Day, *J. Am. Chem. Soc.* **102**, 1725 (1980).
5. C. G. Cooke and M. J. Mays, *J. Organomet. Chem.* **88**, 231 (1975).
6. M. R. Churchill, J. Wormald, J. Knight, and M. J. Mays, *J. Am. Chem. Soc.* **93**, 3073 (1971).
7. M. R. Churchill and J. Wormald, *J. Chem. Soc., Dalton Trans.* p. 2410 (1974).
8. J. S. Bradley, G. B. Ansell, and E. W. Hill, *J. Am. Chem. Soc.* **101**, 7417 (1979).
9. D. Seyferth, J. E. Hallgren, and C. S. Eschbach, *J. Am. Chem. Soc.* **96**, 1730 (1974).
10. See, for example, P. R. Wentreck, B. J. Wood, and H. Wise, *J. Catal.* **43**, 363 (1976); M. Araki and V. Ponec, *ibid.* **44**, 439 (1976); V. Ponec, *Catal. Rev. —Sci. Eng.* **18**, 151

(1978); J. A. Rabo, A. P. Risch, and M. L. Poutsma, *J. Catal.* **53**, 295 (1978); J. W. A. Sachtler, J. M. Kool, and V. Ponec, *ibid.* **56**, 284 (1979); P. Biloen, J. N. Helle, and W. M. H. Sachtler, *ibid.* **48**, 95 (1979); H. H. Nijs and P. A. Jacobs, *ibid.* **66**, 401 (1980).

11. J. S. Bradley, G. B. Ansell, M. E. Leonowicz, and E. W. Hill, *J. Am. Chem. Soc.* **103**, 4968 (1981).
12. J. S. Bradley and E. W. Hill, unpublished observations.
13. G. B. Ansell, M. A. Modrick, and J. S. Bradley, *Acta Crystallogr.* (submitted for publication).
14. J. S. Bradley, *Philos. Trans. R. Soc. London* **A308**, 103 (1982).
15. M. Tachikawa and E. L. Muetterties, *J. Am. Chem. Soc.* **102**, 4541 (1980).
16. J. H. Davis, M. A. Beno, J. M. Williams, J. Zimmie, M. Tachikawa, and E. L. Muetterties, *Proc. Natl. Acad. Sci. U.S.A.* **78**, 668 (1981).
17. E. M. Holt, K. H. Whitmire, and D. F. Shriver, *J. Organomet. Chem.* **213**, 125 (1981).
18. M. A. Beno, J. M. Williams, M. Tachikawa, and E. L. Muetterties, *J. Am. Chem. Soc.* **102**, 4542 (1980).
19. M. A. Beno, J. M. Williams, M. Tachikawa, and E. L. Muetterties, *J. Am. Chem. Soc.* **103**, 1485 (1981).
20. K. Wade, *Inorg. Nucl. Chem. Lett.* **8**, 559 (1972).
21. K. Whitmire and D. F. Shriver, *J. Am. Chem. Soc.* **102**, 1456 (1980).
22. M. Manassero, M. Sansoni, and G. Longoni, *J. Chem. Soc., Chem. Commun.* p. 919 (1976).
23. K. H. Whitmire and D. F. Shriver, *J. Am. Chem. Soc.* **103**, 6754 (1981).
24. J. S. Bradley, E. W. Hill, G. B. Ansell, and M. A. Modrick, *Organometallics* **1**, 1634 (1982).
25. The isolation of $[Fe_5C(CO)_{12}Br_2]^{2-}$ was fortuitous, and attempts at rational synthesis have not been successful. See ref. (*24*).
26. E. W. Hill, unpublished observations.
27. D. Mansuy, J.-P. Lecomte, J.-P. Chottard, and J.-F. Bartoli, *Inorg. Chem.* **20**, 3119 (1981).
27a. V. L. Goedken, M. R. Deakin, and L. A. Bottomley, *J. Chem. Soc., Chem. Commun.* p. 607 (1982).
28. K. Tatsumi, R. Hoffman, and M.-H. Whangbo *J. Chem. Soc., Chem. Commun.* p. 509 (1980).
29. D. H. Farrar, P. F. Jackson, B. F. G. Johnson, J. Lewis, J. N. Nicholls, and M. McPartlin, *J. Chem. Soc., Chem. Commun.* p. 415 (1981).
30. C. R. Eady, B. F. G. Johnson, J. Lewis, and T. Matheson, *J. Organomet. Chem.* **57**, C82 (1973).
31. P. F. Jackson, B. F. G. Johnson, J. Lewis, J. N. Nicholls, M. McPartlin, and W. J. H. Nelson, *J. Chem. Soc., Chem. Commun.* p. 564 (1980).
32. I. A. Oxton, D. B. Powell, D. H. Farrar, B. F. G. Johnson, J. Lewis, and J. N. Nicholls, *Inorg. Chem.* **20**, 4302 (1981).
33. J. M. Fernandez, B. F. G. Johnson, J. Lewis, P. R. Raithby, and G. M. Sheldrick, *Acta Crystalogr., Sect. B* **B34**, 1994 (1978).
34. J. M. Fernandez, B. F. G. Johnson, J. Lewis, and P. R. Raithby, *J. Chem. Soc., Dalton Trans.* p. 2250 (1981).
35. A. G. Orpen and G. M. Sheldrick, *Acta Crystallogr.,* **B34,** 1992 (1978).
36. B. F. G. Johnson, R. D. Johnston, and J. Lewis, *Chem. Commun.* p. 1057 (1967).
37. B. F. G. Johnson, R. D. Johnston, and J. Lewis, *J. Chem. Soc. A* p. 2865 (1968).
38. A. Sirigu, M. Bianchi, and E. Benedetti, *Chem. Commun.* p. 596 (1969).
39. R. Mason and W. R. Robinson, *Chem. Commun.* p. 468 (1968).
40. B. F. G. Johnson, J. Lewis, S. W. Sankey, K. Won, M. McPartlin, and W. J. H. Nelson, *J. Organomet. Chem.* **191,** C3 (1980).

41. J. S. Bradley, G. B. Ansell, and E. W. Hill, *J. Organomet. Chem.* **184,** C33 (1980).
42. B. F. G. Johnson, J. Lewis, and I. G. Williams, *J. Chem. Soc. A* p. 901 (1970).
43. B. F. G. Johnson, J. Lewis, P. Raithby, G. J. Will, M. McPartlin, and W. J. H. Nelson, *J. Organomet. Chem.* **185,** C17 (1980).
44. S. C. Brown, J. Evans, and M. Bebster, *J. Chem. Soc., Dalton Trans.* p. 2263 (1981).
45. J. S. Bradley, G. Yang, and D. VanEngen, *Organometallics,* (submitted for publication).
46. D. VanEngen, J. S. Bradley, and G. Yang, *Am. Crystallogr. Assoc., 1982* Abstract PD10.
47. G. B. Ansell and J. S. Bradley, *Acta Crystallogr. Sect. B* **B36,** 1939 (1980).
48. P. F. Jackson, B. F. G. Johnson, J. Lewis, P. R. Raithby, G. J. Will, M. McPartlin, and W. J. H. Nelson, *J. Chem. Soc., Chem. Commun.* p. 1190 (1980).
49. B. F. G. Johnson, J. Lewis, K. Wong, and M. McPartlin, *J. Organomet. Chem.* **185,** C17 (1980).
50. J. S. Bradley, R. L. Pruett, E. Hill, G. B. Ansell, M. E. Leonowicz, and M. A. Modrick, *Organometallics* **1,** 748 (1982).
51. J. S. Bradley and E. W. Hill, U.S. Patent 4,301,086 (1981).
52. G. B. Ansell, M. A. Modrick, and J. S. Bradley, *Am. Crystallogr. Assoc., 1982,* Abstract PD12.
53. C. R. Eady, J. M. Fernandez, B. F. G. Johnson, J. Lewis, P. R. Raithby, and G. M. Sheldrick, *J. Chem. Soc., Chem. Commun.* p. 421 (1978).
54. C. R. Eady, B. F. G. Johnson, and J. Lewis, *J. Organomet. Chem.* **57,** C84 (1973).
55. C. R. Eady, B. F. G. Johnson, and J. Lewis, *J. Chem. Soc., Dalton Trans.* p. 2606 (1975).
56. P. F. Jackson, B. F. G. Johnson, J. Lewis, M. McPartlin, and W. J. H. Nelson, *J. Chem. Soc., Chem. Commun.* p. 224 (1980).
57. I. A. Oxton, S. F. A. Kettle, P. F. Jackson, B. F. G. Johnson, and J. Lewis, *J. Mol. Struct.* **71,** 117 (1981).
58. D. H. Farrar, P. G. Jackson, B. F. G. Johnson, J. Lewis, W. J. H. Nelson, M. D. Vargas, and M. McPartlin, *J. Chem. Soc., Chem. Commun.* p. 1009 (1981).
59. V. G. Albano, M. Sansoni, P. Chini, and S. Martinengo, *J. Chem. Soc., Dalton Trans.* p. 651 (1973).
60. V. G. Albano, P. Chini, S. Martinengo, M. Sansoni, and D. Strumolo, *J. Chem. Soc., Chem. Commun.* p. 299 (1974).
61. V. G. Albano, D. Braga, S. Martinengo, P. Chini, M. Sansoni, and D. Strumolo, *J. Chem. Soc., Dalton Trans.* p. 52 (1980).
62. V. G. Albano, D. Braga, and S. Martinengo, *J. Chem. Soc., Dalton Trans.* p. 717 (1981).
63. V. G. Albano, P. Chini, G. Ciani, M. Sansoni, and S. Martinengo, *J. Chem. Soc., Dalton Trans.* p. 163 (1980).
64. V. G. Albano, P. Chini, G. Ciani, M. Sansoni, D. Strumolo, B. T. Heaton, and S. Martinengo, *J. Am. Chem. Soc.* **98,** 5027 (1976).
65. J. S. Bradley and E. W. Hill, unpublished observation.
66. V. G. Albano, P. Chini, G. Ciani, S. Martinengo, and M. Sansoni, *J. Chem. Soc., Dalton Trans.* p. 463 (1978).
67. V. G. Albano, D. Braga, G. Ciani, and S. Martinengo, *J. Organomet. Chem.* **213,** 293 (1981).
68. V. G. Albano, D. Braga, P. Chini, G. Ciani, and S. Martinengo, *J. Chem. Soc., Dalton Trans.* p. 645 (1982).
69. G. Bör, G. Gervasio, R. Rossetti, and P. L. Stanghellini, *J. Chem. Soc., Chem. Commun.* p. 841 (1978).
70. G. Bör and P. L. Stanghellini, *J. Chem. Soc., Chem. Commun.* p. 886 (1979).
71. G. Bör, U. K. Kietler, P. L. Stanghellini, G. Gervasio, R. Rossetti, G. Sbrignadello, and G. A. Battiston, *J. Organomet. Chem.* **213,** 277 (1981).

58   J. S. BRADLEY

72. P. L. Stanghellini and G. Bör, EUCHEM, personal communication.
73. V. G. Albano, M. Sansoni, P. Chini, S. Martinengo, and D. Strumolo, *J. Chem. Soc., Dalton Trans.* p. 305 (1975).
74. V. G. Albano, P. Chini, S. Martinengo, M. Sansoni, and D. Strumolo, *J. Chem. Soc., Dalton Trans.* p. 459 (1978).
75. V. G. Albano, M. Sansoni, P. Chini, S. Martinengo, and D. Strumolo, *J. Chem. Soc., Dalton Trans.* p. 970 (1976).
76. G. Ciani, G. D'Alfonso, M. Freni, P. Romiti, and A. Sironi, *J. Chem. Soc., Chem. Commun.* p. 339 (1982).
77. G. Ciani, G. D'Alfonso, M. Freni, P. Romiti, and A. Sironi, *J. Chem. Soc., Chem. Commun.* p. 705 (1982).
78. I. A. Oxton, D. B. Powell, R. J. Goudsmit, B. F. G. Johnson, J. Lewis, W. J. H. Nelson, J. N. Nichols, M. J. Rosales, M. D. Vargas, and K. H. Whitmire, *Inorg. Chim. Acta Lett.* **64,** L259 (1982).
79. J. A. Creighton, R. Della Pergola, B. T. Heaton, S. Martinengo, L. Strona, and D. A. Willis, *J. Chem. Soc., Chem. Commun.* p. 864 (1982).
80. V. G. Albano, P. Chini, S. Martenengo, D. J. A. McCaffrey, D. Strumolo, and B. T. Heaton, *J. Am. Chem. Soc.* **96,** 8106 (1974).
81. B. T. Heaton, L. Strona, S. Martinengo, D. Strumolo, R. J. Goodfellow, and I. H. Sadler, *J. Chem. Soc., Dalton Trans.* p. 1499 (1982).
82. B. T. Heaton, L. Strona, and S. Martinengo, *J. Organomet. Chem.* **215,** 415 (1981).
83. C. R. Eady, B. F. G. Johnson, and J. Lewis, *J. Organomet. Chem.* **39,** 329 (1972).
84. T. Ya. Kosolapova, "Carbides," pp. 180–181. Plenum, New York, 1971.
85. A. C. Sievert, D. S. Strickland, J. R. Shapley, G. R. Steinmetz, and G. L. Geoffroy, *Organometallics* **1,** 214 (1982).
86. J. S. Bradley and E. W. Hill, unpublished observation.
87. J. L. Vidal and W. E. Walker, *Inorg. Chem.* **19,** 896 (1980).
88. D. M. P. Mingos, *Inorg. Chem.* **21,** 464 (1982).
89. B. F. G. Johnson, J. Lewis, W. J. H. Nelson, J. H. Nicholls, and M. D. Vargas, *J. Organomet. Chem.* **249,** 255 (1983).
90. G. Longoni, A. Cerriotti, R. Della Pergola, M. Manassero, M. Perego, G. Piro, and M. Sansoni, *Philos. Trans. R. Soc. London* **A308,** 47 (1982).
91. C-M. T. Hayward and J. R. Shapley, *Inorg. Chem.* **21,** 3816 (1982).
92. C-M. T. Hayward, J. R. Shapley, M. R. Churchill, and C. Bueno, *J. Am. Chem. Soc.* **104,** 7347 (1982).
93. J. W. Kolis, E. M. Holt, M. Drezdzon, K. H. Whitmire, and D. F. Shriver, *J. Am. Chem. Soc.* **104,** 6134 (1982).

ADVANCES IN ORGANOMETALLIC CHEMISTRY, VOL. 22

Vinylidene and Propadienylidene (Allenylidene) Metal Complexes

MICHAEL I. BRUCE and
A. GEOFFREY SWINCER

Jordan Laboratories
Department of Physical and Inorganic Chemistry
University of Adelaide
Adelaide, Australia

I

INTRODUCTION

The rapid development of the chemistry of transition metal complexes containing terminal carbene (**A**) or carbyne (**B**) ligands (*1*) has been followed more recently by much research centered on bridged methylene compounds (**C**) (*2*). The importance of μ-methylidyne complexes, whether in recently established binuclear examples (**D**), the well-known trinuclear derivatives (**E**), or the unusual μ_4-η^2-CH complexes (**F**), has also become apparent. All are based on one-carbon (C_1) fragments, and considerable interest is centered on their possible significance as models for intermediates in surface-catalyzed reactions between carbon monoxide and hydrogen (Fischer–Tropsch reactions) and related processes. These topics have been extensively reviewed (*3*).

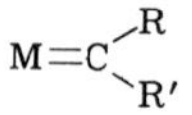

$$M{=}C\overset{\displaystyle R}{\underset{\displaystyle R'}{}}$$

(A)

$$M{\equiv}C{-}R$$

(B)

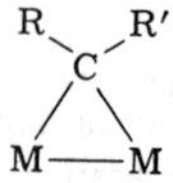

$$\underset{M{-\!\!-}M}{\overset{R\diagup C \diagdown R'}{}}$$

(C)

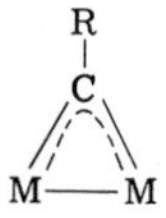

(D)

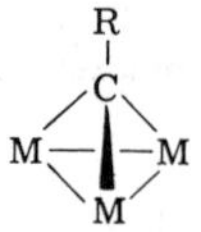

(E)

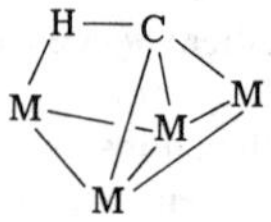

(F)

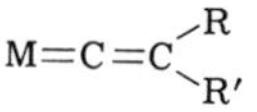

$$M{=}C{=}C\overset{\displaystyle R}{\underset{\displaystyle R'}{}}$$

(G)

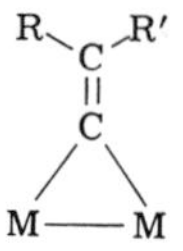

(H)

$$M{=}C{=}C{=}C\overset{\displaystyle R}{\underset{\displaystyle R'}{}}$$

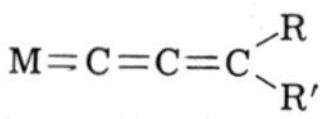

(I)

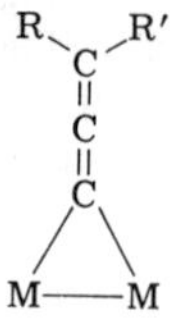

(J)

Complexes containing the unsaturated carbenes vinylidene (**G and H**) and propadienylidene (allenylidene) (**I and J**), and their cluster-bound analogs, are also known, although general routes to these reactive compounds have been discovered only within the last five years or so. Vinylidene, the simplest unsaturated carbene, has never been observed experimentally, as it undergoes an extremely fast 1,2-hydrogen shift to give acetylene:

$$\begin{array}{c} H \\ \diagdown \\ \diagup \\ H \end{array} C = C: \longrightarrow HC \equiv CH \qquad (1)$$

Several theoretical studies have addressed the problem of the relative stabilities of vinylidene and acetylene, one of the most recent concluding that the classical barrier to Eq. (1) is 4 kcal; zero-point energy effects lower this to 2.2 kcal (*4*). Tunneling through this barrier is extremely rapid: the calculated lifetime of vinylidene is ca. 10^{-11} sec, which agrees with a value of ca. 10^{-10} sec deduced from trapping experiments (*5*).

Unsaturated carbenes, such as vinylidenes and cumulenylidenes, have been shown to be intermediates in certain organic reactions, such as α-eliminations from vinyl halides and related compounds. Initial studies were carried out in the 1960s, and an excellent review is available (*6*). As with many other reactive organic intermediates, it has proved possible to stabilize vinylidenes and cumulenylidenes as ligands in transition metal complexes, the first being obtained from diphenyl ketene and $Fe_2(CO)_9$ in 1966 (*7*). This product contained a bridging diphenylvinylidene ligand, and a diiron complex containing the dicyanovinylidene group was described six years later (*8*). At about this time also, mononuclear complexes were obtained from reactions between tertiary phosphines or phosphites and chlorodicyanovinyl molybdenum complexes (*9*). The intermediacy of metal-stabilized carbonium ions was recognized as a contributing factor in the reactions of certain cationic platinum(II)–alkyne complexes; in these, 1-alkynes were considered to rearrange to the platinum–vinylidene moiety, $Pt-\overset{+}{C}=CHR$, but no stable complexes were isolated (*10*).

Rapid development of this area followed the discovery of routes to these complexes, either by ready conversion of terminal alkynes to vinylidene complexes in reactions with manganese, rhenium, and the iron-group metal complexes (*11–14*) or by protonation or alkylation of some metal σ-acetylides (*14*). Recent work has demonstrated the importance of vinylidene complexes in the metabolism of some chlorinated hydrocarbons (DDT) using iron porphyrin-based enzymes (*15*). Interconversions of alkyne and vinylidene ligands occur readily on multimetal centers. Several reactions involving organometallic reagents may proceed via intermediate vinylidene complexes.

The first propadienylidene complexes were reported in 1976 (*16, 17*), and examples of terminal and bridging ligands of this type are now known; their chemistry is now beginning to be understood. This article is concerned with the synthesis, properties, and chemistry of transition metal complexes containing these unsaturated carbene ligands.

II

NOMENCLATURE *

As often occurs in new and developing areas of chemistry, some confusion about the nomenclature of these complexes has arisen. Most workers have used the name "vinylidene" for the $:C{=}CRR'$ ligand depicted in **G** or **H**, and "allenylidene" for those in **I** and **J**. In a recent review, the terms "alkylidene carbene" and "vinylidene carbene," respectively, were employed, whereas yet another group refers to "vinylalkylidene" for the $:C{=}CRR'$ ligand. Current usage refers to the ligand $\equiv CR$ [in $Co_3(\mu_3\text{-}CR)(CO)_9$ or $Ta(CR)(Cl)(PMe_3)_2(\eta\text{-}C_5Me_5)$, for example] as "alkylidyne."

IUPAC rules allow the term "vinylidene," although the *Chemical Abstracts* preferred name is "ethenylidene," whereas for ligands of type $C{=}C{=}CR_2$, the name "propadienylidene" is probably less likely to be misunderstood. Systematic names based on "enylidene" or "dienylidene" roots can be adapted to take account of the longest carbon chain, e.g., $C{=}C{=}CEtPh$ is 3-phenylpenta-1,2-dienylidene. In this review, the name "vinylidene" will be used for **G** and **H** and "propadienylidene" for **I** and **J**.

III

MONONUCLEAR VINYLIDENE METAL COMPLEXES

A. *Preparative Methods*

1. *From Alkynes*

Terminal acetylenes (1-alkynes) undergo a 1,2-hydrogen shift in reactions with many metal centers to give vinylidene complexes. These reactions may proceed via an intermediate η^2-alkyne complex, which has been isolated or detected spectroscopically in some cases (*11–14, 18–20*).

---

* M.I.B. is pleased to acknowledge helpful correspondence with R. Schoenfeld (CSIRO Editorial and Publications Section).

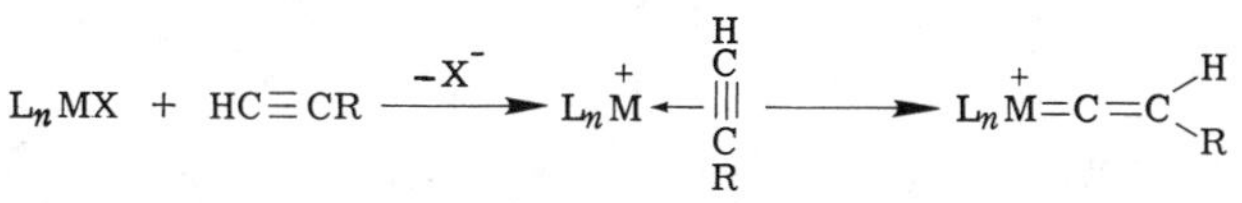

$$L_nMX + HC\equiv CR \xrightarrow{-X^-} L_n\overset{+}{M}\leftarrow \overset{\overset{H}{\overset{|}{C}}}{\underset{\underset{R}{\overset{|}{C}}}{|||}} \longrightarrow L_n\overset{+}{M}=C=C\overset{H}{\underset{R}{<}}$$

$L_nM = Mn(CO)_2(\eta\text{-}C_5H_5)$, $Re(CO)_2(\eta\text{-}C_5H_5)$, $FeCl(depe)_2$, $FeL_2(\eta\text{-}C_5H_5)$, $RuL_2(\eta\text{-}C_5H_5)$, $OsL_2(\eta\text{-}C_5H_5)$; X = solvent, CO or halide; the cyclopentadienyl ring may also be substituted with one or more Me groups.

Recently, the complex $RhCl(\eta^2\text{-}HC_2Ph)(PPr_3^i)_2$ has been found to undergo isomerization to $HRhCl(C_2Ph)(PPr_3^i)_2$ (by intramolecular oxidative addition); with NaC_5H_5, the η^2-alkyne complex $Rh(\eta^2\text{-}HC_2Ph)(PPr_3^i)(\eta\text{-}C_5H_5)$ is formed. In contrast, reaction with pyridine affords $HRhCl(C_2Ph)(py)(PPr_3^i)_2$ (**1**, 90%), and this complex gives the isomeric orange phenylvinylidene complex $Rh(=C=CHPh)(PPr_3^i)(\eta\text{-}C_5H_5)$ (**2**, 65%) on treatment with NaC_5H_5. By working at 0°C, the intermediate $Rh(C_2Ph)(py)(PPr_3^i)_2$ can be isolated; this reacts with cyclopentadiene to give (**2**) (*17a*).

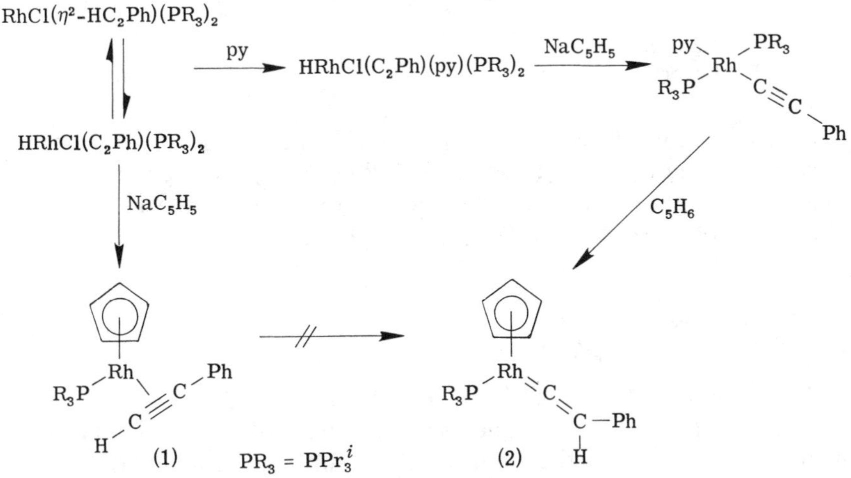

Conversion of $Mn[\eta^2\text{-}HC\equiv C(CO_2Me)](CO)_2(\eta\text{-}C_5H_5)$ to $Mn[=C=CH(CO_2Me)](CO)_2(\eta\text{-}C_5H_5)$ occurs on reaction with organolithium reagents at low temperatures, suggesting that the hydrogen transfer is a base-catalyzed reaction (*21*); this vinylidene complex is obtained in only 1% yield in the absence of base.

In some cases, more than one alkyne is incorporated into the vinylidene ligand:

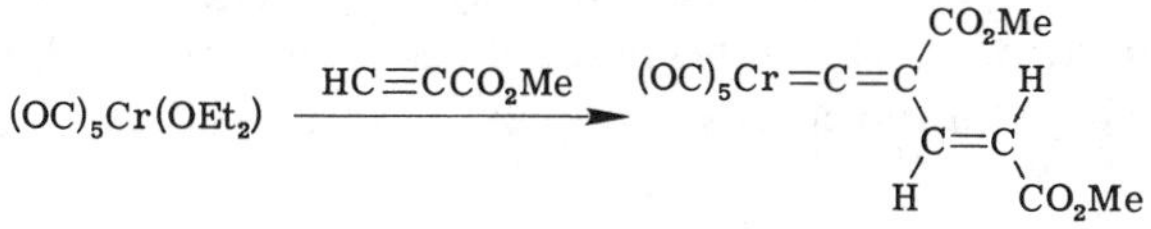

(Ref. *22*)

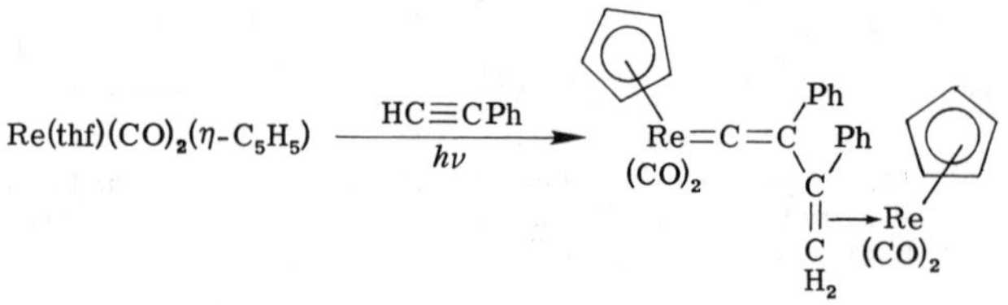

(Ref. 19)

Vinylidene complexes have been obtained from reactions between $PhC{\equiv}CEPh_3$ (E = Si, Ge, or Sn) and manganese complexes (11, 18); solvolysis of the C–E bond is followed by transfer of a proton from the solvent. The yields (E = Si, 0%; Ge, 1%; Sn, 15%) are inversely proportional to the stability of the intermediate η^2-alkyne complex.

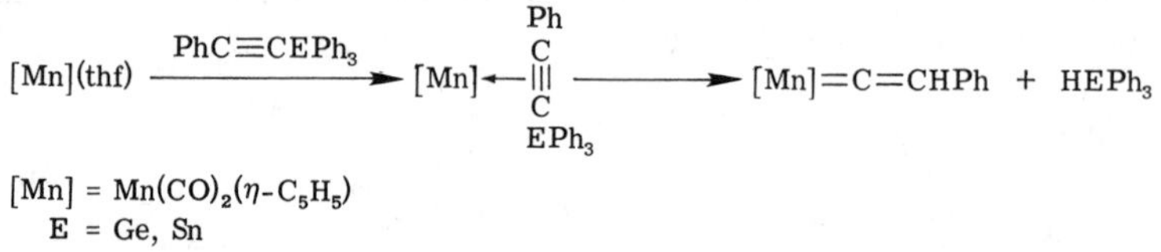

$$[Mn] = Mn(CO)_2(\eta\text{-}C_5H_5)$$
$$E = Ge, Sn$$

A similar reaction between $RuCl(PMe_3)_2(\eta\text{-}C_5H_5)$ and $PhC{\equiv}CSiMe_3$ in the presence of $AgBF_4$ gives $[Ru(=C=CHPh)(PMe_3)_2(\eta\text{-}C_5H_5)]^+$ (85%); an intermediate yellow coloration may indicate formation of the alkyne complex $[Ru(\eta^2\text{-}PhC{\equiv}CSiMe_3)(PMe_3)_2(\eta\text{-}C_5H_5)]^+$ (23).

2. From Metal Acetylides

Protonation or alkylation of several ethynyl–metal derivatives gives the corresponding vinylidene complexes in high yield (14, 24). This is a convenient route to the disubstituted vinylidene complexes, as well as the parent compounds, which cannot be obtained from 1-alkynes; they are formed if $MeOSO_2F$ or $[R_3O]^+$ (R = Me, Et) is used as alkylating agent:

Methylation of $Fe(C{\equiv}CH)(dppe)(\eta\text{-}C_5H_5)$ (3) affords an equimolecular mixture of the vinylidene (4) and dimethylvinylidene (5) complexes, together with spectroscopic amounts of the expected methylvinylidene derivative (6), implying that the ethynyl complex is more basic than the propynyl derivative (25):

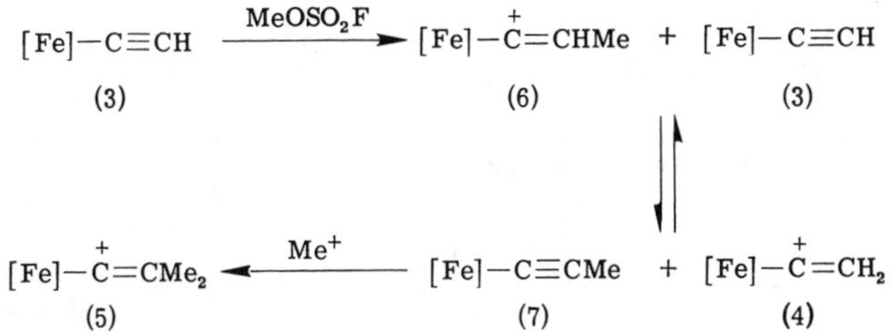

$$[Fe]-C\equiv CH \xrightarrow{\text{MeOSO}_2\text{F}} [Fe]-\overset{+}{C}=CHMe + [Fe]-C\equiv CH$$

(3)   (6)   (3)

$$[Fe]-\overset{+}{C}=CMe_2 \xleftarrow{\text{Me}^+} [Fe]-C\equiv CMe + [Fe]-\overset{+}{C}=CH_2$$

(5)   (7)   (4)

$[Fe] = Fe(dppe)(\eta\text{-}C_5H_5)$

The formation of $Mn(=C=CMe_2)(CO)_2(\eta\text{-}C_5H_5)$ by reaction of the η^2-methyl propiolate complex with $LiNPr_2^i$ proceeds via an insoluble yellow anionic alkynylmanganese complex, which is protonated or methylated to give the vinylidene complexes (26):

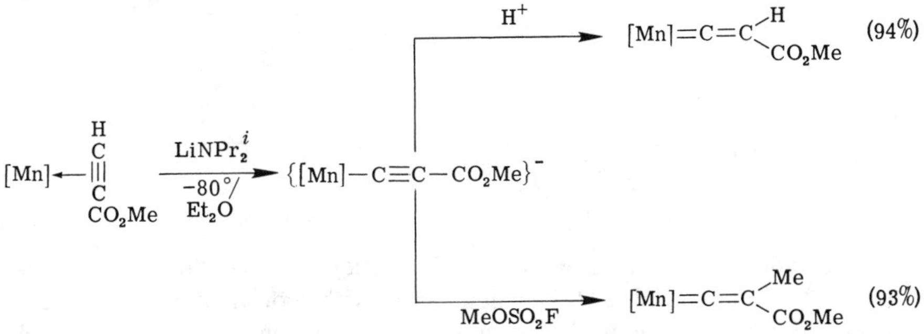

$$[Mn]\leftarrow\overset{\displaystyle H}{\underset{\displaystyle CO_2Me}{\overset{\displaystyle C}{\underset{\displaystyle C}{|||}}}} \xrightarrow[\substack{-80\%\\ \text{Et}_2\text{O}}]{\text{LiNPr}_2^i} \{[Mn]-C\equiv C-CO_2Me\}^-$$

$$\xrightarrow{\text{H}^+} [Mn]=C=C\overset{H}{\underset{CO_2Me}{<}} \quad (94\%)$$

$$\xrightarrow[\text{MeOSO}_2\text{F}]{} [Mn]=C=C\overset{Me}{\underset{CO_2Me}{<}} \quad (93\%)$$

If LiMe is used as the base, addition to the ester function, followed by elimination of acetone, affords the dianion (8), which gives the dimethylvinylidene complex (9) on treatment with methyl fluorosulfonate (27):

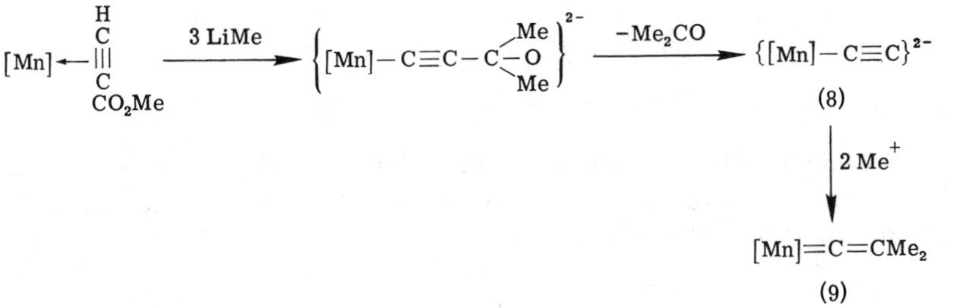

$$[Mn]\leftarrow\overset{\displaystyle H}{\underset{\displaystyle CO_2Me}{\overset{\displaystyle C}{\underset{\displaystyle C}{|||}}}} \xrightarrow{\text{3 LiMe}} \left\{[Mn]-C\equiv C-C\overset{Me}{\underset{Me}{<}}O\right\}^{2-} \xrightarrow{-\text{Me}_2\text{CO}} \{[Mn]-C\equiv C\}^{2-}$$

(8)

$$\downarrow 2\,\text{Me}^+$$

$$[Mn]=C=CMe_2$$

(9)

A similar sequence of reactions occurs with $LiBu^t$, although in this instance methylation or protonation of the intermediates formed after addition of either two or three equivalents of alkyllithium reagent, respectively, allow the corresponding acyl– or hydroxyalkyl–vinylidene complexes to be isolated (26):

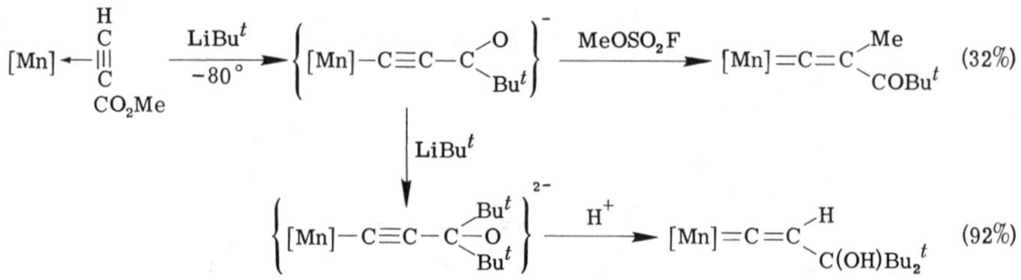

$$[Mn] \leftarrow \overset{\overset{\displaystyle H}{\overset{\displaystyle \|}{\underset{\displaystyle CO_2Me}{C}}}}{\underset{\displaystyle C}{\|}} \quad \xrightarrow[-80°]{LiBu^t} \quad \left\{ [Mn]-C\equiv C-C \overset{O}{\underset{Bu^t}{<}} \right\}^- \quad \xrightarrow{MeOSO_2F} \quad [Mn]=C=C \overset{Me}{\underset{COBu^t}{<}} \quad (32\%)$$

$$\Bigg\downarrow LiBu^t$$

$$\left\{ [Mn]-C\equiv C-C\overset{Bu^t}{\underset{Bu^t}{<}}O^- \right\}^{2-} \quad \xrightarrow{H^+} \quad [Mn]=C=C\overset{H}{\underset{C(OH)Bu_2^t}{<}} \quad (92\%)$$

Related to these reactions is that of $Cr(CO)_6$ with a dilithium alkynolate (28); the intermediate acyl dianion loses CO on irradiation to give **10**, which reacts with acetyl chloride in the presence of PPh_3 to give a low yield of **11**:

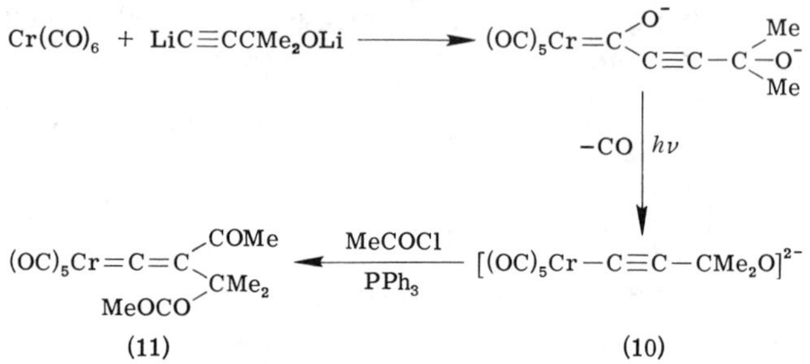

$$Cr(CO)_6 + LiC\equiv CCMe_2OLi \longrightarrow (OC)_5Cr=C\overset{O^-}{\underset{C\equiv C-C\overset{-O^-}{\underset{Me}{<}}}{<}}{}^{Me}$$

$$-CO \Big| h\nu$$

$$(OC)_5Cr=C\overset{COMe}{\underset{MeOCO}{<}}{}^{CMe_2} \quad \xleftarrow[PPh_3]{MeCOCl} \quad [(OC)_5Cr-C\equiv C-CMe_2O]^{2-}$$

$$\qquad\qquad (11) \qquad\qquad\qquad\qquad\qquad\qquad (10)$$

Cycloaddition of CS_2 to $Fe(C\equiv CMe)(dppe)(\eta\text{-}C_5H_5)$ affords the σ-2H-thiete-2-thione complex (**12**), which reacts with iodomethane to give **13**, which contains perhaps the most strongly electron-withdrawing ligand known (29):

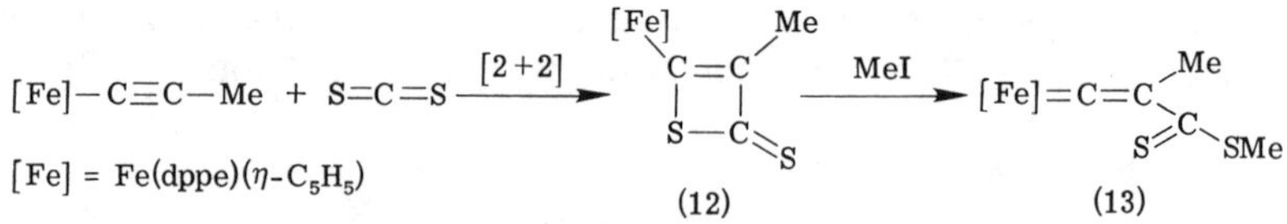

$$[Fe]-C\equiv C-Me + S=C=S \xrightarrow{[2+2]} \overset{[Fe]}{\underset{S-C\underset{S}{\searrow}}{\overset{|}{C}=C\overset{Me}{|}}} \xrightarrow{MeI} [Fe]=C=C\overset{Me}{\underset{S}{\searrow}}\overset{}{\underset{C}{\searrow}}SMe$$

$$[Fe] = Fe(dppe)(\eta\text{-}C_5H_5)$$

$$\qquad\qquad\qquad\qquad\qquad\qquad (12) \qquad\qquad\qquad (13)$$

3. From σ-Vinyl–Metal Complexes

Several vinylidene complexes of the group VI metals have been obtained by heating σ-chlorovinyl derivatives with tertiary phosphines, phosphites, arsines, or stibines (9, 30):

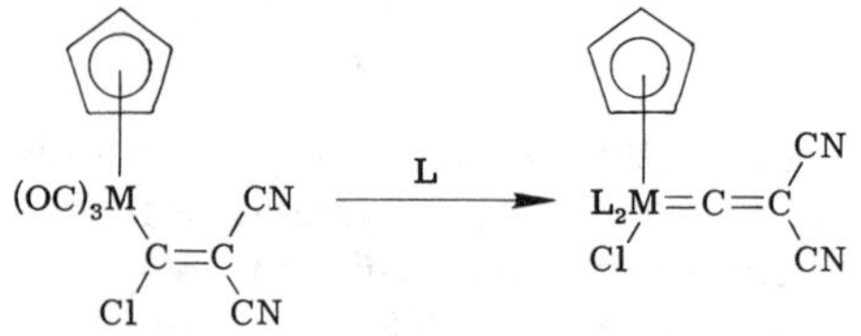

$$(OC)_3M\overset{CN}{\underset{Cl}{\overset{|}{C}=C\overset{}{\underset{CN}{<}}}} \xrightarrow{L} L_2M=C=C\overset{CN}{\underset{CN}{<}}$$

M = Mo or W
L = PPh_3, PMe_2Ph, $P(OMe)_3$, $P(OEt)_3$, $P(OPh)_3$,
$\quad AsPh_3$, $SbPh_3$; L_2 = dppe

A similar reaction with $PhP(CH_2CH_2PPh_2)_2$ (L_3) afforded the cationic complex $[Mo\{C\!=\!C(CN)_2\}(L_3)(\eta\text{-}C_5H_5)]^+$ (61%). The reaction of the tungsten analog with PPh_3 is more sluggish, and also gives $WCl(CO)_2(PPh_3)(\eta\text{-}C_5H_5)$ by loss of the dicyanovinylidene ligand. The transfer of halide from the α-carbon to the metal is similar to that suggested to occur in the easy hydrolysis of the platinum(II) complex $trans\text{-}PtCl(CCl\!=\!CH_2)(PMe_2Ph)_2$, which is supposed to proceed via an intermediate vinylidene (31). It is of some interest that the α-chlorovinyl group in the platinum derivative is already distorted to allow transfer of the chlorine atom, with a long $C-Cl$ bond and the chlorine displaced toward the metal.

4. From Carbene or Carbyne Complexes

Iron(II) porphyrins react readily with haloalkanes in the presence of reducing agents, e.g., excess iron powder, to give chlorocarbene complexes (15, 32). With the insecticide DDT [2,2-bis(4-chlorophenyl)-1,1,1-trichloroethane, $(ClC_6H_4)_2CHCCl_3$], the reaction proceeds one step further, by elimination of HCl from the carbene complex, to give diarylvinylidene complexes (33):

$$Fe(por) + Cl_3CCH(C_6H_4Cl)_2 \rightarrow (por)Fe[CClCH(C_6H_4Cl)_2]$$

por = tpp, ttp, tap, oep, ppix, dpdme $\qquad \Big\downarrow -HCl$

$$(por)Fe[C\!=\!C(C_6H_4Cl)_2]$$

Deprotonation of some molybdenum–carbyne complexes affords vinylidene derivatives; suitable reagents are aryldiazonium salts, trifluoroiodomethane (34), or n-butyllithium (35):

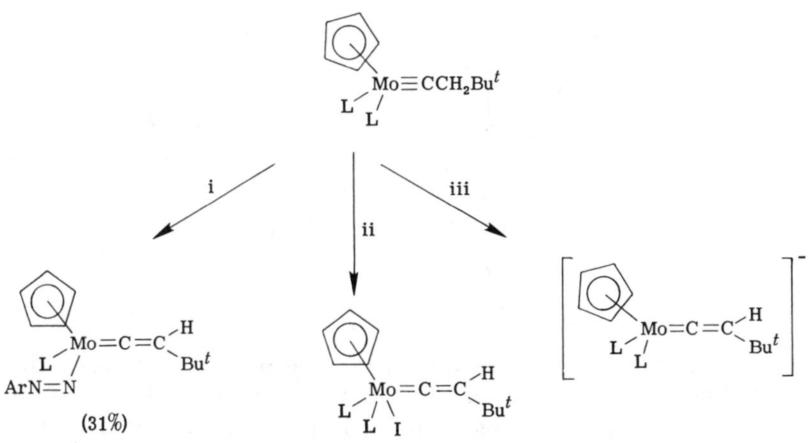

L = $P(OMe)_3$
Reagents: (i) ArN_2^+; (ii) CF_3I; (iii) LiBu.

In the latter case, the red anionic vinylidene complex is formed at $-78°C$; quenching with D_2O gives a $1:1$ mixture of the mono- and dideuterocarbyne complexes, as a result of a primary kinetic isotope effect by which the anion selectively abstracts a proton from the monodeuterocarbyne complex before quenching is complete. Similar observations were made in the tungsten series (35).

The reaction between $Li[HBEt_3]$ and $[Os\{C(tol)\}(CO)(L)(PPh_3)_2]^+$ [14; $L = CO$ or $CN(tol)]$ gives 15, formed by attack of hydride on the vinylidene resonance hybrid of 14 (36):

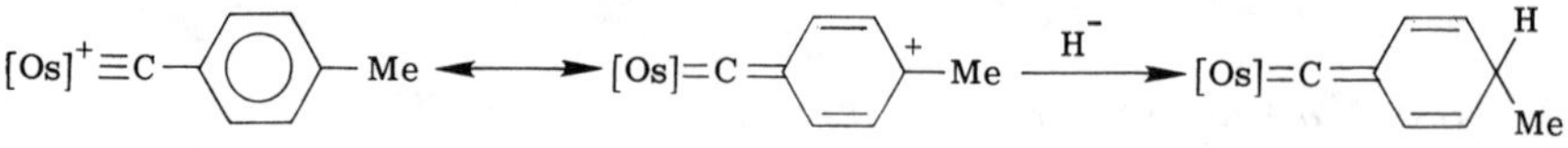

(14)   (15)

$[Os] = Os(CO)(L)(PPh_3)_2$, $L = CO$ or $CN(tol)$

5. From Acyl Complexes

Treatment of some iron–acyl complexes with trifluoromethanesulphonic anhydride (Tf_2O) affords vinylidene derivatives directly (37, 38). The reaction is envisaged as a nucleophilic attack on Tf_2O by the acyl, followed by deprotonation to the vinyl ether complex. A combination of an excellent leaving group (TfO^-) with a good electron-releasing substituent on the same carbon atom facilitates the subsequent formation of the vinylidene:

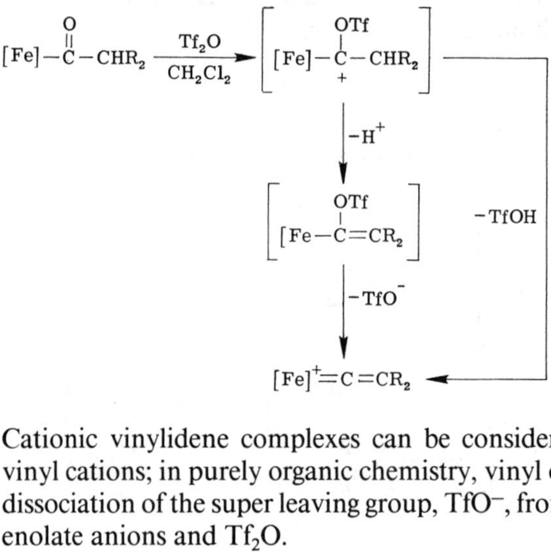

R	L
H	PPh_3, PMe_2Ph, PCy_3
Me	PPh_3

Cationic vinylidene complexes can be considered to be metal-stabilized vinyl cations; in purely organic chemistry, vinyl cations can be obtained by dissociation of the super leaving group, TfO^-, from the esters obtained from enolate anions and Tf_2O.

The acylate anion from $Mn(CO)_3(\eta\text{-}C_5H_5)$ and LiMe reacts similarly; reaction with 1,8-bis(dimethylamino)naphthalene (proton sponge) affords the vinylidene, although in this reaction, it is the binuclear complex which is isolated as the final product (*38*):

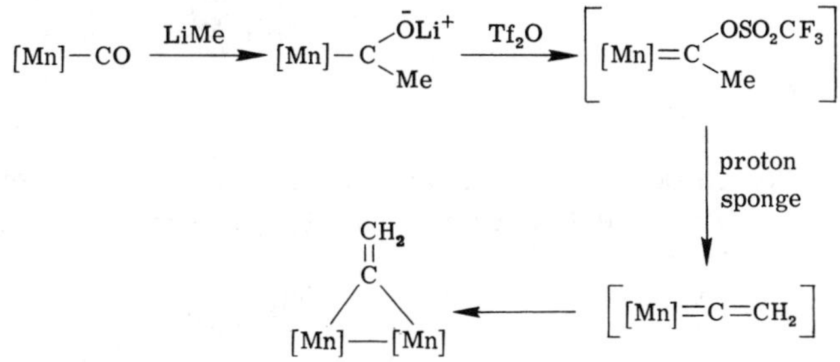

$$[Mn]-CO \xrightarrow{\text{LiMe}} [Mn]-C\underset{Me}{\overset{\bar{O}Li^+}{<}} \xrightarrow{Tf_2O} \left[[Mn]=C\underset{Me}{\overset{OSO_2CF_3}{<}} \right]$$

proton sponge

$$\underset{[Mn]---[Mn]}{\overset{\overset{CH_2}{\underset{||}{C}}}{\bigtriangleup}} \longleftarrow \left[[Mn]=C=CH_2 \right]$$

B. *Reactions*

The electron distribution in the vinylidene ligand, which results in a high electron deficiency on the α-carbon, but with considerable electron density on the β-carbon (see Section VII,A), renders this ligand subject to nucleophilic attack on the former, and electrophilic attack on the latter.

1. *Nucleophilic Addition to the α-Carbon*

Those vinylidene complexes which are not readily deprotonated by bases undergo attack at the α-carbon by anions such as H^-, MeO^-, NH_2^- to give vinyl derivatives

$$[Fe(=C=CMe_2)(dppe)(\eta\text{-}C_5H_5)]^+ \xrightarrow{H^-} Fe(CH=CMe_2)(dppe)(\eta\text{-}C_5H_5)$$

(Ref. *25*)

$$Mn[=C=CH(CO_2Me)](CO)_2(\eta\text{-}C_5H_5) \xrightarrow{B^-} [Mn\{CB=CH(CO_2Me)\}(CO)_2(\eta\text{-}C_5H_5)]^-$$
$$B = MeO, NH_2$$
$$\downarrow H^+$$
$$[Mn]=CBCH_2(CO_2Me) \qquad \text{(Ref. } 26)$$

In the case of the anionic manganese complexes, subsequent protonation affords the neutral alkylcarbene complexes. Intramolecular addition of deprotonated dppe to the α-carbon occurs on reaction of $[Fe(=C=CMe_2)(dppe)(\eta\text{-}C_5H_5)]^+$ with base [NaN(SiMe_3)_2 or KOH in tetrahydrofuran] to give **16** (*39*):

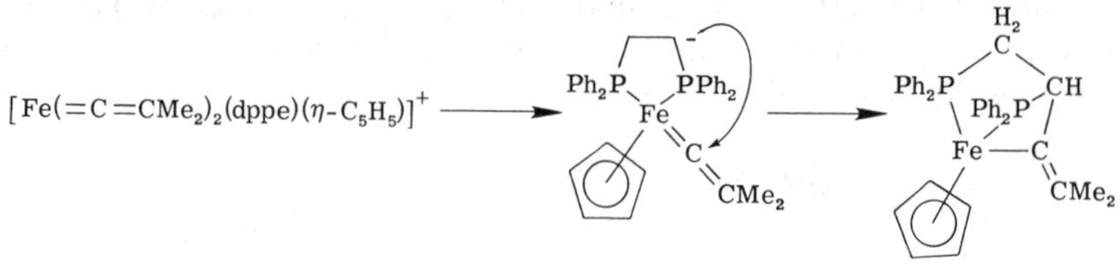

(16)

Alkoxy(alkyl)carbene complexes are formed more or less rapidly from alcohols and cationic vinylidene complexes (*14, 40–43*):

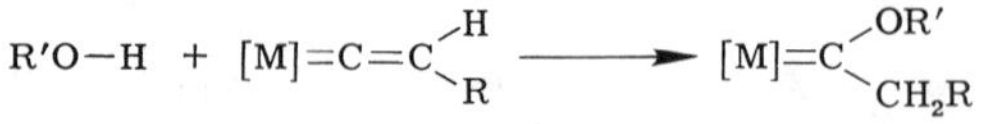

$[M] = Mn(CO)_2(\eta\text{-}C_5H_5), [Fe(CO)(L)(\eta\text{-}C_5H_5)]^+, [RuL_2(\eta\text{-}C_5H_5)]^+$

The addition follows the expected direction to give conventional Fischer-type carbene complexes. Some qualitative experiments have shown that the observed variation in reactivity results from a combination of steric and electronic factors. Bulky ligands protect the α-carbon from attack, and the rates of formation of the alkoxycarbene complexes are inversely proportional to the cone angle of L in $Ru(L)(PPh_3)(\eta\text{-}C_5H_5)$: CO (cone angle ca. 95°) ⁓ CNBu^t (ca. 95°) > P(OMe)$_3$ (107°) > PMe$_3$ (118°) ⁓ PPh$_3$ (145°). Broad agreement of reactivity is also found with Tolman's electronic factors (*44*), which would predict an order of reactivity

$$CO > P(OMe)_3 > CNBu^t > PPh_3 > PMe_3$$

An increase in the electron-withdrawing power of the vinylidene substituent, C=CHR, would be expected to increase the reaction rate, as is found with the series:

$$R = CO_2Me > Ph > Me$$

On the other hand, the *t*-butylvinylidene complex was unchanged after 24 hr in refluxing methanol, suggesting that the bulky CMe$_3$ group exerts a predominantly steric control over the reaction. A limited range of analogous complexes were available for comparison within the Fe–Ru–Os triad. Whereas the reaction of $[Fe(=C=CHPh)(CO)(PPh_3)(\eta\text{-}C_5H_5)]^+$ with methanol is fast (*43*), $[Fe(=C=CHMe)(dppe)(\eta\text{-}C_5H_5)]^+$ does not react (*24, 25*),* nor do any of the osmium complexes that were investigated. As the alcohol is changed, the reactivity decreases

$$MeOH > EtOH > Pr^iOH$$

---

* Reinvestigation of the chemistry of the iron complexes has shown that reactions with alcohols or water are extremely fast (S. G. Davies, private communication, 1983).

It is noteworthy that the methylplatinum(II)–vinylidene intermediates are four-coordinate square-planar complexes, with essentially no hindrance to attack from above or below the square-plane. Only alkoxycarbene complexes have ever been isolated from the reactions in alcohol solvents (10).

Rapid intramolecular addition of the alcohol function to a supposed intermediate vinylidene complex occurs in reactions of $HC\equiv C(CH_2)_nOH$ with metal halide complexes (45); the cyclic carbene complexes (17) are isolated instead:

$[M] = [Ru(PPh_3)_2(\eta\text{-}C_5H_5)]^+$; $n = 2,3$ (17)

The conventional vinylidene complex could be isolated when the hydroxyl group was protected as the tetrahydropyranyl ether derivative; reaction of this with acid immediately gave the cyclic carbene complex, even under mild conditions. The reaction is related to the formation of similar nickel(II)– and platinum(IV)–carbene complexes from the ω-hydroxyalkyne in the presence of silver(I) ion (46, 47):

$$trans\text{-}NiCl(C_6Cl_5)(PMe_2Ph)_2 \xrightarrow[\text{AgClO}_4]{HC\equiv C(CH_2)_2OH} [trans\text{-}Ni(C_6Cl_5)\{\overline{C(CH_2)_3O}\}(PMe_2Ph)_2]^+$$

$$PtIMe_2(CF_3)(PMe_2Ph)_2 \xrightarrow[\text{AgPF}_6]{HC\equiv C(CH_2)_2OH} \{PtMe_2(CF_3)[\overline{C(CH_2)_3O}](PMe_2Ph)_2\}^+$$

The iron complexes $[Fe(=C=CRR')(dppe)(\eta\text{-}C_5H_5)]^+$ (R = R' = H, Me; R = H, R' = Me) and $[Fe(=C=CMe_2)(PMe_3)_2(\eta\text{-}C_5H_5)]^+$ are inert to alcohols and water (39).* In contrast, the α-carbon in $[Fe(=C=CH_2)(CO)(PPh_3)(\eta\text{-}C_5H_5)]^+$ is attacked by a range of nucleophiles (Scheme 1) (43). Water affords the acetyl complex (18), probably via a hydroxy(methyl)carbene intermediate; some alcohols give alkoxy(methyl)carbene complexes (19), with the hydroxylic proton adding to the β-carbon as shown by labeling studies. t-Butanol does not react. Hydrogen sulfide and methanethiol react to give methyl(thio)carbene derivatives (20), the former being deprotonated to the unstable thioacyl $Fe(CSMe)(CO)(PPh_3)(\eta\text{-}C_5H_5)$. According to their basicity, amines add to give aminocarbene complexes, or deprotonate the vinylidene to $Fe(C\equiv CH)(CO)(PPh_3)(\eta\text{-}C_5H_5)$ (21). Thus, benzylamine (pK_a 9.33) gives 22, whereas methylamine (pK_a 10.66) and dimethylamine (pK_a 10.77) give 23 and 24, respectively, accompanied by about 10% 21. In the absence of solvent, the stronger base NMe_3 (pK_a 9.81) gives only 21. Dry

* See footnote on facing page.

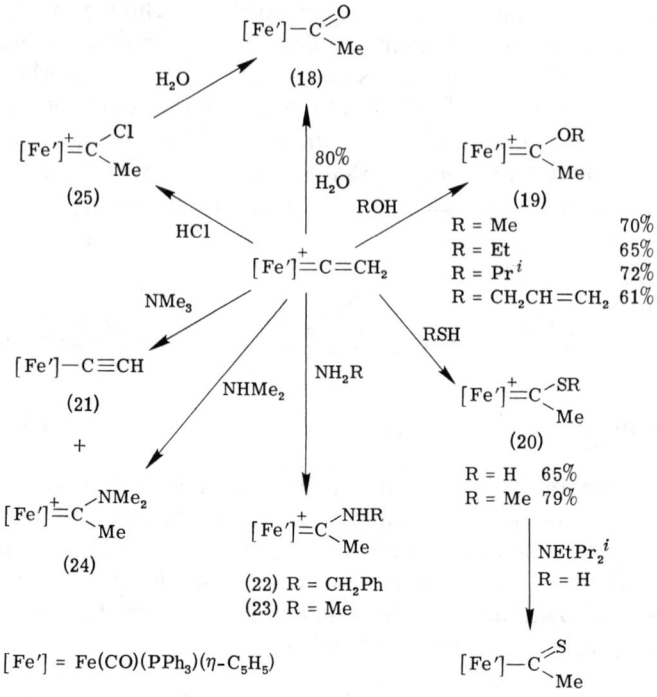

SCHEME 1. Some reactions of [Fe(=C=CH₂)(CO)(PPh₃)(η-C₅H₅)]⁺ (43).

HCl also adds to the α-carbon to give the chloro(methyl)carbene complex (25), which is rapidly hydrolyzed to the acetyl derivative (18).

The reaction of tertiary phosphines with alkyne complexes to give α-metallated vinylphosphonium derivatives has been taken to indicate the intermediacy of vinylidene complexes (48):

$$[Mn] \leftarrow \overset{\overset{\displaystyle H}{\overset{\displaystyle |}{C}}}{\underset{\underset{\displaystyle CO_2Me}{\displaystyle |}}{C}} \quad \xrightarrow[\substack{20-40°/3\ hr \\ pentane}]{PPh_3} \quad \underset{\underset{+}{Ph_3P}}{\overset{[Mn]}{\diagdown}} C=CH(CO_2Me)$$

(85%)

However, substitution of CO by tertiary phosphine or phosphite occurs in similar reactions of Mn(=C=CHPh)(CO)₂(η-C₅H₅) (18):

Mn=C=CHPh with OC and CO ligands → PR₃ → Mn=C=CHPh with OC and PR₃ ligands

R = Ph, OEt, OPh

Addition of PPh_3 to $[Fe(=C=CH_2)(CO)(PPh_3)(\eta\text{-}C_5H_5)]^+$ proceeds in dichloromethane to give $[Fe[C(PPh_3)=CH_2](CO)(PPh_3)(\eta\text{-}C_5H_5)][BF_4]$, which was structurally characterized (38). Similar adducts were obtained with PMe_2Ph, pyridine, and 4-methylpyridine.

Reaction of ethylvinyl ether (ethoxyethylene) to $[Fe(=C=CH_2)(CO)$ $(PPh_3)(\eta\text{-}C_5H_5)]^+$ affords the ethoxy(methyl)carbene complex (26) with elimination of acetylene (43):

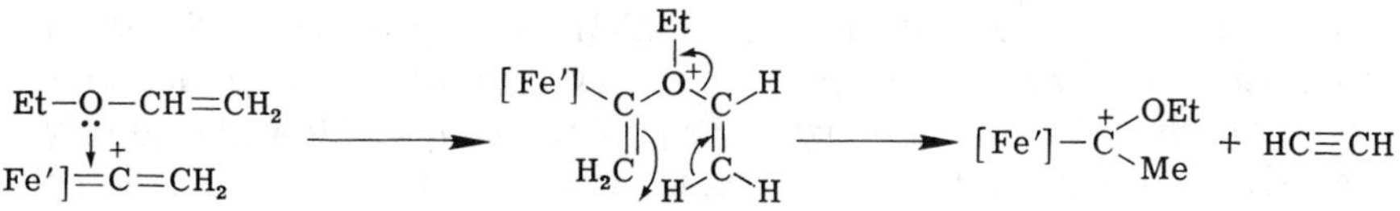

Addition of hydride (using $Na[HB(OMe)_3]$ in thf) to $[Fe(=C=CMe_2)$ $(CO)(PPh_3)(\eta\text{-}C_5H_5)]^+$ gives a $4:1$ ratio of $Fe(CH=CMe_2)(CO)(PPh_3)$ $(\eta\text{-}C_5H_5)$ and $HFe(CO)(PPh_3)(\eta\text{-}C_5H_5)$ (25).

2. Deprotonation

Cationic vinylidene complexes bearing a β-hydrogen are readily deprotonated to give the corresponding neutral σ-acetylide complexes:

$$[M]^+=C=CHR \rightleftharpoons [M]-C\equiv CR + H^+$$

The base used may be hydroxide, alkoxide, carbonate, alkyl lithium, or alumina (13, 20, 24, 41, 49). The reaction is the reverse of the vinylidene synthesis by protonation of the σ-acetylide, and the two complexes form a simple acid–base system. For the iron complexes in Eq. (2), the pK_a has been measured at 7.78 (in $2:1$ thf-H_2O) (24).

$$[Fe]-C\equiv CMe + H^+ \rightleftharpoons [Fe]^+{=}C{=}C\overset{H}{\underset{Me}{<}} \qquad (2)$$

3. Addition to the β-Carbon

Carbyne complexes are formed by reaction of D_2O with 26, in a reversal of the deprotonation reaction used to form it (35):

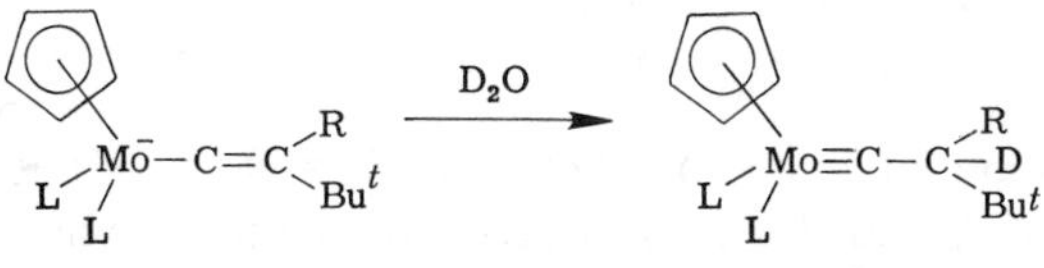

(26) R = H, D

4. *Protonation*

Addition of HCl to the osmium–vinylidene complex (**15**, L = CO) gave the 4-methylbenzyl complex Os(CH$_2$C$_6$H$_4$Me-*p*)(CO)$_2$(PPh$_3$)$_2$ (*36*).

5. *Ligand-Exchange Reactions*

Several simple ligand-exchange reactions of CO for PR$_3$ (R = Ph, OEt, OPh) have been reported for Mn(C=CHPh)(CO)$_2$(*η*-C$_5$H$_5$) (*18*, *50*). In contrast the cationic complexes [Ru(C=CHR)(PPh$_3$)$_2$(*η*-C$_5$H$_5$)]$^+$ are stable toward such exchanges, although the complexes can be obtained readily by ligand exchange on the corresponding acetylides, followed by protonation of the latter to give the vinylidene complexes (*42*).

6. *Miscellaneous Reactions*

The complex [Fe(=C=CH$_2$)(CO)(PPh$_3$)(*η*-C$_5$H$_5$)][BF$_4$] is soluble in dichloromethane or chloroform without change, solutions in thf rapidly afford the binuclear complex (**28**) which exists as a 3 : 1 ratio of the meso and *R,R:S,S* isomers (*43*). The analogous PCy$_3$ complex is obtained as the pure *R,R:S,S* isomer. The formation of (**28**) occurs via partial deprotonation to the ethynyl complex, followed by cycloaddition of this to the remaining vinylidene complex. Confirmation of this route is provided by the synthesis of **28** by addition of half an equivalent of HBF$_4$·OEt$_2$ to the ethynyl complex. Either a stepwise (A) or concerted [$\pi^2_s + \pi^2_a$] (B) intermediate may be involved:

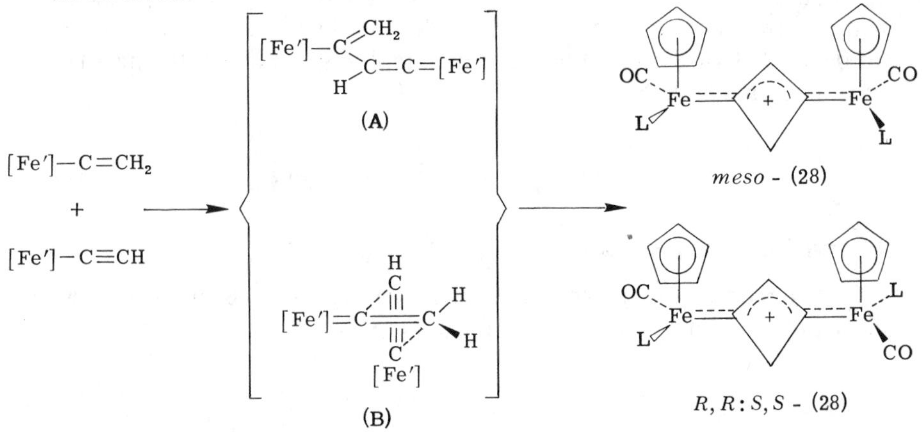

Similar cyclobutenylidene complexes (**29**; R = H) have been obtained by addition of strong acids (HClO$_4$ or HBF$_4$ in acetic anhydride or dichloromethane) to Fe(C≡CPh)(CO)$_2$(*η*-C$_5$H$_5$); in the presence of PPh$_3$ the vinylphosphonium complex $[Fe\{C(PPh_3)=CHPh\}(CO)_2(\eta\text{-}C_5H_5)]^+$ is formed (13, 51). The methyl derivative (29; R = Me) was obtained from the reaction with $MeOSO_2F$.

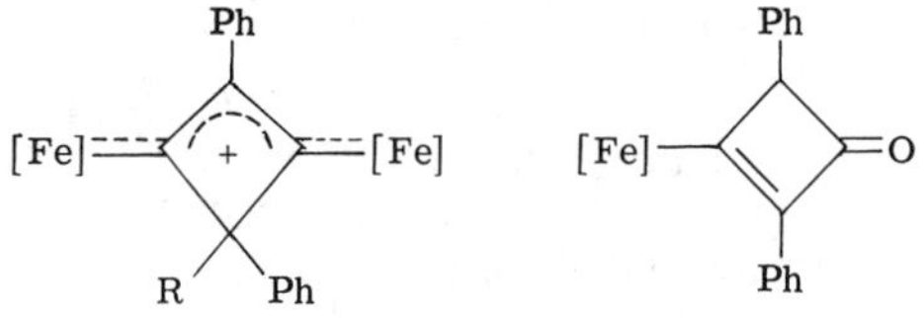

(29)   (30)

Further reaction of (29; R = H) with triethylamine has given the mononuclear cyclobutenone complex (30) (52). Complexes of this type can also be obtained by cycloaddition of ketenes to the σ-acetylide (53):

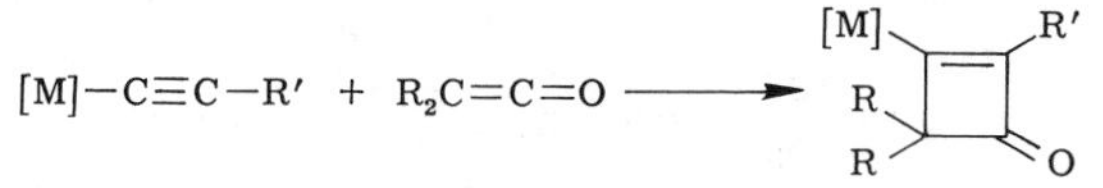

$[M] = Ni(PPh_3)(\eta\text{-}C_5H_5),\ Fe(CO)(L)(\eta\text{-}C_5H_5)$

One of the most interesting reactions of cationic vinylidene complexes is that with dioxygen, which affords the carbonyl cation and the corresponding aldehyde (41):

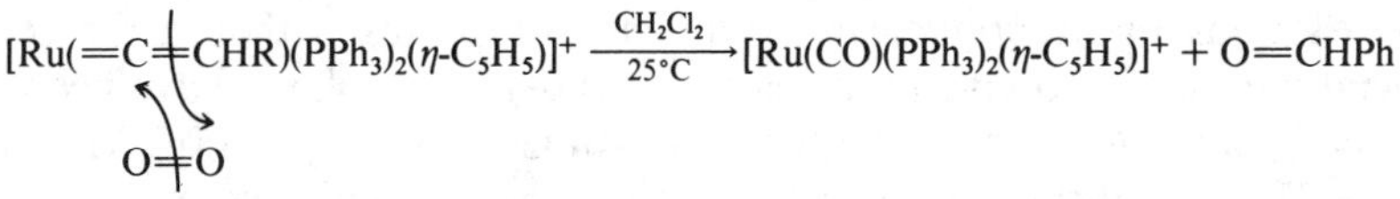

This reaction is a formal dismutation of the multiply bonded components, but no detailed mechanism has been elucidated.

IV

BINUCLEAR VINYLIDENE COMPLEXES

A. Preparative Methods

1. From Mononuclear Complexes

Several complexes containing μ-C=CHR ligands have been obtained directly from 1-alkynes and two equivalents (or excess) of an appropriate precursor (*11, 18, 54–59*):

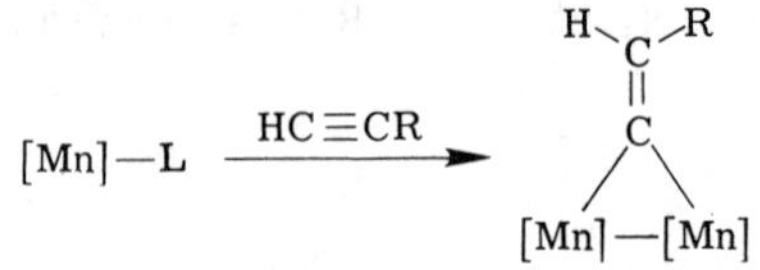

$$[Mn]—L \xrightarrow{HC\equiv CR} \begin{array}{c} H{\diagdown}_{C}{\diagup}^{R} \\ \parallel \\ C \\ \diagup \diagdown \\ [Mn]—[Mn] \end{array}$$

L = CO or thf
R = H, Me, Ph, CO_2Me, COPh
[Mn] = $Mn(CO_2)(\eta\text{-}C_5H_5)$

These reactions undoubtedly proceed via the mononuclear complexes described in Section III, and, accordingly, reactions of the latter with metal complexes afford the binuclear products. Mixed-metal derivatives can also be obtained by this route (*18, 26*):

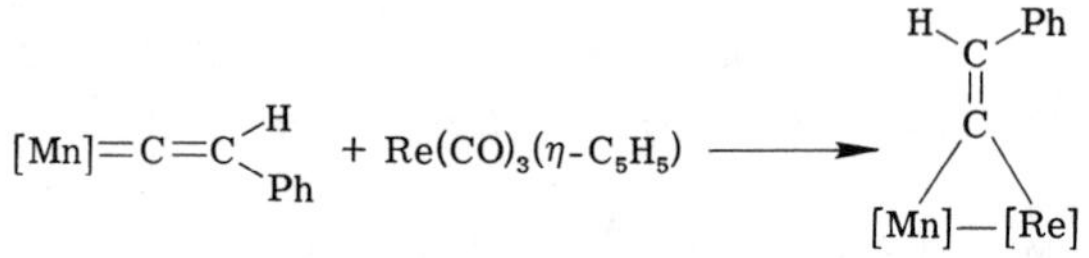

$$[Mn]{=}C{=}C{\diagdown}^{H}_{Ph} + Re(CO)_3(\eta\text{-}C_5H_5) \longrightarrow \begin{array}{c} H{\diagdown}_{C}{\diagup}^{Ph} \\ \parallel \\ C \\ \diagup \diagdown \\ [Mn]—[Re] \end{array}$$

The mononuclear complexes can also be converted into the binuclear analogs by heating, e.g., with $Mn(=C=CHPh)(CO)_2(\eta\text{-}C_5H_5)$ (*18*), or by reaction with ethanolic KOH solution, e.g., $Re(=C=CHPh)(CO)_2(\eta\text{-}C_5H_5)$ (*19*).

Reactions of mononuclear vinylidene complexes with other reactive metal complexes to give binuclear μ-vinylidene complexes have been described above. Addition of $Fe_2(CO)_9$ to $Mn(C=CHPh)(CO)_2(\eta\text{-}C_5H_5)$ also gives **31**, by addition of a CO group to the α-carbon; structural data are consistent with the delocalized formulation (**31b**), with its obvious resemblances to trimethylenemethane (*60*):

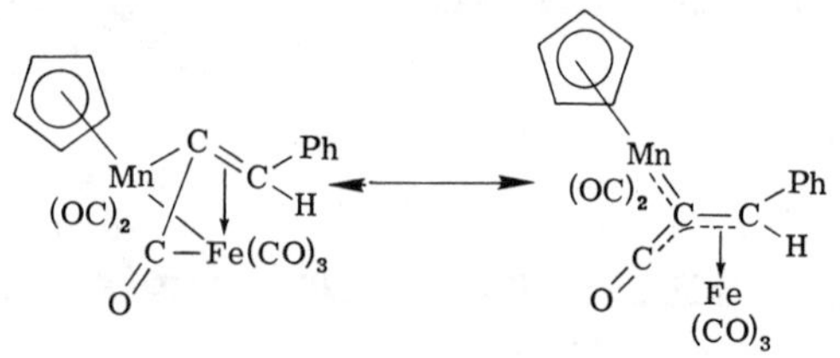

(a)   (b)

(31)

2. From Diphenylketene

Irradiation of mixtures of $Ph_2C=C=O$ and iron carbonyls afforded **32**, the first vinylidene complex to be characterized (*7*):

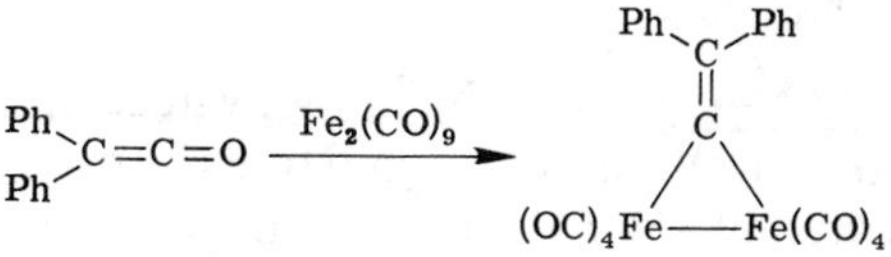

(32)

3. From Alkynes

The reaction between acetylene and $Rh(CO)(\eta\text{-}C_2H_4)(\eta^5\text{-}C_9H_7)$ [which acts as a source of the $Rh(CO)(\eta^5\text{-}C_9H_7)$ fragment] affords 33 in 50% yield (61). The reaction is supposed to proceed via oxidative addition of the alkyne to the rhodium fragment, followed by isomerization to the vinylidene complex which then interacts with a second rhodium fragment:

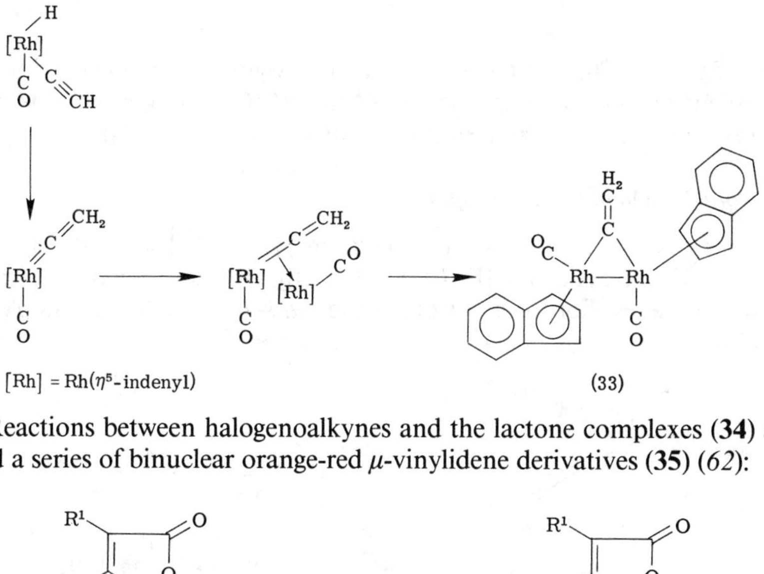

[Rh] = Rh(η^5-indenyl)

(33)

Reactions between halogenoalkynes and the lactone complexes (34) afford a series of binuclear orange-red μ-vinylidene derivatives (35) (62):

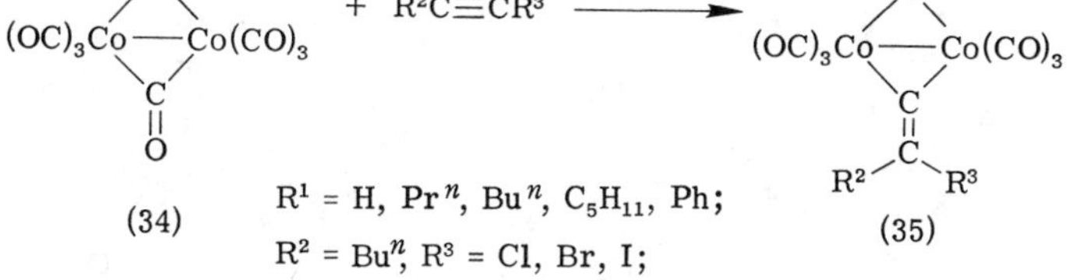

(34)

$+ \ R^2C{\equiv}CR^3 \longrightarrow$

$R^1 = H, Pr^n, Bu^n, C_5H_{11}, Ph;$
$R^2 = Bu^n, R^3 = Cl, Br, I;$
$R^2 = Ph, R^3 = Br; R^2 = R^3 = Br, I$

(35)

The formation of these compounds is unusual in that it involves 1,2-migration of the halogen atom.

4. *From 1,1-Dichloroolefins*

The reaction of $Cl_2C{=}C(CN)_2$ with the highly nucleophilic anionic metal carbonyl $[Fe(CO)_2(\eta\text{-}C_5H_5)]^-$ affords cis and trans isomers of **36** separable by column chromatography (*8, 63*). These red complexes are analogs of $[Fe(CO)_2(\eta\text{-}C_5H_5)]_2$, although in the latter case, facile interconversion of isomers occurs at room temperature. A further difference is that the Fe–Fe bond in **36** resists cleavage with iodine to give a mononuclear product such

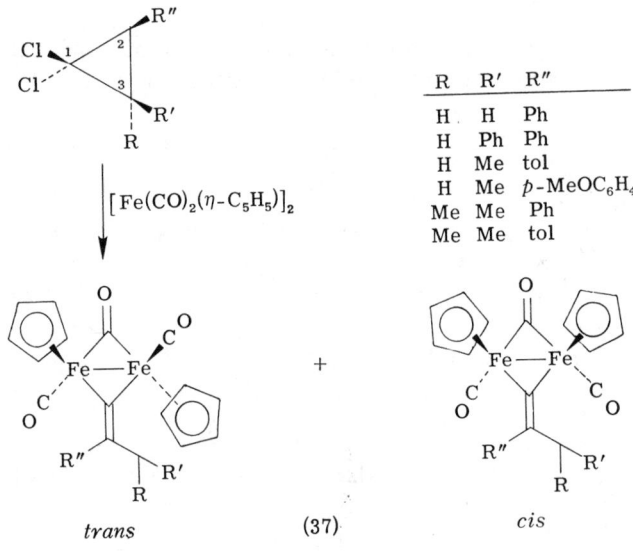

cis   (36)   trans

as $FeI(CO)[C{=}C(CN)_2](\eta\text{-}C_5H_5)$, even under vigorous reaction conditions.

Less nucleophilic anions react with $Cl_2C{=}C(CN)_2$ to give chlorodicyanovinyl derivatives, such as $Mo[CCl{=}C(CN)_2](CO)_3(\eta\text{-}C_5H_5)$.

5. *From 1,1-Dichlorocyclopropanes*

Under normal phase-transfer conditions [NaOH–H_2O (50:50), thf, $(NBu_4)HSO_4$], reactions of $[Fe(CO)_2(\eta\text{-}C_5H_5)]_2$ with substituted 1,1-dichlorocyclopropanes afford a variety of *cis*- and *trans*-alkylvinylidene complexes (**37**) (*64*).

R	R'	R"
H	H	Ph
H	Ph	Ph
H	Me	tol
H	Me	p-MeOC$_6$H$_4$
Me	Me	Ph
Me	Me	tol

trans   (37)   cis

The trans form isomerizes on heating: for R = Ph, the half-life is ca. 5 hr at 80°C, the equilibrium constant of ca. 20 favoring the cis form. The reaction is general for 2-arylcyclopropanes which also have hydrogen attached to C(2); 1,1-dichloro-2-methyl-2-phenylcyclopropane does not react. In the formation of 37, none of the C=CR(CH$_2$Ph) isomer is obtained, showing that it is the C(1)–C(3) bond that is cleaved. The initial step is base-promoted elimination of HCl from the dichlorocyclopropane to give the cyclopropene; these compounds also react with [Fe(CO)$_2$(η-C$_5$H$_5$)]$_2$ to give the bridged vinylidene complexes under these conditions.

6. From Other Binuclear Complexes by Modification of Bridging Ligands

Nucleophilic attack by methyllithium on the μ-CO group in 38 or 39, followed by acidification, affords the μ-vinylidene complexes 40 and 41, respectively (65, 66):

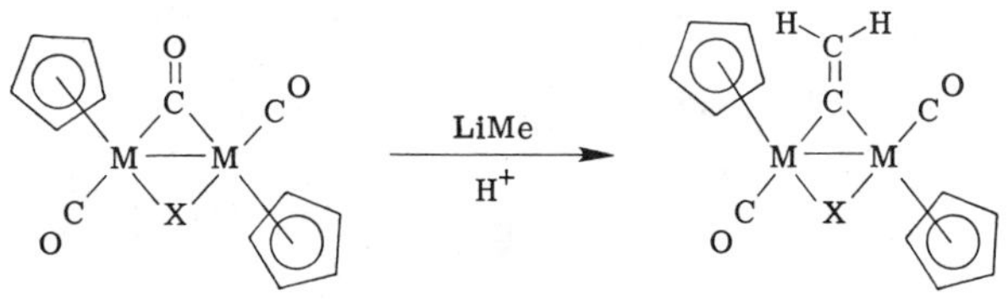

(38) M = Fe, X = CO
(39) M = Ru, X = CMe$_2$

(40) M = Fe, X = CO
(41) M = Ru, X = CMe$_2$

The reaction is assumed to proceed by addition of methyl to the bridging carbon atom to give an acylate anion. Protonation, followed by loss of water from the resulting hydroxycarbene complex, gives the neutral vinylidene complexes (Scheme 2). Further protonation affords a cationic μ-methylcarbyne complex, which was first described from the successive reactions between [Fe(CO)$_2$(η-C$_5$H$_5$)]$_2$ and methyllithium and tetrafluoroboric acid (67); in the synthesis of 40, the weaker trifluoroacetic acid is used in the protonation step. The ruthenium complex [Ru$_2$(μ-CMe)(μ-CMe$_2$)(CO)$_2$(η-C$_5$H$_5$)$_2$]$^+$ is deprotonated on shaking with a water–dichloromethane mixture to give yellow 41. An alternative route is by thermolysis of 42 in refluxing toluene; the latter complex is formed from [Ru(CO)$_2$(η-C$_5$H$_5$)]$_2$ and acetylene (68). Complex 43 is formed as a mixture of the cis and trans isomers, which can be separated by chromatography, but slowly interconvert in solution. Labeling studies with the phenyl-substituted complexes show that the rearrangement of 42 to 43 is an intramolecular process:

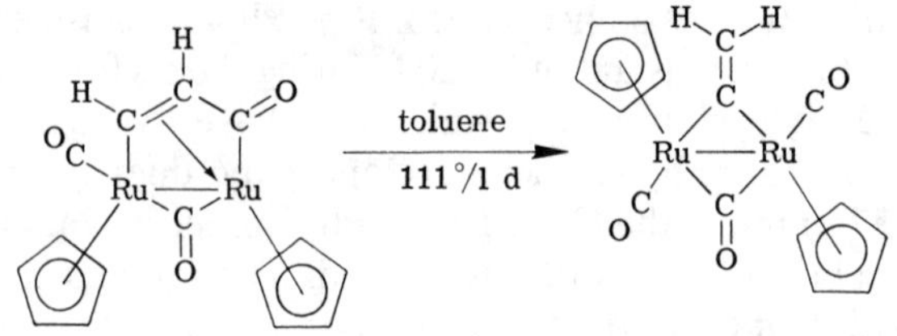

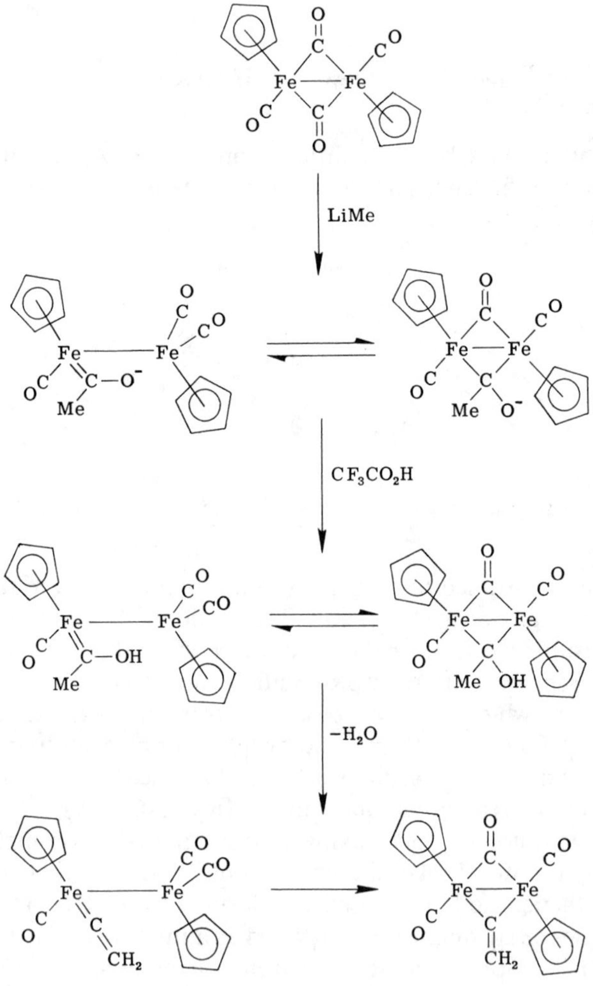

SCHEME 2. Synthesis of a μ-vinylidene complex.

Complex **44** has been detected by mass spectrometry among the products of the reaction between $[Fe(CO)_2(\eta\text{-}C_5H_5)]_2$ and $Ph_3P{=}CH_2$, being formed by a Wittig reaction on the μ-CO groups (69):

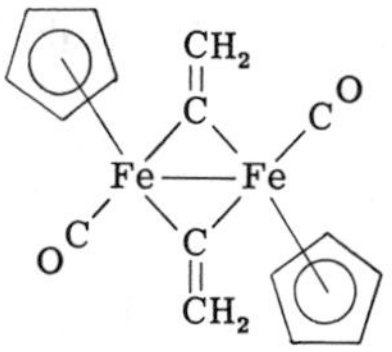

(44)

B. *Reactions*

Electrophilic attack on μ-vinylidene complexes can occur either on the methylene carbon, or at the metal–metal bond. With the manganese complexes (**45**, R = H or Me), protonation affords the μ-carbyne complexes (**46**), which in the case of R = Me, exist in the stereoisomeric forms shown (57). Interconversion of the two forms is slow at room temperature:

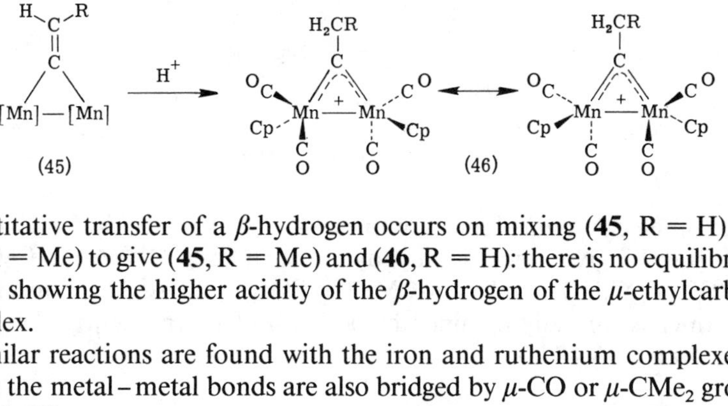

Quantitative transfer of a β-hydrogen occurs on mixing (**45**, R = H) and (**46**, R = Me) to give (**45**, R = Me) and (**46**, R = H): there is no equilibrium set up showing the higher acidity of the β-hydrogen of the μ-ethylcarbyne complex.

Similar reactions are found with the iron and ruthenium complexes in which the metal–metal bonds are also bridged by μ-CO or μ-CMe$_2$ groups (65, 66, 68):

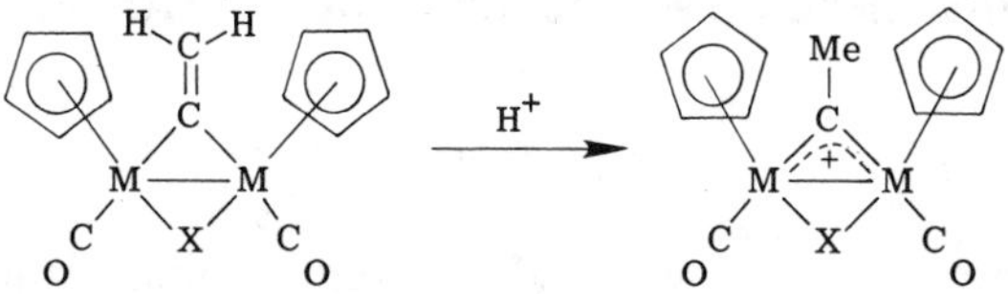

M = Fe, X = CO
X = Ru, X = CO or CMe$_2$

Reaction of the $[cis\text{-}Ru_2(\mu\text{-}CMe)(\mu\text{-}CO)(CO)_2(\eta\text{-}C_5H_5)_2]^+$ cation with sodium borohydride gives the μ-methylcarbene complex $cis\text{-}Ru_2(\mu\text{-}CHMe)(\mu\text{-}CO)(CO)_2(\eta\text{-}C_5H_5)_2$ (68).

In contrast, addition of $HBF_4 \cdot OEt_2$ to the dirhodium complex (33) affords the $\mu\text{-}\eta^1,\eta^2$-vinyl complex (47), probably via initial attack on the Rh–Rh bond (61):

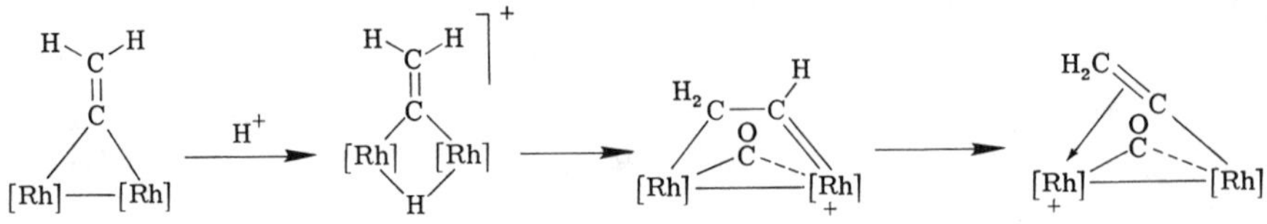

(33)

(47)

[Rh] = Rh(CO)(η^5-indenyl)

Nucleophilic attack on 45 (R = H) by four equivalents of $Li[HBEt_3]$, followed by addition of iodomethane, affords the μ-allene complex (48) in a reaction which involves reduction of a coordinated CO and linkage to the vinylidene within the coordination sphere (57, 70):

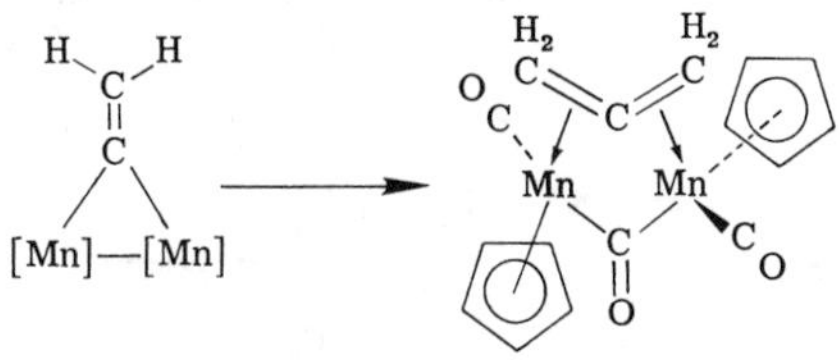

(45)

(48)

The binuclear μ-vinylidene complex (40) reacts with $H_3Mn_3(CO)_{12}$ or $Fe_2(CO)_9$ to give the μ_3-ethylidyne complexes $Fe_2Mn(\mu_3\text{-}CMe)(\mu\text{-}CO)_3(CO)_3(\eta\text{-}C_5H_5)_2$ and $Fe_3(\mu_3\text{-}CMe)(\mu_3\text{-}CO)_2(CO)_6(\eta\text{-}C_5H_5)$, respectively; the former is formally obtained by addition of $HMn(CO)_4$ to the $Fe_2(\mu\text{-}C{=}CH_2)$ moiety (70a).

V

PROPADIENYLIDENE (ALLENYLIDENE) COMPLEXES

A. Preparation

The first complexes to be described (16) were obtained from Fischer-type carbene complexes by successive reactions with a Lewis acid and a weak base, which in this case was the solvent tetrahydrofuran:

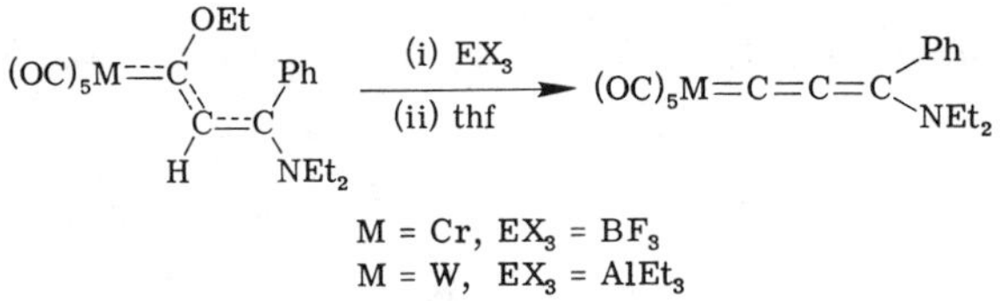

$$(OC)_5M \overset{OEt}{\underset{\underset{\underset{NEt_2}{C \cdots C}}{\overset{Ph}{|}}}{=\!\!=\!\!C}} \quad \underset{\text{(ii) thf}}{\overset{\text{(i) } EX_3}{\longrightarrow}} \quad (OC)_5M\!\!=\!\!C\!\!=\!\!C\!\!=\!\!C\overset{Ph}{\underset{NEt_2}{\diagdown}}$$

$$M = Cr, \ EX_3 = BF_3$$
$$M = W, \ \ EX_3 = AlEt_3$$

A general route to complexes containing propadienylidene ligands is by loss of water or alcohols from suitable carbene or vinylidene precursors, or of oxo or alkoxy functions from ynolate anions. The latter are generally obtained from reactions of alkyne complexes containing an ester group, for example, methyl propiolate. Vinylidene complexes containing hydroxy groups on the γ-carbon can be readily converted to the propadienylidene derivative. These reactions are promoted by base, or by use of $COCl_2$ or $CSCl_2$:

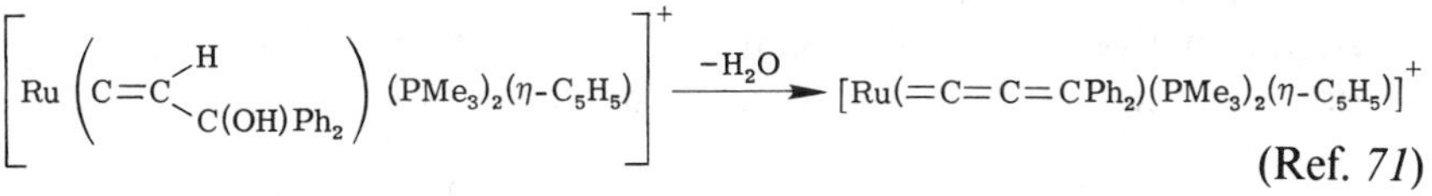

$$\left[Ru \left(C\!\!=\!\!C \overset{H}{\underset{C(OH)Ph_2}{\diagdown}} \right) (PMe_3)_2(\eta\text{-}C_5H_5) \right]^+ \xrightarrow{-H_2O} [Ru(\!=\!C\!\!=\!\!C\!\!=\!\!CPh_2)(PMe_3)_2(\eta\text{-}C_5H_5)]^+$$

(Ref. 71)

Attempts to prepare the dimethyl analog from $HC\!\!\equiv\!\!CCMe_2(OH)$ led instead to the binuclear vinylidene–carbene complex (49) (72):

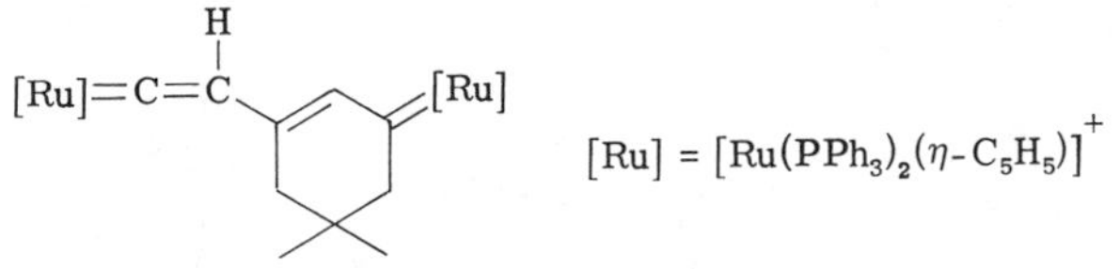

$$[Ru] = [Ru(PPh_3)_2(\eta\text{-}C_5H_5)]^+$$

(49)

The complex $Mn[\!=\!C\!\!=\!\!CHC(OH)Bu_2{}^i](CO)_2(\eta\text{-}C_5H_5)$ dehydrates in the mass spectrometer to give the parent ion of the corresponding propadienylidene complex, $[Mn(\!=\!C\!\!=\!\!C\!\!=\!\!CBu_2{}^i)(CO)_2(C_5H_5)]^+$ (26).

The acylynolate anion (50) can be decarbonylated on irradiation, and subsequent deoxygenation (formal loss of O^{2-}) affords the propadienylidene complex (22). In some cases, this is not isolable or spontaneously polymerizes, but its intermediacy can be inferred from products obtained when the reaction is run in the presence of PPh_3:

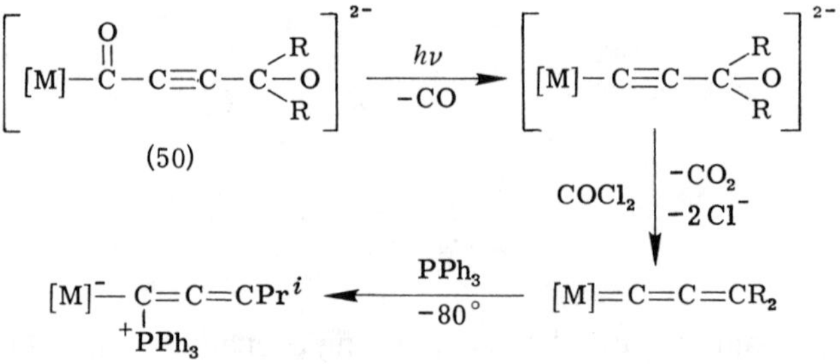

$$\left[[M]-\overset{\overset{O}{\|}}{C}-C{\equiv}C-C\overset{R}{\underset{R}{\diagdown}}O \right]^{2-} \xrightarrow[-CO]{h\nu} \left[[M]-C{\equiv}C-C\overset{R}{\underset{R}{\diagdown}}O \right]^{2-}$$

(50)

$$\text{COCl}_2 \Bigg\downarrow \begin{array}{l} -CO_2 \\ -2\,Cl^- \end{array}$$

$$[M]^-\underset{\overset{|}{{}^+PPh_3}}{C}{=}C{=}CPr^i \xleftarrow[-80°]{PPh_3} [M]{=}C{=}C{=}CR_2$$

$[M] = Cr(CO)_5, W(CO)_5; \quad R = Pr^i, Bu^t$

The anionic complexes (50) are obtained from the group VI carbonyl and the dilithium derivatives of the hydroxyalkynes $HC{\equiv}CCR_2(OH)$; subsequent protonation or acylation affords propadienylidene (51) or carbene (52) complexes by complex cyclization and addition reactions (Scheme 3) (28). Attempts to obtain the dimethyl complex by reaction of the chromium alkynolate dianion with $COCl_2$ gave only polymeric material. More stable complexes were obtained with R = aryl (73).

An alternative route to the dianionic ynolato complex is from propiolic ester complexes and excess organolithium reagents (21, 26, 27):

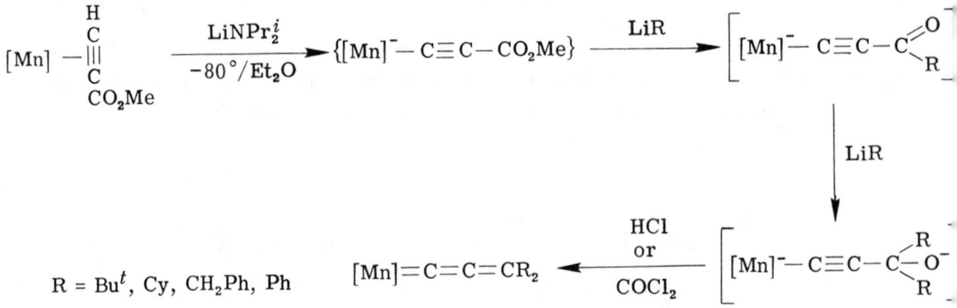

$$[Mn]-\overset{\overset{H}{C}}{\underset{\underset{CO_2Me}{C}}{\overset{|||}{C}}} \xrightarrow[-80°/Et_2O]{LiNPr^i_2} \{[Mn]^--C{\equiv}C-CO_2Me\} \xrightarrow{LiR} \left[[Mn]^--C{\equiv}C-C\overset{O}{\underset{R}{\diagdown}} \right]^-$$

$$\Bigg\downarrow LiR$$

$$R = Bu^t, Cy, CH_2Ph, Ph \quad [Mn]{=}C{=}C{=}CR_2 \xleftarrow[COCl_2]{\begin{array}{c} HCl \\ or \end{array}} \left[[Mn]^--C{\equiv}C-C\overset{R}{\underset{R}{\diagdown}}O \right]^-$$

If methyllithium is used, only polymeric products result in the final step; neutralization with $MeOSO_2F$ affords the dimethylvinylidene complex (Section III,A,2).

The oxapropatrienylidene complex (54) was obtained from $[CrI(CO)_5]^-$ and $AgC{\equiv}CCO_2Na$ in the presence of $AgBF_4$, probably via the η^2-alkyne intermediate (53) (17):

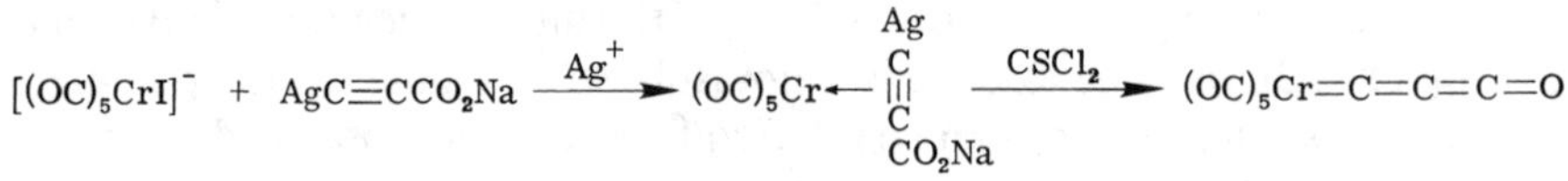

$$[(OC)_5CrI]^- + AgC{\equiv}CCO_2Na \xrightarrow{Ag^+} (OC)_5Cr{\leftarrow}\overset{\overset{Ag}{\underset{C}{\overset{|||}{C}}}}{\underset{CO_2Na}{}} \xrightarrow{CSCl_2} (OC)_5Cr{=}C{=}C{=}C{=}O$$

(53)   (54)

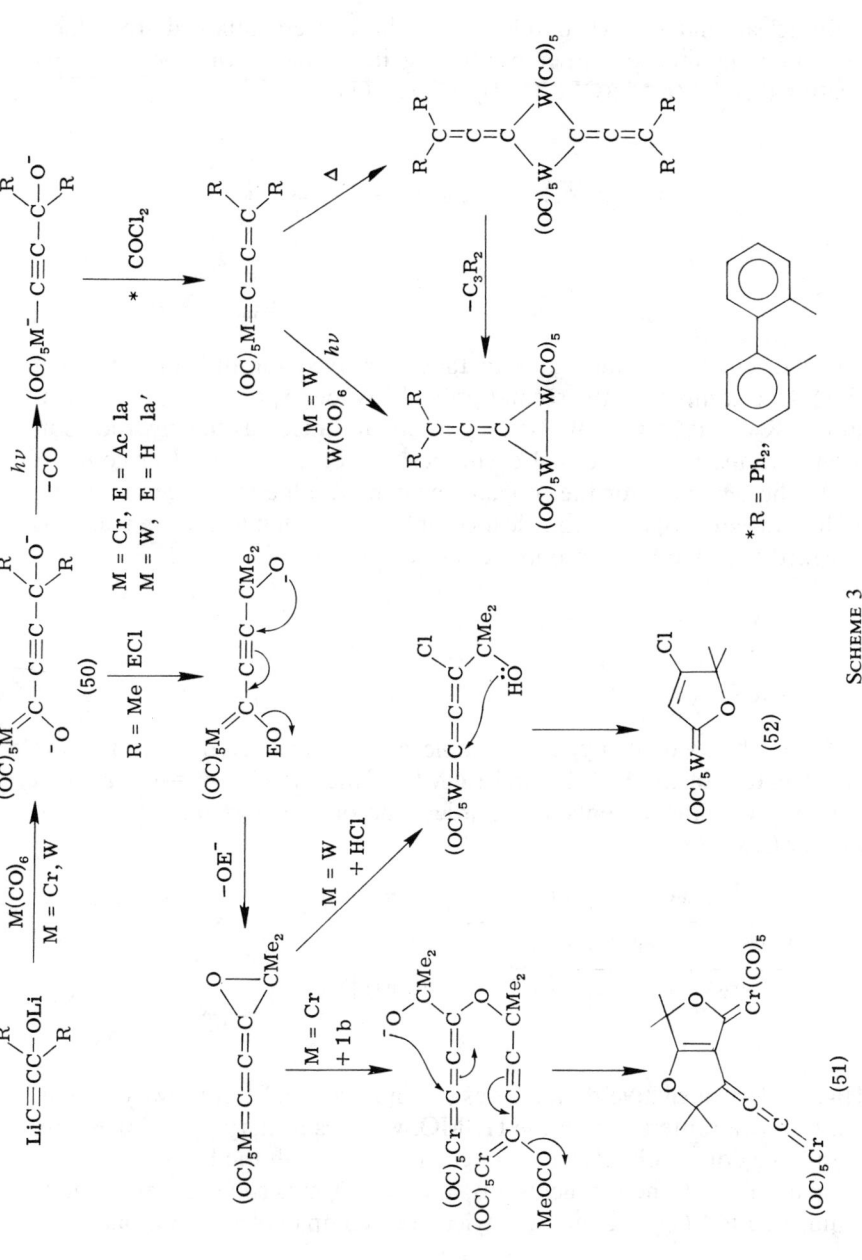

SCHEME 3

Binuclear manganese complexes (55) have been obtained from their mononuclear analogs, either by heating in an inert atmosphere, or by treating them with $Mn(OEt_2)(CO)_2(\eta-C_5H_5)$ (74):

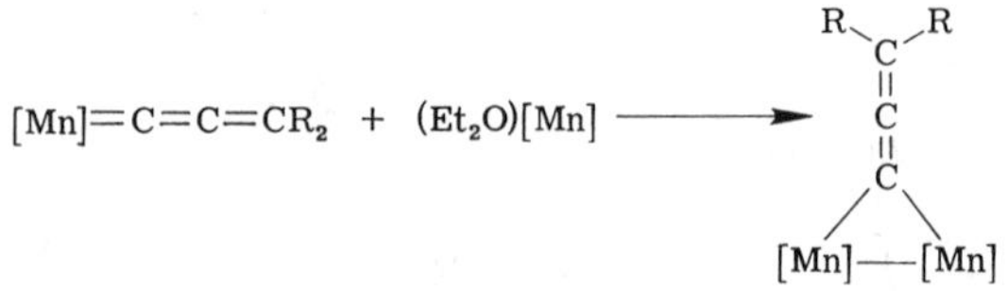

$$[Mn]=C=C=CR_2 \;+\; (Et_2O)[Mn] \longrightarrow$$

R = But, CH₂Ph, Ph, Cy   (55)

These reactions are analogous to those observed for the corresponding vinylidene complexes; the second propadienylidene group is lost, presumably as $R_2C=(C=C)_2=CR_2$; only in one instance has this hydrocarbon been detected in the thermal decomposition of a mononuclear complex (17). The tendency for the propadienylidene residue to bridge two metal atoms is so great that the binuclear complexes are often formed in reactions designed to generate the mononuclear derivatives (73).

B. Reactions

The reactivity of allenylidene complexes is rationalized by consideration of the nature of the HOMO and LUMO of the $M=C=C=C$ fragment (75). Nucleophilic reagents can be ordered according to whether they add to C(1) or C(3):

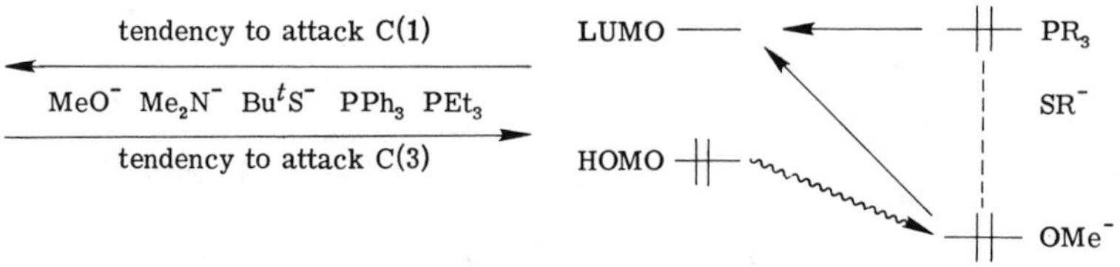

tendency to attack C(1)

MeO⁻ Me₂N⁻ BuᵗS⁻ PPh₃ PEt₃

tendency to attack C(3)

LUMO ——   ——||— PR₃

——||— SR⁻

HOMO ——||—   ——||— OMe⁻

Highly electronegative donor bases (methoxide, etc.) have low-lying lone pairs and can interact with the HOMO, whereas tertiary phosphines have high-lying donor orbitals which interact directly with the LUMO.

In accord with these conclusions, anionic reagents such as methoxide or amide add to C(1), and afford vinylcarbene complexes after protonation:

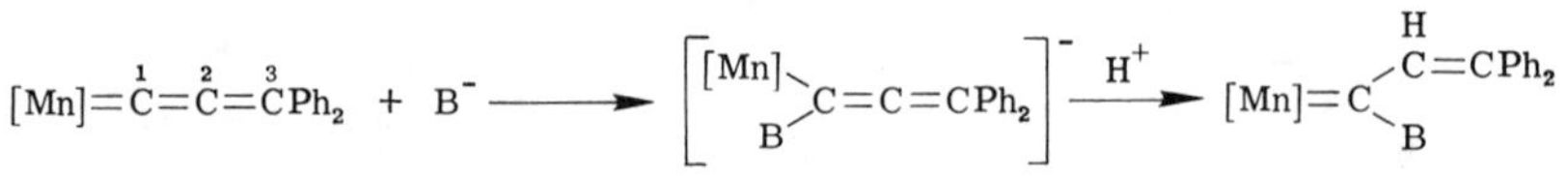

$$[Mn]=\overset{1}{C}=\overset{2}{C}=\overset{3}{C}Ph_2 \;+\; B^- \longrightarrow \left[[Mn]\underset{B}{\overset{}{C}}=C=CPh_2 \right]^- \overset{H^+}{\longrightarrow} [Mn]=C\underset{B}{\overset{C=CPh_2}{\diagup}}$$

Reactions of tertiary phosphines afford zwitterionic phosphonium salts by addition to C(3):

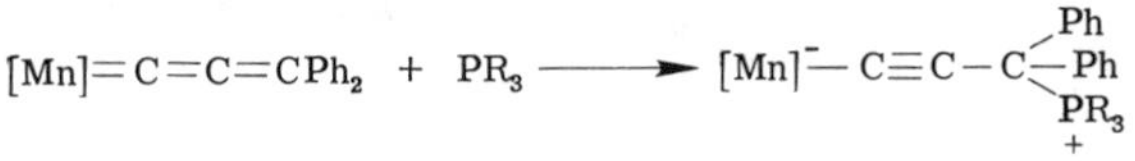

$$[Mn]\!=\!C\!=\!C\!=\!CPh_2 \ + \ PR_3 \longrightarrow [Mn]^- \!-\! C\!\equiv\!C\!-\!C\!\!\begin{array}{l} \nearrow Ph \\ \!\!-\!Ph \\ \searrow PR_3 \\ \ \ + \end{array}$$

The *t*-butylthiolate anion exhibits ambient behavior, giving both types of product, complexes **56** and **57**:

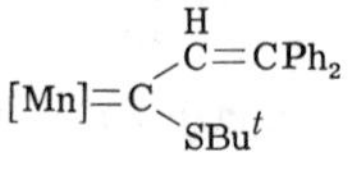

$$[Mn]\!=\!C\!\!\begin{array}{l} \nearrow \overset{\displaystyle H}{C\!=\!CPh_2} \\ \searrow SBu^t \end{array} \qquad\qquad [Mn]\!=\!C\!=\!C\!\!\begin{array}{l} \nearrow H \\ \searrow C(SBu^t)Ph_2 \end{array}$$

$$(56) \qquad\qquad\qquad\qquad (57)$$

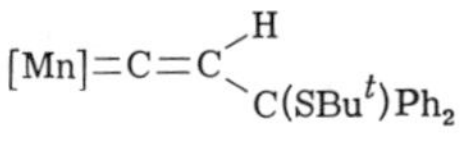

In the reactions of group VI allenylidene complexes, however, tertiary phosphines add to C(1) to give yellow **58**, which was structurally characterized (*22*):

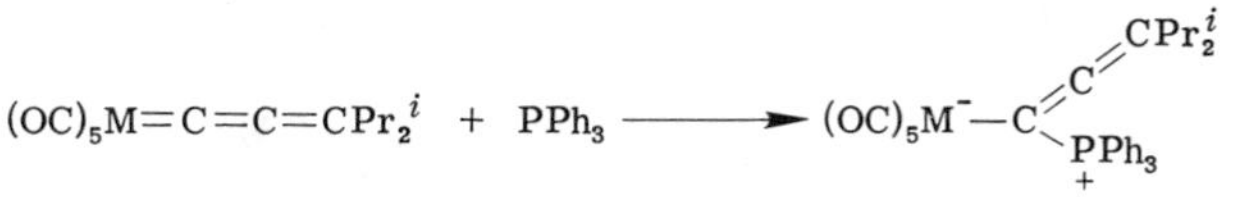

$$(OC)_5M\!=\!C\!=\!C\!=\!CPr_2^i \ + \ PPh_3 \longrightarrow (OC)_5M^-\!-\!C\!\!\begin{array}{l} \nearrow C\!\!\begin{array}{l}\nearrow CPr_2^i\end{array} \\ \searrow PPh_3 \\ \ \ \ + \end{array}$$

$$(58)$$

Addition of *t*-butyllithium, followed by neutralization with HCl or methyl fluorosulfonate, affords vinylidene complexes (**59**) by addition of the Bu^t residue to C(3):

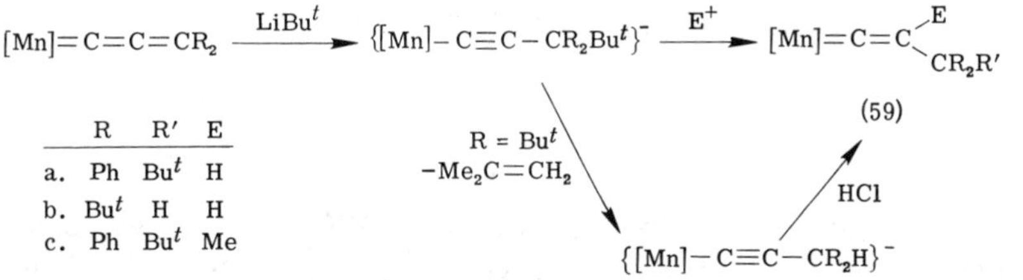

$$[Mn]\!=\!C\!=\!C\!=\!CR_2 \ \xrightarrow{\ LiBu^t\ } \ \{[Mn]\!-\!C\!\equiv\!C\!-\!CR_2Bu^t\}^- \ \xrightarrow{\ E^+\ } \ [Mn]\!=\!C\!=\!C\!\!\begin{array}{l}\nearrow E \\ \searrow CR_2R'\end{array}$$

$$(59)$$

	R	R'	E
a.	Ph	Bu^t	H
b.	Bu^t	H	H
c.	Ph	Bu^t	Me

$$R = Bu^t \searrow \qquad\qquad \swarrow HCl$$
$$-Me_2C\!=\!CH_2$$

$$\{[Mn]\!-\!C\!\equiv\!C\!-\!CR_2H\}^-$$

While the diphenyl complex affords **59a** and **59c**, the di-*t*-butyl derivative undergoes loss of 3-methylbut-l-ene from the intermediate alkynyl complex to give **59b** after protonation (*76*).

VI

SOME CHEMISTRY OF VINYLIDENE COMPLEXES

The majority of the chemistry of vinylidene and propadienylidene complexes is concerned with their synthesis and reactions, which have been described in Sections III–V. Some individual complexes have received more attention, however, and these results are summarized in this section.

A. *Chromium, Molybdenum, and Tungsten*

Although no simple vinylidene complex was obtained from reactions between $Cr(CO)_5(OEt_2)$ and methyl propiolate, complexes **60**, **61**, and **62** were formed in the ratio $4:5:1$ (*22*); the latter two each contain two molecules of the alkyne, and may be formed from an intermediate $Cr[=C=CH(CO_2Me)](CO)_5$ complex (*22*).

(60)   (61)

In nonpolar solvents, **62** changes from red to violet, suggesting formation of the tautomer **62b**:

(a)   (b)

(62)

A low yield of $Cr[=C=C(COMe)CMe_2(OCOMe)](CO)_5$ was obtained from the reaction of $[Cr(C≡CMe_2O)(CO)_5]^{2-}$ with acetyl chloride in the presence of triphenylphosphine (*28*).

Protonation of $W(C≡CPh)(CO)_3(\eta-C_5H_5)$ does not afford a vinylidene complex, the cationic **63** being obtained instead (*77, 78*). However, in the presence of triphenylphosphine, $[W\{C(PPh_3)=CHPh\}(CO)_3(\eta-C_5H_5)]BF_4$ is formed, suggesting the inital formation of a vinylidene derivative (*79*). If the reaction is carried out in the presence of diphenylacetylene, $[W(CO)(\eta^2-HC_2Ph)(\eta^2-C_2Ph_2)(\eta-C_5H_5)]BF_4$ is obtained in 83% yield. Similar molybdenum complexes have also been described. In this reaction, it is suggested that the usual rearrangement of a coordinated 1-alkyne to the analogous vinylidene is reversed, i.e., protonation of the acetylide moiety to the phenylvi-

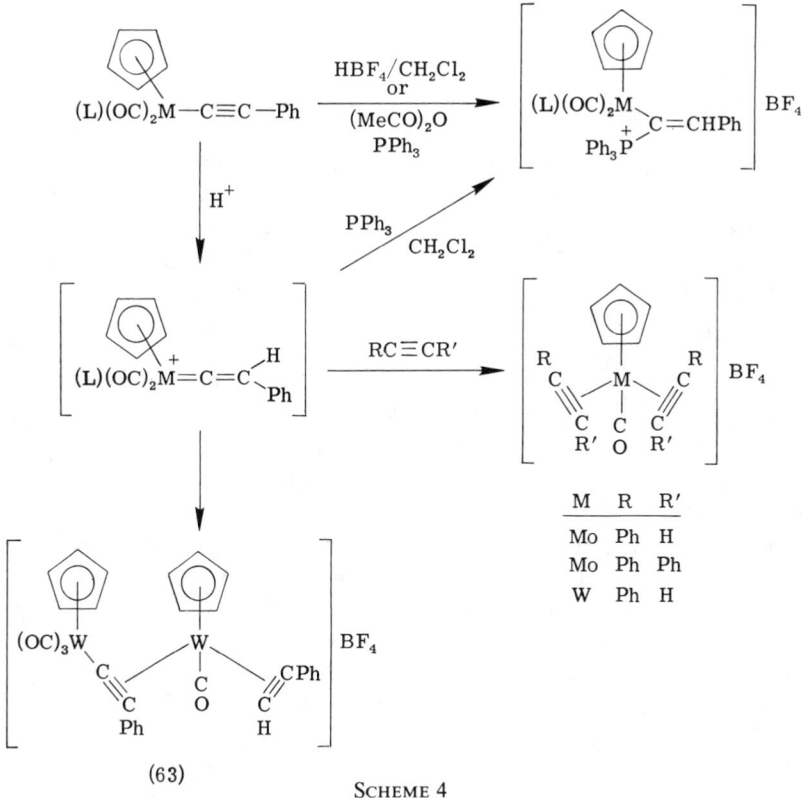

SCHEME 4

nylidene ligand is followed by isomerisation to the η^2-HC$_2$Ph complex (Scheme 4).

B. Manganese and Rhenium

Reactions between Mn(CO)$_3$(η-C$_5$H$_5$) and 1-alkynes with electron-withdrawing substituents, such as CO$_2$Me and COPh, generally stop with formation of the η^2-alkyne complex. Further reaction with base, typically an organolithium compound, affords the vinylidene complex in up to 50% yield, but care must be taken to avoid reaction of excess LiR with the ester or acyl group (21, 59). Unstable Mn(η^2-HC$_2$Ph)(CO)$_2$(η-C$_5$H$_5$) reacts with aqueous methanolic KOH to give light-red Mn$_2$(C$_{16}$H$_{10}$)(CO)$_4$(η-C$_5$H$_5$)$_2$, which may contain a divinylidene ligand :C=CPhCPh=C: (18); the rhenium analog forms a η^1,η^2-2,3-diphenylbutadienylidene complex (19).

C. *Iron*

1. *The Chemistry of Alkyne–Fe(CO)₂(η-C₅H₅) Complexes*

The cationic complex $[Fe(\eta^2\text{-}HC_2Me)(CO)_2(\eta\text{-}C_5H_5)]^+$ has been reported from the protonation (HPF₆) of $Fe(CH=C=CH_2)(CO)_2(\eta\text{-}C_5H_5)$ (*80*). It is described as very air sensitive, and reacts with water rapidly to give a 2:1 mixture of $Fe(CH_2COMe)(CO)_2(\eta\text{-}C_5H_5)$ and $Fe(COEt)(CO)_2(\eta\text{-}C_5H_5)$. The same products were obtained from the allenyl complex and ethanolic HCl, although only the propionyl complex was obtained from $Fe(C\equiv CMe)(CO)_2(\eta\text{-}C_5H_5)$, while aqueous HClO₄ or HBF₄ afforded $[Fe(CO)_3(\eta\text{-}C_5H_5)]^+$ (*81*). Similar reactions occur with $Fe(C\equiv CPh)(CO)_2(\eta\text{-}C_5H_5)$ either aqueous HCl (*82*) or HBF₄·OMe₂ in methanol (*13*) giving the phenylacetyl complex $Fe(COCH_2Ph)(CO)_2(\eta\text{-}C_5H_5)$, and between aqueous HBF₄ and $Fe(C\equiv CPh)(CO)(PPh_3)(\eta\text{-}C_5H_5)$ to give $Fe(COCH_2Ph)(CO)(PPh_3)(\eta\text{-}C_5H_5)$ (*13*).

These reactions have been interpreted as proceeding via attack on an intermediate vinylidene complex $[Fe(=C=CHR)(CO)_2(\eta\text{-}C_5H_5)]^+$, which with water gives the hydroxycarbene complex $[Fe\{C(OH)CH_2R\}(CO)_2(\eta\text{-}C_5H_5)]^+$ (Section III,B,1). As already described, the methoxy(benzyl)carbene complex can be obtained from the phenylacetylide, and an unstable phenyl-vinylidene complex was formed with the triphenylphosphine-substituted complex. Further information on these reactions came from reactions of 1-alkynes, $HC\equiv CR$, with $[Fe(\eta^2\text{-}CH_2=CMe_2)(CO)_2(\eta\text{-}C_5H_5)]^+$, from which cation the isobutylene is readily displaced. In ethanol/dichlorometh-ane mixtures, the ethoxycarbene complexes $[Fe\{C(OEt)R\}(CO)_2(\eta\text{-}C_5H_5)]^+$ were obtained. Phenylacetylene gave a low yield of 2-phenylnaphthalene in a reaction described as "mildly catalytic" in the isobutylene complex; labeling experiments excluded a pathway involving the phenylvinylidene complex. Methyl propiolate afforded a mixture of lactones by trans addition of isobutylene to $[Fe(\eta^2\text{-}HC\equiv CCO_2Me)(CO)_2(\eta\text{-}C_5H_5)]^+$, and in ethanol the formation of *trans*-$CH(OEt)=CH(CO_2Me)$ is catalyzed by the isobuty-lene complex. These products appear to be derived from reactions of the η^2-alkyne complexes which are competitive with their rearrangements to vinylidene complexes and subsequent reactions (*83*).

This is further indicated in the reactions of 3-butyn-1-ol with $[Fe(\eta^2\text{-}CH_2=CMe_2)(CO)_2(\eta\text{-}C_5H_5)]^+$, which afford a mixture of dihydrofuran complex (**64**) and the oxacyclopentylidene complex (**65**) (*84*). The forma-tion of these two derivatives involves a common η^2-alkyne intermediate, which either forms **64** directly by internal nucleophilic attack of the oxygen on the complexed $C\equiv C$ triple bond, or rearranges to the vinylidene. This forms **65** by a similar attack of the hydroxy group on the α-carbon, followed

by hydrogen transfer. The proposed mechanisms are supported by deuteration studies:

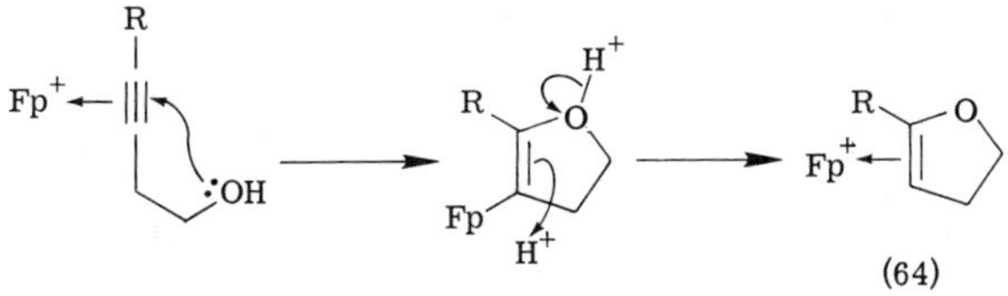

(64)

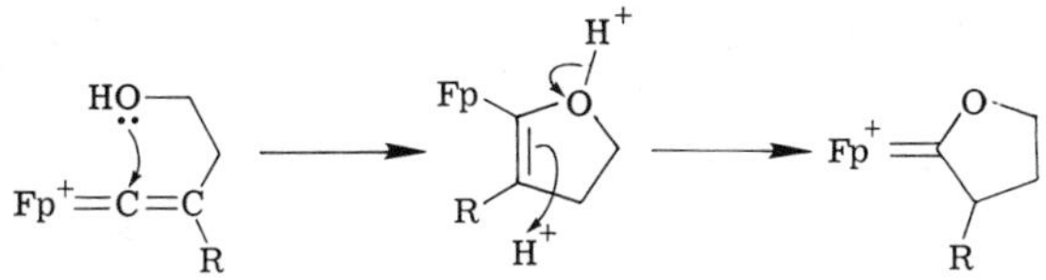

Fp = Fe(CO)$_2$(η-C$_5$H$_5$)

(65)

The reaction with 4-pentyn-1-ol gave only [Fe{η^2-CH$_2$=C(CH$_2$)$_3$O}
(CO)$_2$(η-C$_5$H$_5$)]$^+$, and 3-hexyn-1-ol afforded (64, R = Et) (84); no evidence
for the participation of the vinylidene tautomers was found. With ruthenium (45) and platinum (47) complexes, on the other hand, rearrangement
to the vinylidene is faster than internal attack on the η^2-alkyne, and only the
cyclic carbene complex is formed.

2. Iron–Porphyrin Complexes

Vitamin B$_{12}$ is the only authenticated example of a naturally occurring
organotransition metal complex. The observation of a change in the spectrum of a porphyrin bound to an undecapeptide derived from cytochrome c
after reaction with DDT [1,1-bis(4-chlorophenyl)-3,3,3-trichloroethane,
CCl$_3$CH(C$_6$H$_4$Cl-p)$_2$] in the presence of excess sodium dithionite (85) has
been linked to the intriguing chemistry of iron–diarylvinylidene complexes
uncovered by Mansuy and co-workers (15), suggesting that new organoiron
complexes from natural sources may await characterization.

The "active oxygen"–cytochrome P$_{450}$ complex is a powerful oxidizing
agent towards almost any organic compound, including alkanes, and the
central iron–porphyrin complex has been depicted as

$$(\text{porph})\text{Fe}^{IV}\!\!-\!\text{O}\cdot \quad \text{or} \quad (\text{porph})\text{Fe}^{V}\!\!=\!\text{O}$$

In reduced cytochrome P$_{450}$, the iron(II) is electron-rich by virtue of the
porphyrin ring and the endogenous thiolate, and can reduce a variety of
organic substrates, including nitroarenes, amine oxides, and halomethanes
(15). The metabolism of CCl$_4$ or CF$_3$CHClBr ("halothane" anesthetic)

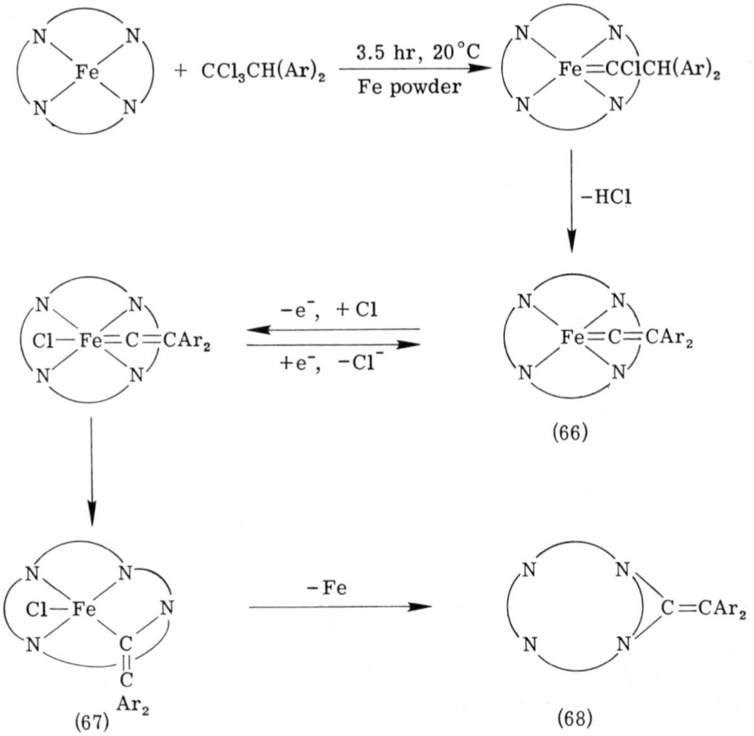

SCHEME 5. Synthesis and reactions of iron–porphyrin vinylidene complexes.

proceeds via the formation of iron–carbene complexes containing CCl_2 or $CH(CF_3)$ ligands. Studies of model compounds produced in reactions between reduced iron porphyrins and CCl_4 afforded complexes such as [(tpp)FeCCl$_2$],[1] but with DDT in the presence of excess iron powder as a reducing agent the purple vinylidene complex (66) was obtained in 50% yield via loss of HCl from the intermediate chlorocarbene complex (Scheme 5) (33). This complex is remarkably stable to oxygen (compared with the dichlorocarbene complex, which has a half-life of 4 hr), and can be purified by normal column chromatographic or TLC methods. Similar compounds were obtained with other iron porphyrin systems containing ttp, tap, oep, dpdme, or ppix.[1] Several ligands (MeOH, pyridine, N-methylimidazole) can be bound in the trans site, although excess pyridine slowly reacts to give Fe(py)$_2$(tpp). The electronic spectra are analogous to those of the oxygen complexes mentioned above, which are probably present in catalase and horseradish peroxidase.

[1] tpp = tetraphenylporphyrin; ttp = tetra-p-tolylporphyrin; tap = tetra-p-anisylporphyrin; oep = octaethylporphyrin; dpdme = deuteroporphyrin IX dimethyl ester; ppix = protoporphyrin IX (all dianions).

Chemical oxidation [Cl_2, Br_2, $CuCl_2$, $FeCl_3$ or $Fe(ClO_4)_3$] results in formation of the chloro complex (67), in which the vinylidene ligand has inserted into an Fe–N bond; complex 66 is regenerated by reaction with iron powder or sodium dithionite, whereas on heating, FeCl(tpp) and the alkyne $C_2(C_6H_4Cl-4)_2$ are formed (86). Confirmation of this unusual structure was obtained from the ^1H and ^{13}C NMR spectra (87), and by two independent X-ray studies (87, 87a), and a similarity with reactions of cobalt porphyrins had been noted (88). Excess oxidant affords the deep green iron-free porphyrin ligand (68) in which the vinylidene bridges two nitrogen atoms; addition of trifluoroacetic acid to 67 affords to corresponding N-vinylporphyrin (89). It is of some interest that (67) is the first example of an iron(III) porphyrin complex found to be in a "pure" intermediate $S = \frac{3}{2}$ spin state (90).

Electrochemical studies of the reversible reduction of 66 have been interpreted in terms of a two-step mechanism:

$$(tpp)Fe^{II}=C=CAr_2 \underset{-e}{\overset{+e}{\rightleftarrows}} [(tpp)Fe^{II}=C=CAr_2]^- \underset{-e}{\overset{+e}{\rightleftarrows}} [(tpp)Fe^{II}=C=CAr_2]^-$$

(lifetime ca 1 sec)

$-H^+ \big\updownarrow +H^+ \qquad\qquad -H^+ \big\updownarrow +H^+$

$$[(tpp)Fe^{III}(CH=CAr_2)] \underset{-e}{\overset{+e}{\rightleftarrows}} [(tpp)Fe^{III}(CH=CAr_2)]^-$$

corresponding to a ($2e + H^+$)-reduction of the vinylidene to a σ-alkyliron(II) complex, which can be reoxidized to a σ-alkyliron(III) derivative (91).

D. Reactions Thought to Proceed via Vinylidene Intermediates

The isolation of stable vinylidene complexes and elucidation of many of their reactions have given substance to speculations concerning their intermediacy in many reactions. Indeed, the reactions of many alkynes with a series of platinum(II) complexes were explained several years ago by considering the formation of metal-stabilized carbonium ions as nonisolable intermediates (10). Summarized below are several reactions that may reasonably be assumed to proceed via vinylidene complexes.

1. Synthesis of σ-Acetylide Complexes

Several metal complexes are known to react with 1-alkynes in the presence of base to give moderate to high yields of the corresponding σ-acetylide. In many cases, similar reactions of main group acetylide reagents, e.g.,

lithium or magnesium derivatives, do not proceed so efficiently. The intermediate formation of a vinylidene complex which is rapidly deprotonated
by the base allows one interpretation of this result:

$$[M]X \xrightarrow{HC\equiv CR} \left\{ [M]=C=C\diagdown_R^H \right\}^+ \xrightarrow{base} [M]-C\equiv C-R$$

[M] = NiR′(PMe$_2$Ph)$_2$ $(46,92,93)$, Ni(NCS)(PR′$_3$)$_2$ (94), PdX(PEt$_3$)$_2$ (95),
PtCl(PMe$_2$Ph)$_2$ (96), Rh(CO)(PR′$_3$)$_2$ (97), Ir(CO)$_2$PPh$_3$)$_2$ (98)

$$[M]X_2 \xrightarrow[\text{stepwise}]{2\,HC\equiv CR} \left\{ [M]\left(=C=C\diagdown_R^H\right)_2 \right\}^{2+} \xrightarrow{base} [M]-(C\equiv C-R)_2$$

[M] = Ni(PR′$_3$)$_2$ $(94, 99)$, Pd(PEt$_3$)$_2$ (100), Pt(PR′$_3$)$_2$ $(96,101)$.

2. Reactions of σ-Acetylide Complexes

a. Protonation. The protonation of several σ-acetylides in the presence
of triphenylphosphine to give vinylphosphonium complexes has been described already. Addition of acetic acid to Fe(CH=C=CH$_2$)(CO)$_2$(η-C$_5$H$_5$)
may proceed by initial protonation to give the methylvinylidene cation,
followed by attack of the α-carbon by acetate to give
Fe[C(OAc)=CHMe](CO)$_2$(η-C$_5$H$_5$) (81).

b. Formation of Alkoxy(alkyl)carbene Complexes. Extensive studies
$(10, 31, 102)$ of the reactions of platinum complexes with 1-alkynes in
alcoholic solvents, of platinum acetylides in alcohols, or of the alcoholysis of
α-chlorovinyl–platinum complexes, all of which afford alkoxy(alkyl)car-
bene complexes, are consistent with the intermediate formation of cationic
platinum–vinylidene complexes. Addition of anhydrous HCl to *trans*-
Pt(C≡CH)$_2$(PMe$_2$Ph)$_2$ affords acetylene and *trans*-PtCl(C≡CH)(L)$_2$ via
α-chlorovinyl complexes, and a stepwise addition–elimination sequence
(Scheme 6) also involving the cationic platinum–vinylidene complexes has
been proposed (31). It is of interest that the heterolytic cleavage of the C–Cl
bond in **69** is apparently facilitated by the unusually large separation
[1.809(6) Å] of the carbon and chlorine atoms. Oxidative addition of HCl to
the acetylide to give a platinum(IV) complex is followed by elimination of
acetylene to generate the chloroplatinum(II) compound.
Further support for the presence of these intermediates is given by the
finding that H/D exchange of *trans*-Pt(C≡CR)$_2$(PMe$_2$Ph)$_2$ (R = H or D)
with MeOD proceeds only in the presence of weak acids; protonation of the
ethynyl group results in exchange because of the greater acidity of the
vinylidene proton.

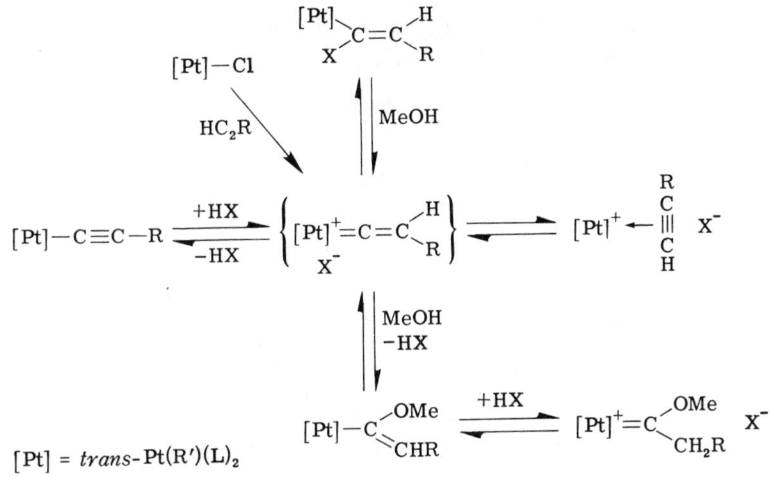

SCHEME 6. Reaction of platinum(II) acetylides with HX.

Several other carbene complexes have been isolated from similar reactions, or from metal halides and 1-alkynes in the presence of alcohols and strong acids ($HClO_4$, HBF_4, HPF_6):

$$[M]-C{\equiv}C-R \xrightarrow[R'OH]{H^+} \left\{ [M]-C{\overset{OR'}{\underset{CH_2R}{\diagup}}} \right\}^+$$

$[M] = trans\text{-}M(C_6Cl_5)(PMe_2Ph)_2$ (M = Ni, Pd, Pt) $(46,92,103)$

$$Pd_2X_4L_2 \xrightarrow[R'OH]{HC{\equiv}CR} cis\text{-}PdX_2L[C(OR')CH_2R] \qquad (ref. 104)$$

X = Cl, Br, I; L = PMe_2Ph, PEt_3; R = Me, Et, Ph; R' = Me, Et, Pr

$$H_2PtCl_6 \cdot 6\,H_2O \begin{cases} \xrightarrow[Pr^iOH]{HC{\equiv}CBu^t} [PtCl_2\{C(OPr^i)CH_2Bu^t\}]_2 \\[2ex] \xrightarrow[Pr^iOH]{HC{\equiv}CSiMe_3} PtCl_2\{C(OPr^i)Me\}_2 \end{cases} \qquad (ref. 105)$$

$$trans\text{-}PtCl(C{\equiv}CFc)(PMe_2Ph)_2 \xrightarrow[MeOH]{H^+} [trans\text{-}PtCl\{C(OMe)CH_2Fc\}(PMe_2Ph)_2]^+ \quad (ref. 106)$$

The last example suggests that the platinum–vinylidene intermediate is more stable than the corresponding ferrocenyl-stabilized carbonium ion.

3. Miscellaneous Reactions

Cycloaddition of dicyclohexylcarbodiimide to an intermediate

$Cr(C=CH_2)(CO)_5$ complex is suggested (*107*) as a possible route for the formation of **70** from $Cr[C(OH)Me](CO)_5$:

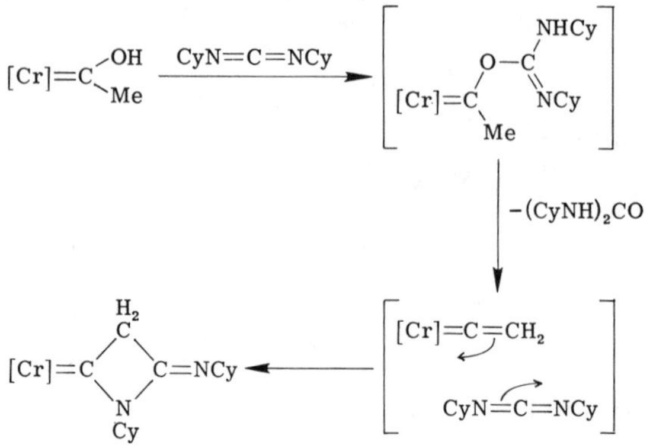

(70)

[Cr] = Cr(CO)$_5$

Mechanistic studies (*108*) have suggested that the formation of the binuclear complex (**71**) on acidification of the product of the reaction between methyllithium and $W(CO)_5[C(OEt)Me]$ proceeds via the intermediate formation of dimethylcarbene– and vinylidene–tungsten complexes. These then cycloadd and rearrange as shown:

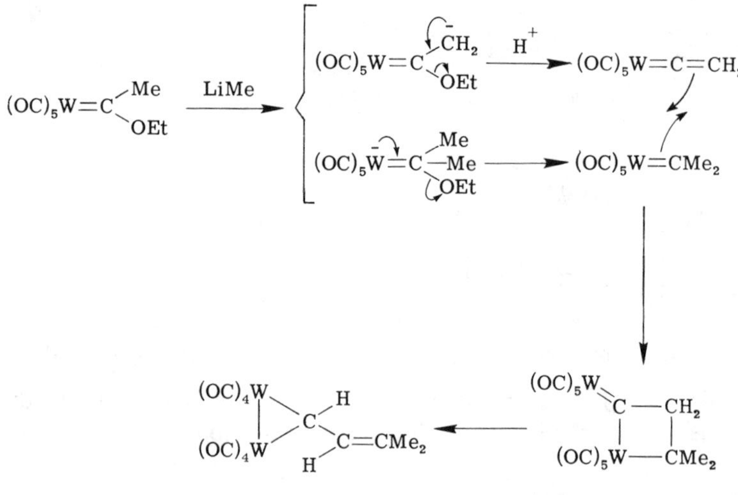

(71)

VII

BONDING AND STRUCTURE IN MONONUCLEAR
AND BINUCLEAR VINYLIDENE COMPLEXES

A. *Bonding*

In principle, free vinylidenes may have structures **K, L,** or **M,** where **K** and

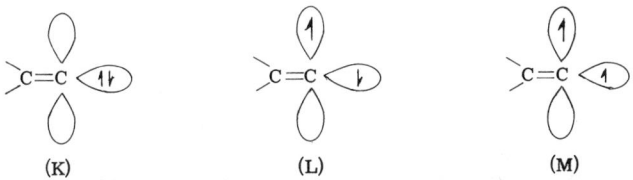

(K)   (L)   (M)

L are singlet states, the latter having two unpaired but spin-paired electrons; form **M** is a triplet with the two unpaired electrons having parallel spins. In contrast with CH_2, which has a triplet ground state, theoretical calculations and experimental results are both consistent with a singlet ground state for $H_2C=C:$. Propadienylidene, $H_2C=C=C:$, is considered to be a hybrid:

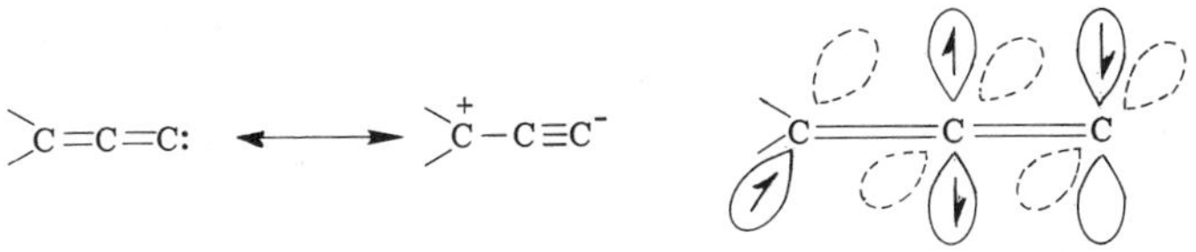

the nature of the substituents and environment, e.g., solvent, influencing the relative contribution from each form. The ground state is a singlet.

In their metal complexes, bonding of either species to the metal atom is via a ligand $\rightarrow$ metal σ donor bond and a metal $\rightarrow$ ligand π bond, enabling back donation of electron density to the π^* orbitals of the C–C multiple bond system to take place. Vinylidene is one of the best π-acceptors known, and is exceeded only by SO_2 and CS in this respect; the relationship between phenylvinylidene and other common ligands has been determined (*18*) from the CO force constants exhibited by a series of $Mn(CO)_2(\eta\text{-}C_5H_5)$ complexes, which increase in this order:

$$PPh_3 < CPh_2 < PhC\equiv CPh < PF_3 < CO < AsF_3 < C=CHPh < CS < SO_2$$

The extreme electron deficiency of the metal-bonded carbon atom is shown by the very low field ^{13}C chemical shifts (ca. 300–450 ppm), and the short

$M-C$ bond lengths testify to efficient backbonding resulting in a bond order of between 2 and 3. This back-donation is stronger with phosphine-substituted complexes.

Calculations of orbital interactions of CH_2, $C=CH_2$ and $C=C=CH_2$ ligands with $Mn(CO)_2(\eta-C_5H_5)$, $Fe(CO)_2(\eta-C_5H_5)^+$, and $Fe(PH_3)_2(\eta-C_5H_5)^+$ moieties (*109, 110*) show that the preferred orientation of the carbene ligand will result in optimal interaction of the empty carbon p orbital with the a'' orbital of the $ML_2(\eta-C_5H_5)$ fragment:

$$a'' \qquad\qquad p$$

Since this p orbital is constrained to lie in the same plane as the substituents on the β-carbon atom, these will thus be in the symmetry plane of the $ML_2(\eta-C_5H_5)$ fragment. Introduction of further carbon atoms will result in successive 90° twists:

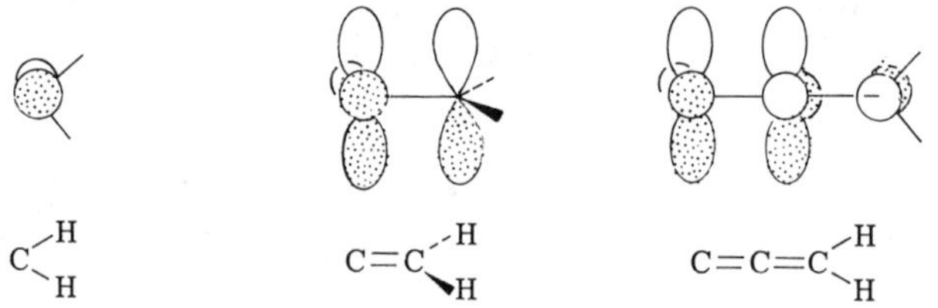

Barrier to rotation in $Fe(CO)_2(\eta-C_5H_5)$ complex:
    6.2   3.6   2.7   kcal mol^{-1}

Calculated barriers to rotation are shown for the iron complexes; for $Mn(=C=C=CH_2)(CO)_2(\eta-C_5H_5)$ the calculated barrier is 3.2 kcal mol^{-1}, which is consistent with the NMR results for the di-*t*-butyl complex (*17, 27*), in which both substituents are magnetically equivalent, probably as a result of free rotation about the $Mn-C$ bond.

In the complexes for which calculations have been performed (*110*), the HOMO is between 25 and 30% localized on C_β, whereas the LUMO, which is π-antibonding between C_α and the metal, is 50–60% localized on C_α. The calculations also indicate that C_β is the most negative ligand atom in the iron complexes; this is consistent with the observed addition to HCl to $[Fe(=C=CH_2)(CO)(PPh_3)(\eta-C_5H_5)]^+$ to give $[Fe(CClMe)(CO)(PPh_3)(\eta-C_5H_5)]^+$. Electrophilic attack on neutral complexes is largely controlled by charge distribution, and addition to C_β usually occurs; however, addition of electrophiles (H^+, Me^+) to cationic complexes has not been described so far. Generally it is found that nucleophiles add to C_α, and this process is frontier – orbital controlled, the addend donating electron density, e.g., from the lone pair HOMO, to the LUMO, breaking the $M–C$ π bond (110). Similar arguments have been advanced to account for the regioselectivity of nucleophilic attack on manganese – propadienylidene complexes. Bases which have high-lying donor orbitals on polarizable atoms, e.g., PR_3 or SBu^t, attack the γ-carbon, whereas bases with more electronegative donor atoms, such as OMe^- and NMe_2^-, add to the α-carbon (75).

B. Structure

Tables I and II summarize the structural studies of mononuclear and binuclear vinylidene complexes, and Table III those of propadienylidene complexes which had been reported to mid-1982. As can be seen, the $C{=}C$ bond lengths range from 1.29 to 1.38 Å, and the $M–C$ bond (1.7–2.0 Å) is considerably shorter than those found in alkyl or simple carbene complexes. Both observations are consistent with the theoretical picture outlined above, and in particular, the short $M–C$ bonds confirm the efficient transfer of electron density to the π^* orbitals. In mononuclear complexes, the $M{-}C{=}C$ system ranges from strictly linear to appreciably bent, e.g., $167°$ in $MoCl[C{=}C(CN)_2][P(OMe_3)_2]_2(\eta\text{-}C_5H_5)$; these variations have been attributed to electronic rather than steric factors. In the molybdenum complex cited, the vinylidene ligand bends towards the cyclopentadienyl ring (111).

The major orbital interactions of the μ-vinylidene ligand in binuclear complexes are with (i) an unperturbed π_{xy} orbital of the Rh_2 fragment, (ii) a bonding linear combination of the Rh_2 π_{xy}^* and C p orbitals, and (iii) a bonding linear combination of the Rh_2 σ orbitals. The small rotation ($0° – 14°$) which is often found between the CR_2 and the CM_2 planes serves to optimize these orbital overlaps.

The pattern of $C{=}C$ bond lengths in propadienylidene complexes is a short $C(1)–C(2)$ bond (ca. 1.25 Å) and a normal $C(2)–C(3)$ bond (ca. 1.37 Å).

C. Spectroscopic Properties

All known vinylidene and propadienylidene complexes are listed in Tables IV – VII, together with some salient physical properties. In their IR spectra, $\nu(CC)$ generally occurs between 1590 and 1660 cm^{-1}; in their Raman spectra, strong $\nu(CC)$ bands are found at 1590 and 1594 cm^{-1} for

TABLE I

MONONUCLEAR VINYLIDENE COMPLEXES: SOME STRUCTURAL PARAMETERS[a]

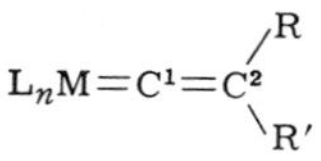

$$L_nM=C^1=C\begin{smallmatrix}R\\R'\end{smallmatrix}$$

L_nM	R	R'	$M-C^1$	C^1-C^2	$M-C^1-C^2$	Reference
MoCl[P(OMe)₃]₂(η-C₅H₅)	CN	CN	1.833(6)	1.378(8)	166.6(4)	141
Mn(CO)₂(η-C₅H₅)	H	Ph	1.68(2)	1.34(3)	174(2)	142
Mn(CO)₂(η-C₅H₄Me)	Me	Me	1.79(2)	1.33(2)	176(2)	76
Re(CO)₂(η-C₅H₅)	Ph	(η²-CPh=CH₂)-Re(CO)₂(η-C₅H₅)	1.90(2)	1.33(3)	179(3)	19
[Fe(dppe)(η-C₅H₅)]⁺ I⁻	Me	CS₂Me	1.74(2)	1.31(2)	176(1)	29
[Ru(PMe₃)₂(η-C₅H₅)]⁺ [PF₆]⁻	H	Me	1.845(7)	1.313(10)	180(2)	49
[Ru(PPh₃)₂(η-C₅H₅)]⁺ [PF₆]⁻	Me	Me	1.88	1.29	168.1	143
Os(CO)₂(PPh₃)₂	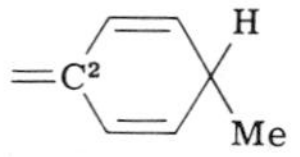 (=C²⟨cyclohexadienyl with H, Me⟩)		1.90(1)	1.33(1)	169(3)	36

[a] Distances are in angstroms, angles in degrees.

TABLE II

MONONUCLEAR PROPADIENYLIDENE COMPLEXES: SOME STRUCTURAL PARAMETERS[a]

$$L_nM=C^1=C^2=C^3\begin{smallmatrix}R\\ \diagdown\\ \diagup\\ R'\end{smallmatrix}$$

ML_n	R	R'	$M—C^1$	$C^1—C^2$	$C^2—C^3$	$M—C^1—C^2$	$C^1—C^2—C^3$	Reference
Cr(CO)$_5$	Ph	NMe$_2$	2.015(15)	1.236(22)	1.372(21)			16
Cr(CO)$_5$	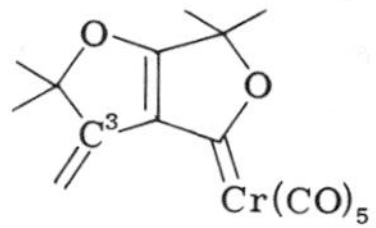 (cyclic structure with Cr(CO)$_5$, C^3)	Cr(CO)$_5$	1.913(7)	1.26(1)	1.359(9)	173.4(6)	174.3(7)	28
Mn(CO)$_2$(η-C$_5$H$_4$Me)	Cy	Cy	1.806(6)	1.2528(8)	1.342(8)	177.9(5)	175.1(6)	75
[Ru(PMe$_3$)$_2$(η-C$_5$H$_5$)]$^+$ [PF$_6$]$^-$	Ph	Ph	1.884(5)	1.255(8)	1.329(9)	175.9(5)	175.1(7)	71

[a] Distances are in angstroms, angles in degrees.

TABLE III

BINUCLEAR VINYLIDENE AND PROPADIENYLIDENE COMPLEXES: SOME STRUCTURAL PARAMETERS[a]

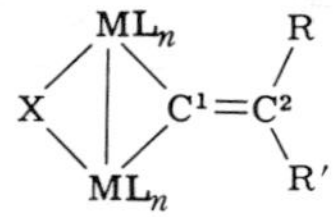

(A) 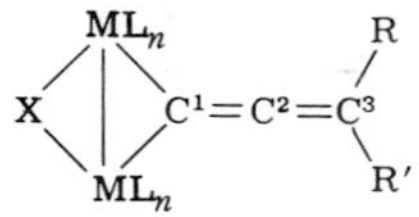   (B)

$[ML_n]_2$	X	R	R'	M—C	C^1—C^2	M—M	M—C^1—M	Torsion angle[b]	Reference
(A) Vinylidene complexes									
$[Mn(CO)_2(\eta\text{-}C_5H_5)]_2$	—	H	H	1.979(7), 1.971(6)	1.308(10)	2.759(2)	88.6(3)	11	56
$[Mn(CO)_2(\eta\text{-}C_5H_5)]_2$	—	H	Ph	1.94(1), 1.99(1)	1.35(2)	2.734(2)	88.0(5)	7	144
$Mn(CO)_2(\eta\text{-}C_5H_5)$ ⎫ $Fe(CO)_4$ ⎬	—	H	CO_2Me	1.95(2), 1.94(2)	1.30(2)	2.703(4)	88.1(7)	7	139
$[Fe(CO)_4]_2$	—	Ph	Ph	1.98(1), 1.98(1)	1.33(1)	2.635(3)	83(1)	13.9	7

cis-[Fe(CO)(η-C₅H₅)]₂	CO	Ph	CH₂Ph	1.936(2), 1.944(2)	1.326(3)	2.5104(5)	80.6(1)	24.8	145
cis-[Fe(CO)(η-C₅H₅)]₂[c]	CO	CN	CN	1.84(2), 1.90(2)	1.38(3)	2.511(4)	84(2)	0.3	146
cis-[Ru(CO)(η-C₅H₅)]₂	CO	H	H	2.025(7), 2.033(7)	1.325(11)	2.695(1)	83.2(3)		68
[Co(CO)₃]₂	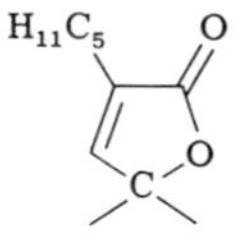 (structure: $H_{11}C_5$–dimethyl furanone ring)	I	I	1.89, 1.91	1.32	2.38			62
[Rh(CO)(η⁵-C₉H₇)]₂	—	H	H	1.982(3), 1.988(3)	1.304(5)	2.691(1)	85.4(1)	7	61
(B) Propadienylidene complex [W(CO)₅]₂[c]	—	Ph	Ph	2.21(5), 2.23(4)	1.28(4)	3.16(6)	91(1)	8.13	73

[a] Distances are in angstroms, angles in degrees.
[b] Angle between M—C¹—M and CRR′ planes.
[c] Average values for 2 molecules in asymmetric unit; C²–C³ 1.33(4) Å.

103

TABLE IV

$$L_nM=C^1=C^2\begin{smallmatrix}R\\ \\R'\end{smallmatrix}$$

ML_n	R	R'	Color
$Cr(CO)_5$	COMe	$CMe_2(OCOMe)$	
$Cr(CO)_5$	CO_2Me	$CH=CH(CO_2Me)$	Orange-red
$MoCl(PPh_3)_2(\eta\text{-}C_5H_5)$	CN	CN	Red-orange
$MoCl(PMe_2Ph)_2(\eta\text{-}C_5H_5)$	CN	CN	Orange-yellow
$MoCl[P(OMe)_3]_2(\eta\text{-}C_5H_5)$	CN	CN	Orange-yellow
$MoCl[P(OEt)_3]_2(\eta\text{-}C_5H_5)$	CN	CN	Yellow
$MoCl[P(OPh)_3]_2(\eta\text{-}C_5H_5)$	CN	CN	Orange-yellow
$MoCl(\eta^1\text{-dppe})_2(\eta\text{-}C_5H_5)$	CN	CN	Red-orange
$MoCl(dppe)(\eta\text{-}C_5H_5)$	CN	CN	
$MoCl(cis\text{-}Ph_2PCH=CHPPh_2)(\eta\text{-}C_5H_5)$	CN	CN	Deep green
$MoCl(AsPh_3)_2(\eta\text{-}C_5H_5)$	CN	CN	Orange-brown
$MoCl(SbPh_3)_2(\eta\text{-}C_5H_5)$	CN	CN	Red-brown
$[Mo\{PhP(CH_2CH_2PPh_2)_2\}(\eta\text{-}C_5H_5)]^+$	CN	CN	Yellow-orange[b]
$[Mo\{P(OMe)_3\}_2(\eta\text{-}C_5H_5)]^-$	H	Bu^t	Red
$MoI[P(OMe)_3]_2(\eta\text{-}C_5H_5)$	H	Bu^t	Red
$Mo(N_2C_6H_4F\text{-}4)[P(OMe)_3](\eta\text{-}C_5H_5)$	H	Bu^t	Dark red
$WCl(PPh_3)_2(\eta\text{-}C_5H_5)$	CN	CN	Yellow-orange
$WCl[P(OMe)_3]_2(\eta\text{-}C_5H_5)$	CN	CN	Yellow
$WCl[P(OEt)_3]_2(\eta\text{-}C_5H_5)$	CN	CN	Yellow
$WCl(AsPh_3)_2(\eta\text{-}C_5H_5)$	CN	CN	Orange-yellow
$Mn(CO)_2(\eta\text{-}C_5H_5)$	H	Ph	Dark red
$Mn(CO)_2(\eta\text{-}C_5H_5)$	H	CO_2Me	Orange-brown
$Mn(CO)_2(\eta\text{-}C_5H_5)$	H	$C(OH)Bu^t_2$	Pale red
$Mn(CO)_2(\eta\text{-}C_5H_5)$	Me	Me	Red-orange
$Mn(CO)_2(\eta\text{-}C_5H_5)$	Me	$COBu^t$	Orange-red
$Mn(CO)_2(\eta\text{-}C_5H_5)$	Me	CO_2Me	Orange
$Mn(CO)(PPh_3)(\eta\text{-}C_5H_5)$	H	Ph	Dark orange
$Mn(CO)[P(OEt)_3](\eta\text{-}C_5H_5)$	H	Ph	Red
$Mn(CO)[P(OPh)_3](\eta\text{-}C_5H_5)$	H	Ph	Orange-red
$Mn(CO)_2(\eta\text{-}C_5H_4Me)$	H	$CHBu^t_2$	Pale red
$Mn(CO)_2(\eta\text{-}C_5H_4Me)$	H	CBu^tPh_2	Red-orange
$Mn(CO)_2(\eta\text{-}C_5H_4Me)$	H	$CPh_2(SBu^t)$	Yellow
$Mn(CO)_2(\eta\text{-}C_5H_4Me)$	Me	CBu^tPh_2	Red-orange
$Mn(CO)_2(\eta\text{-}C_5H_4Et)$	H	Ph	Red
$Re(CO)_2(\eta\text{-}C_5H_5)$	H	Ph	Red
$Re(CO)_2(\eta\text{-}C_5H_5)$	Ph	$(\eta^2\text{-}CPh=CH_2)\text{-}Re(CO)_2(\eta\text{-}C_5H_5)$	Light red
$FeCl(depe)_2$	H	Ph	Green[b]
$[Fe(CO)(PPh_3)(\eta\text{-}C_5H_5)]^+$	H	H	Yellow-gold[c]
$[Fe(CO)(PPh_3)(\eta\text{-}C_5H_5)]^+$	H	Ph	Pink[c]
$[Fe(CO)(PPh_3)(\eta\text{-}C_5H_5)]^+$	Me	Me	Yellow[c]
$[Fe(CO)(PMe_2Ph)(\eta\text{-}C_5H_5)]^+$	H	H	Lime green[c]

MONONUCLEAR VINYLIDENE COMPLEXES

Yield (%)	mp (°C)	v (C=C)	δ (C^1)^a	δ (C^2)^a	Reference
7		1697			*28*
8.6					*22*
90	162–164d				*9,30*
11	181–183				*30*
61	135–136d				*30*
49	86–87				*30*
67	123–124d				*30*
50	123–124				*30*
60	199–202d				*30*
68	252–254				*30*
76	130–132d				*30*
34	120d				*30*
85	dec > 275				*9,30*
		1602	322.8t(17)	121.8s	*35*
		1605	326.4d(51)	132.7t(12)	*34*
31		1608	348.6d(30)	141.3s	*34*
35	166–168d				*9,30*
26	dec > 149				*30*
17					*30*
10	dec > 131				*30*
10	64–65		379.5	123.5	*18*
94		1610			*26*
92		1652			*26*
41					*27*
32		1595			*26*
93		1631			*26*
24	174d				*18*
24	Oil				*18*
21	110–112	1590			*18*
41	55–56	1658			*76*
62	95–96	1650			*76*
37	101–102·				*75*
46	Oil	1649			*76*
	Oil				*18*
12	75–76	1591	329.5	119.5	*19*
4.3	150–152				*19*
		1572–1609			*12*
90	154–155d	1630	372.4d(29)	107.1	*37,38*
		1675			*13*
90	135–136d	1621			*37,38*
70	130–134d	1633			*37,38*

(*Continued*)

TABLE IV

$$L_nM = C^1 = C^2 \underset{R'}{\overset{R}{<}}$$

ML_n	R	R'	Color
$[Fe(CO)(PCy_3)(\eta\text{-}C_5H_5)]^+$	H	H	Lime green[c]
$[Fe(dppe)(\eta\text{-}C_5H_5)]^+$	H	H	
$[Fe(dppe)(\eta\text{-}C_5H_5)]^+$	H	Me	Pale orange
$[Fe(dppe)(\eta\text{-}C_5H_5)]^+$	Me	Me	Orange[d]
$[Fe(dppe)(\eta\text{-}C_5H_5)]^+$	Me	CS_2Me	Yellow-brown[e]
Fe(tpp)	Ph	Ph	
Fe(tpp)	C_6H_4Cl-4	C_6H_4Cl-4	Purple
Fe(tpp)	C_6H_4Cl-4	C_6H_4Cl-4	Purple
Fe(tap)	C_6H_4Cl-4	C_6H_4Cl-4	Purple
Fe(oep)	C_6H_4Cl-4	C_6H_4Cl-4	Purple
Fe(ppix)	C_6H_4Cl-4	C_6H_4Cl-4	Purple
$[Ru(PMe_3)_2(\eta\text{-}C_5H_5)]^+$	H	H	Yellow[b]
$[Ru(PMe_3)_2(\eta\text{-}C_5H_5)]^+$	H	Me	Orange[b]
$[Ru(PMe_3)_2(\eta\text{-}C_5H_5)]^+$	H	Ph	Pink[b]
$[Ru(PMe_3)_2(\eta\text{-}C_5H_5)]^+$	Me	Ph	Pink[b]
$[Ru(PPh_3)_2(\eta\text{-}C_5H_5)]^+$	H	Me	Orange-red[b]
$[Ru(PPh_3)_2(\eta\text{-}C_5H_5)]^+$	H	Pr	Tan[b]
$[Ru(PPh_3)_2(\eta\text{-}C_5H_5)]^+$	H	CO_2Me	Orange-brown[b]
$[Ru(PPh_3)_2(\eta\text{-}C_5H_5)]^+$	H	Ph	Red-purple[b]
$[Ru(PPh_3)_2(\eta\text{-}C_5H_5)]^+$	H	C_6H_4F-4	Red-brown[b]
$[Ru(PPh_3)_2(\eta\text{-}C_5H_5)]^+$	H	C_6F_5	Tan[b]
$[Ru(PPh_3)_2(\eta\text{-}C_5H_5)]^+$	Me	Me	Deep red[b]
$[Ru(PPh_3)_2(\eta\text{-}C_5H_5)]^+$	Me	Pr	Deep red[b]
$[Ru(PPh_3)_2(\eta\text{-}C_5H_5)]^+$	Me	Ph	Red[b]
$[Ru(PPh_3)_2(\eta\text{-}C_5H_5)]^+$	Me	C_6F_5	Pink[b]
$[Ru(PPh_3)_2(\eta\text{-}C_5H_5)]^+$	Et	Ph	Red[b]
$[Ru(dppm)(\eta\text{-}C_5H_5)]^+$	H	Ph	Buff[b]
$[Ru(dppe)(\eta\text{-}C_5H_5)]^+$	H	Bu	
$[Ru(dppe)(\eta\text{-}C_5H_5)]^+$	H	Ph	Pale orange[b]
$Os(CO)_2(PPh_3)_2$	$=C^2\!\!-\!\!\langle\rangle\!\!<^H_{Me}$		Yellow
$Os(CO)(CNC_6H_4Me\text{-}4)(PPh_3)_2$	$=C^2\!\!-\!\!\langle\rangle\!\!<^H_{Me}$		Yellow
$[Os(PPh_3)_2(\eta\text{-}C_5H_5)]^+$	H	Ph	Light purple

[a] Chemical shifts in ppm; $J(CP)$ in parentheses.
[b] PF_6 salt.
[c] BF_4 salt.
[d] SO_3F salt.
[e] I salt.

(*Continued*)

Yield (%)	mp (°C)	ν (C=C)	δ (C¹)ᵃ	δ (C²)ᵃ	Reference
88	170d	1629			*38*
		1625			*24*
		1658	358.3t(33)	118.0	*24*
	230–232d	1675	363.3t(33)	127.8	*24,39*
64	171–177d	1550	364.5	144.7	*29*
					91
50	dec > 70				*33*
					33
					33
					33
					33
79	dec > 100	1632		127.35	*49*
83	155d	1650			*49*
85	185–190d	1650		114.7	*23,49*
56	195d	1650			*49*
89	120–125	1655	349.4t(17)	109.0s	*20*
86	dec > 98	1660		114.7	*20*
92		1640		113.3	*20*
88	105–110d	1640	358.9t(24)	119.6	*14,20*
84	120	1635			*20*
80	dec > 60	1640			*20*
67	202–207	1678			*20*
77	194–196	1678	352.0m	121.3	*20*
93	205–210d	1665	365.7	125.2	*20*
61	191–194	1638			*20*
87	195–200	1665			*20*
92	208–210	1651			*20*
					138
79	210–212	1646			*20*
		1649			*36*
		1604			*36*
94	136–138	1648		108.2	*20*

TABLE V
BINUCLEAR VINYLIDENE DERIVATIVES

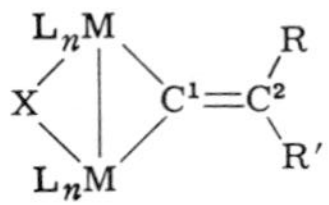

$$\begin{array}{c} L_nM \\ \diagdown \\ X \quad\; C^1\!=\!C^2 \\ \diagup \qquad\quad \diagdown R' \\ L_nM \qquad\quad R \end{array}$$

$[ML_n]_2$	X	R	R'	Color	Yield (%)	mp (°C)	ν(C=C)	Reference
$[Mn(CO)_2(\eta\text{-}C_5H_5)]_2$	—	H	H	Red-purple	42			38
$[Mn(CO)_2(\eta\text{-}C_5H_5)]_2{}^a$	—	H	Ph	Violet	27	144	1550	18
$[Mn(CO)_2(\eta\text{-}C_5H_5)]_2$	—	H	COPh		1–2			58
$[Mn(CO)_2(\eta\text{-}C_5H_5)]_2$	—	H	CO_2Me	Red-violet	70		1525	21,26,29
$Mn(CO)_2(\eta\text{-}C_5H_5)$ $\}$ $Re(CO)_2(\eta\text{-}C_5H_5)$ $\}$	—	H	Ph	Dark orange	4	161–163	1552	149
$Mn(CO)_2(\eta\text{-}C_5H_5)$ $\}$ $Fe(CO)_4$ $\}$	—	H	CO_2Me	Dark red	73	110–111d	1545	139
$Mn(CO)_2(\eta\text{-}C_5H_5)$ $\}$ $Fe(CO)_4$ $\}$	—	CO_2Me	H	Red	8.7	100–102	1555	139
$[Re(CO)_2(\eta\text{-}C_5H_5)]_2$	—	H	Ph	Yellow	3	193–194	1555	19
$[Fe(CO)_4]_2$	CO	Ph	Ph	Orange	93			7
$[Fe(CO)_2(\eta\text{-}C_5H_5)]_2{}^b$	CO	H	H	Orange-red				65,147
$[Fe(CO)(\eta\text{-}C_5H_5)]_2$ cis	CO	Ph	CH_2Ph	Red	17			64
$[Fe(CO)(\eta\text{-}C_5H_5)]_2$ trans	CO	Ph	CH_2Ph	Purple				64

Compound		R	R'	Color		mp (°C)	ν(C=C) (cm^{-1})	Ref.
[Fe(CO)(η-C$_5$H$_5$)]$_2$	CO	Me	CH$_2$(tol)					64
[Fe(CO)(η-C$_5$H$_5$)]$_2$	CO	Me	CHMePh					64
[Fe(CO)(η-C$_5$H$_5$)]$_2$	CO	Me	CHMe(tol)					64
[Fe(CO)(η-C$_5$H$_5$)]$_2$	CO	Me	CH$_2$C$_6$H$_4$OMe-4					64
[Fe(CO)(η-C$_5$H$_5$)]$_2$	CO	Me	Ph					64
[Fe(CO)(η-C$_5$H$_5$)]$_2$ cis	CO	CN	CN	Red-orange	2–7	dec 260	1480	8,63
[Fe(CO)(η-C$_5$H$_5$)]$_2$ trans	CO	CN	CN	Red-violet	0.3	dec 255	1480	8,63
[Ru(CO)(η-C$_5$H$_5$)]$_2$ cis^c	CO	H	H	Yellow	30	178–180		68
[Ru(CO)(η-C$_5$H$_5$)]$_2$ trans	CO	H	H	Yellow	35	177		68
[Ru(CO)(η-C$_5$H$_5$)]$_2$ trans^d	CO	H	H	Yellow	66			66
[Co(CO)$_3$]$_2$^{e}	CMe$_2$	I	I	Orange-red				62
[Rh(CO)(η^5-C$_9$H$_7$)]$_2$^{f}	H	—	H	Red	50			61

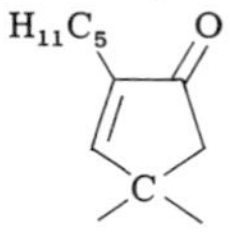

(Structure drawn in the [Co(CO)$_3$]$_2$ / CMe$_2$ entry: a cyclopentenone ring bearing H$_{11}$C$_5$, O, and CMe$_2$ substituents.)

^{a} δ(C$_1$) 284.2, δ(C$_2$) 125.2 ppm.

^{b} δ(C$_1$) + δ(μ-CO) 271.3, 277.2; δ(C$_2$) 125.6 ppm.

^{c} δ(C$_1$) 249.1, δ(C$_2$) 122.7 ppm.

^{d} δ(C$_1$) 244.5, δ(C$_2$) 122.2 ppm.

^{e} Several complexes with R = Cl, R' = Bu; R = Br, R' = Br, Bu, Ph; R = I, R' = Bu and a variety of similar μ-lactone–carbene ligands described: ν(C=C) 1550–1590 cm^{-1}.

^{f} δ(C$_2$) 111.2 ppm (−90°C).

TABLE VI

MONONUCLEAR PROPADIENYLIDENE COMPLEXES

$$L_nM=C^1=C^2=C^3{\overset{R}{\underset{R'}{}}}$$

ML_n	R	R'	Color	Yield (%)	mp (°C)	ν(C=C)	δ(C^1)	δ(C^2)	δ(C^3)	Reference
Cr(CO)₅	CHMe₂	CHMe₂	Deep red	34	dec	1930				22
Cr(CO)₅	Buᵗ	Buᵗ	Red violet	32		1930				22
Cr(CO)₅	Ph	NMe₂	Blue	24	151d	1923	198.9	121.3	157.3	16
Cr(CO)₅	Ph	Ph	Black-violet	36	dec > 10	1930				73
Cr(CO)₅		=O	Red-violet	27	dec 32	2028	440.6		389.9	140
Cr(CO)₅	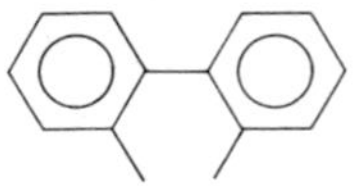 (2,2′-dimethylbiphenyl structure)				dec	1920				73
Cr(CO)₅	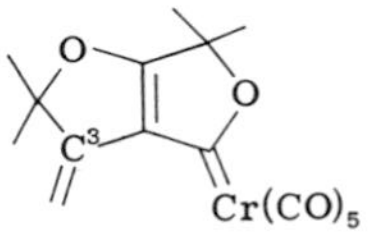 (furanone–C^3 / Cr(CO)₅ structure)		Green	36	>280	1933				28
W(CO)₅	CHMe₂	CHMe₂	Deep red	23	dec	1930				22
W(CO)₅	Buᵗ	Buᵗ		<1		1930				22
W(CO)₅	Ph	Ph	Red			1920				73
Mn(CO)₂(η-C₅H₅)	Buᵗ	Buᵗ	Orange-brown	90	123	1922	331.2	167.0	213.6	17,21,27
Mn(CO)₂(η-C₅H₅)	CH₂Ph	CH₂Ph	Orange-red	25	24	1922	382.4	140.2	229.6	27
Mn(CO)₂(η-C₅H₅)	Cy	Cy	Brown-black	42	125	1925	323.4	169.5	202.3	27
Mn(CO)₂(η-C₅H₅)	Ph	Ph	Black-violet	43	dec 125	1909	304.5	139.8	223.3	21,27
[Ru(PMe₃)₂(η-C₅H₅)]⁺	Ph	Ph	Orange-brown	76		1926	295.8	216.0	153.8	71

^{a} PF₆ salt.

TABLE VII
BINUCLEAR PROPADIENYLIDENE COMPLEXES

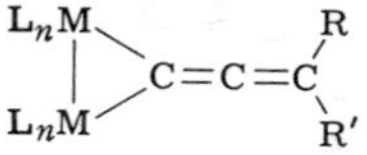

$$L_nM - \overset{\displaystyle R}{\underset{\displaystyle R'}{C=C=C}}$$

(with L_nM bridging)

[ML$_n$]$_2$	R	R'	Color	Yield (%)	mp (°C)	ν(C=C)	Reference
[Cr(CO)$_5$]$_2$	Ph	Ph	Deep blue	24	dec −10		73
[W(CO)$_5$]$_2$	Ph	Ph	Deep blue	10	dec 93	1866	73
[W(CO)$_5$]$_2$	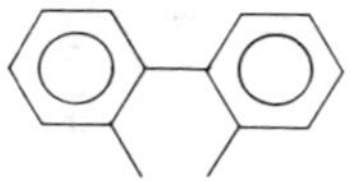 (2,2'-dimethylbiphenyl-diyl)		Red-violet	11		1879	73
[Mn(CO)$_2$(η-C$_5$H$_5$)]$_2$	Bu^t	Bu^t	Black-brown	57		1862	74
[Mn(CO)$_2$(η-C$_5$H$_5$)]$_2$	CH$_2$Ph	CH$_2$Ph	Red-violet	82		1887	74
[Mn(CO)$_2$(η-C$_5$H$_5$)]$_2$	Cy	Cy		54		1878	74
[Mn(CO)$_2$(η-C$_5$H$_5$)]$_2$	Ph	Ph	Black	76		1873	74
Mn(CO)$_2$(η-C$_5$H$_5$) Fe(CO)$_4$ }^{a}	Ph	Ph	Violet	65	99–101		*139*

^{a} δ(C$_1$) 333.25 ppm.

$M(=C=CHPh)(CO)_2(\eta\text{-}C_5H_5)$ (M = Mn and Re, respectively). The $\nu(C=C=C)$ absorptions are found between 1862 and 1887 cm^{-1} for manganese–propadienylidene complexes.

The ^1H NMR spectra are generally consistent with substituent groups and other ligands present, no features being characteristic of the vinylidene ligands. In mononuclear complexes, the α- and β-carbons resonate between δ 320–380 ppm, and between δ 118–142 ppm, respectively, consistent with the electron deficiency of the former. A considerably wider range is found for the α carbons in binuclear complexes, from δ 250 ppm for $Ru_2(\mu\text{-}C=CH_2)(\mu\text{-}CO)(CO)_2(\eta\text{-}C_5H_5)_2$ to δ 329.5 ppm for $Mn_2(\mu\text{-}C=CHPh)(CO)_4(\eta\text{-}C_5H_5)_2$. As expected, the α carbons in propadienylidene complexes are more electron deficient, as indicated by a chemical shift of δ 441 ppm in $Cr(=C=C=C=O)(CO)_5$, although a value of δ 331.2 ppm is found in $Mn(=C=C=CBu_2^t)(CO)_2(\eta\text{-}C_5H_5)$.

Mass spectral fragmentations of mononuclear and binuclear manganese and rhenium complexes have been described by Russian workers (*111a*). Apart from the usual loss of CO ligands, elimination of the vinylidene moiety was noted, and of $M(CO)_2(C_5H_5)$ (M = Mn or Re) groups from binuclear complexes. Perhaps the most unusual feature was found in the spectrum of the olefinic vinylidene complex $(\eta\text{-}C_5H_5)(OC)_2Re(\eta^2\text{-}CH_2=CPhCPh=C=)Re(CO)_2(\eta\text{-}C_5H_5)$, in which the carbonyl-free $[Re_2(C_{26}H_{22})]^+$ ion underwent successive loss of *five* H_2 molecules.

VIII

THE RELATION OF VINYLIDENE TO OTHER

η^1-CARBON-BONDED LIGANDS

The vinylidene ligand occupies an important place in the sequence of reactions linking a variety of well-known η^1-carbon-bonded ligands:

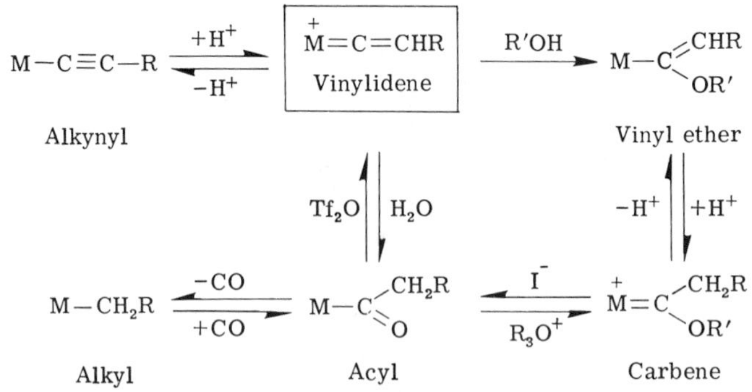

and in the more general conversion of alkynyl to alkyl ligands by a succession of electrophilic (E,E′) and nucleophilic (N,N′) additions:

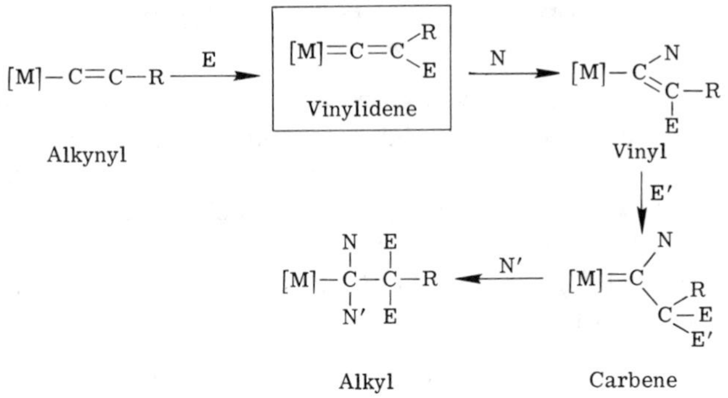

All of these individual conversions are now well documented, or have come to light during investigations of the chemistry of vinylidene complexes. An example is to be found in the reactions of $[Ru(=C=CHPh)(PPh_3)_2(\eta\text{-}C_5H_5)]PF_6$ (42). Deprotonation under mild conditions affords the corresponding σ-phenylacetylide complex, which can be reprotonated with $HPF_6 \cdot OEt_2$. Reaction of the vinylidene with methanol affords the methoxy(benzyl)carbene complex, which can be deprotonated to the vinyl ether with base. With water, on the other hand, the product is $Ru(CH_2Ph)(CO)(PPh_3)(\eta\text{-}C_5H_5)$, this compound forming by spontaneous decarbonylation of the phenylacetyl complex $Ru(COCH_2Ph)(PPh_3)_2(\eta\text{-}C_5H_5)$, with concomitant loss of a triphenylphosphine ligand; the acyl derivative is formed by deprotonation of the unobserved hydroxy(benzyl) carbene complex.

The stepwise reduction of the ethynyl complex $Fe(C\equiv CH)(dppe)(\eta\text{-}C_5H_5)$ to the neopentyl derivative has been achieved by two sequences of methylation and hydride additions, illustrating the propensity for nucleophilic addition to C_α, and electrophilic addition to C_β (25). The initial conversion to $[Fe(=C=CMe_2)(dppe)(\eta\text{-}C_5H_5)]^+$ (5) has been described above (Section III,A,2); further reactions of this complex with $Na[HB(OMe)_3]$, Me_3O^+, and $NaBH_4$ afford the neopentyl complex:

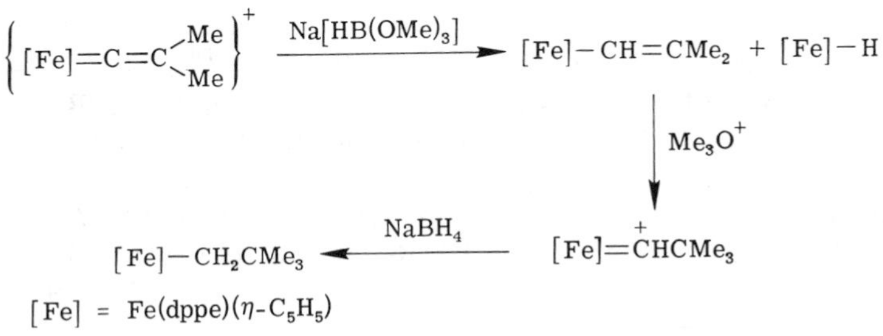

IX

DISPLACEMENT OF VINYLIDENE AND
PROPADIENYLIDENE LIGANDS

Little success has attended experiments designed to use these metal complexes as sources of free vinylidenes or propadienylidenes. Thus, it has not been possible to transfer vinylidenes to olefins in reactions with manganese complexes. As described above (Section IV,A,1), heating mononuclear $Mn(CO)_2(\eta\text{-}C_5H_5)$ complexes containing these ligands affords the binuclear analogs, the second propadienylidene fragment probably being lost as a hexapentaene derivative:

$$2R_2C{=}C{=}C: \longrightarrow R_2C{=}(C{=}C)_2{=}CR_2$$

Indeed, tetra-t-butylhexapentaene was obtained in 52% yield by the thermal decomposition of $Mn({=}C{=}C{=}CBu_2^t)(CO)_2(\eta\text{-}C_5H_5)$ at 100°C/1 atm. (17).

The chromium oxapropadienylidene complex affords dimethyl malonate when oxidized in methanol (17), and the diester is assumed to be produced by addition of the alcohol to the first-formed tricarbon dioxide:

$$(OC)_5Cr{=}C{=}C{=}C{=}O \xrightarrow{[O]} O{=}C{=}C{=}C{=}O \xrightarrow{MeOH} CH_2(CO_2Me)_2$$

If a mixture of diphenylacetylene and $Mo[C(Cl){=}C(CN)_2](CO)_3(\eta\text{-}C_5H_5)$ is heated, part of the cyanocarbon ligand is incorporated into the olive-green 6,6-dicyano-1,2,3,4-tetraphenylfulvene (72), which can be formed by cyclization of a $C{=}C(CN)_2$ fragment with two molecules of alkyne (112):

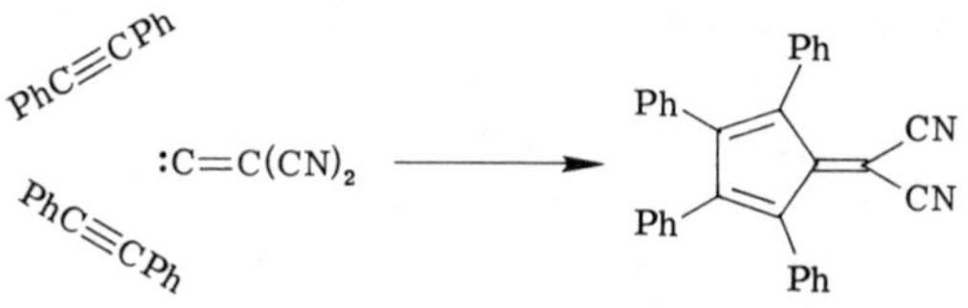

(72)

The analogous tungsten complex reacts with PPh_3 to give not only $WCl[C{=}C(CN)_2](PPh_3)_2(\eta\text{-}C_5H_5)$, but also minor amounts of $WCl(CO)_3(\eta\text{-}C_5H_5)$, which suggests loss of dicyanovinylidene can occur from some intermediate complex ($9, 30$).

Finally, the stereospecific formation of $trans$-1,4-di-t-butylbuta-2,3-triene from $HC{\equiv}CBu^t$, catalyzed by $H_2Ru(CO)(PPh_3)_3$, suggests the intermediacy of a t-butylvinylidene–ruthenium complex which facilitates formation of the C_4 hydrocarbon (113).

X

POLYNUCLEAR COMPLEXES

Metal cluster complexes containing vinylidene ligands have been considered as models of species present when olefins or alkynes are chemisorbed on metal surfaces (114). Vinylidene has been detected in reactions of ethylene or acetylene with Fe(100), Ni(111), and Pt(111) surfaces (115), and was shown to be an intermediate by theoretical studies on a manganese surface (116). The facile cleavage of $C-H$ bonds which occurs in these systems, together with hydrogen addition or abstraction, also occurs on metal clusters. Typical of the reactions considered is the hydrogen transfer reaction

$$3C_2H_4(\text{surface}) \longrightarrow C_2H_6(g) + 2CMe(\text{surface})$$

which may be envisaged to occur via the sequence

$$H_2[M_n] \xrightleftharpoons{\;C_2H_4\;} H_2[M_n](CH_2{=}CH_2) \rightleftharpoons$$

$$HEt[M_n] \xrightleftharpoons{\;C_2H_4\;} HEt[M_n](CH_2{=}CH_2) \xrightleftharpoons{\;-C_2H_6\;} [M_n](CH_2{=}CH_2) \rightarrow$$

$$H[M_n](CH{=}CH_2) \rightarrow H_2[M_n](C{=}CH_2) \xrightarrow{\;H_2\;} H_3[M_n](CMe)$$

Several triosmium carbonyl complexes containing the various C_2 ligands appearing in the sequence have been structurally characterized.

A. Complexes Containing μ_2-Vinylidene Ligands

There are no structurally characterized cluster complexes containing μ_2-vinylidene ligands, although several of the minor products isolated from reactions between $Ru_3(CO)_{12}$ and $NaBH_4$ in refluxing tetrahydrofuran are supposed to have this feature (117): examples are complexes 73–75.

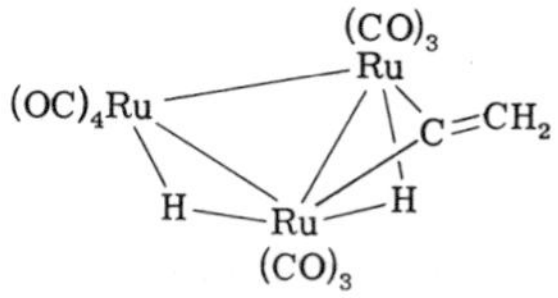

(73)

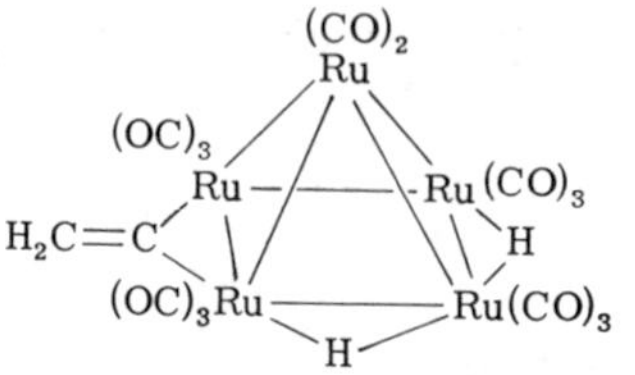

(74)

$H_2Ru_5(\mu\text{-}C{=}CH_2)(CO)_{15}$

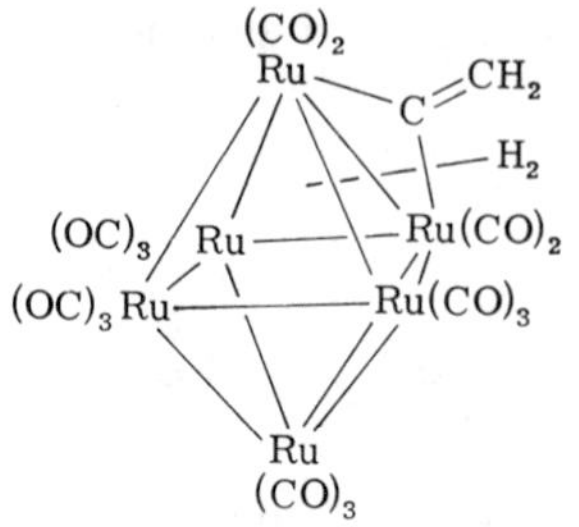

(75)

$H_2Ru_6(\mu\text{-}C{=}CH_2)(CO)_{16}$

B. Complexes Containing μ_3-Vinylidene Ligands

1. Cobalt

Carbonium ions based on the well-known Co_3C cluster can be obtained by dehydration of the hydroxy complexes $Co_3[\mu_3\text{-}CCH(OH)R](CO)_9$ with HPF_6 in propionic anhydride:

$$[Co_3]\,C{-}CH(OH)R \xrightarrow[\text{(EtCO)}_2O]{HPF_6} \{[Co_3]\,C{=}CHR\}^+$$

R = H, Me, Ph

$$[Co_3]\,C \equiv \text{(structure: C bonded to } (OC)_3Co{-}|{-}Co(CO)_3 \text{ and } Co(CO)_3)$$

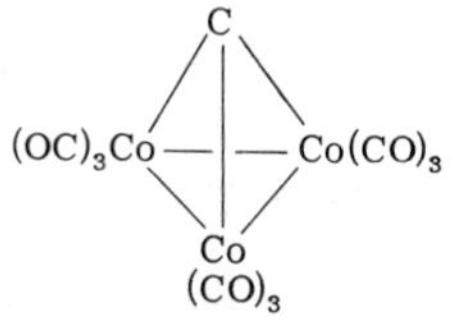

These compounds form black crystalline solids which are stable in the absence of air or moisture (*118*). Similar tertiary carbonium ion derivatives could be obtained from vinylic derivatives, for example, on protonation (*119*):

$$[Co_3]CCMe{=}CH_2 \xrightarrow[\text{(EtCO)}_2O]{HPF_6} \{[Co_3]C{=}CMe_2\}^+PF_6^-$$

These complexes are very reactive electrophiles, and reactions with alcohols, a thiol, or amines generate new functionally substituted cluster compounds:

$$\{[Co_3]{=}C{=}CRR'\}^+ + Nu^- \longrightarrow [Co_3]{-}C{-}\underset{R'}{\overset{R}{C}}{-}Nu$$

The chemistry of these interesting compounds has been well studied, and has been reviewed in this series (*120*).

NMR studies of these compounds confirm the electron deficiency of the

α-carbon atoms, which have chemical shifts in the region 258–286 ppm (*121*). The stability of these species was attributed to overlap of the empty p orbital on the β-carbon with the filled σ orbital of the $Co–C_\alpha$ bond, as in **76**, with C_β bending towards the metal; the threefold opportunity for such overlap suggests that the cations would be fluxional. Subsequently, extended Hückel calculations also showed that a conformation away from the upright is preferred (*122*). A further feature shown by the calculations is the rotation

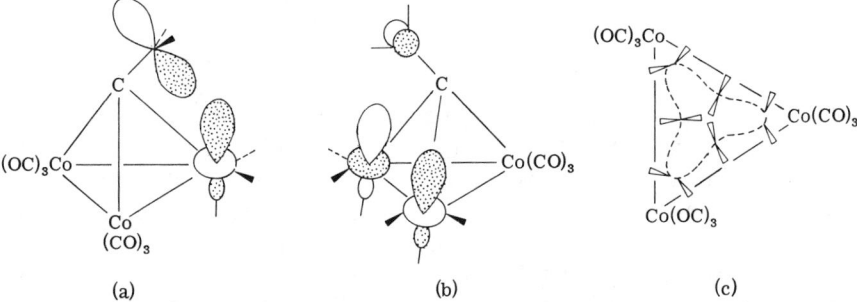

of the CH_2 plane as the vinylidene ligand bends towards a metal–metal bond (CH_2 perpendicular to the M_3 plane) or towards a metal atom (CH_2 rotated to some degree). A consequence of this is rotation of the CH_2 group as the vinylidene group moves around the M_3 cluster (**77**), which is consistent with the orbital interactions shown in Fig. 1, which illustrates interaction of the vinylidene p orbital with (a) a metal d_{z^2} orbital, and (b) with the antisymmetric e-type cluster MO.

2. Iron, Ruthenium, and Osmium

The μ_3-methylidyne anion (**77**), which is obtained from $[HFe_3(CO)_{11}]^-$ and acetylene, is converted to the μ_3-vinylidene cluster complex (**78**) by

FIG. 1. Diagrams showing (a) interaction of vinylidene p orbital with metal d_{z^2} orbital, (b) interaction of vinylidene p orbital with antisymmetric e-type orbital, and (c) circumambulation of vinylidene around Co_3 triangle, with concomitant rotation of the CH_2 group [after B. E. Schilling and R. Hoffmann, *J. Am. Chem. Soc.* **101**, 3456 (1979)].

heating in refluxing acetone (*123*). Under CO, **78** rearranges back to **77**, and this reversible reaction indicates the remarkable lability of the C–H and Fe–H bonds in these derivatives:

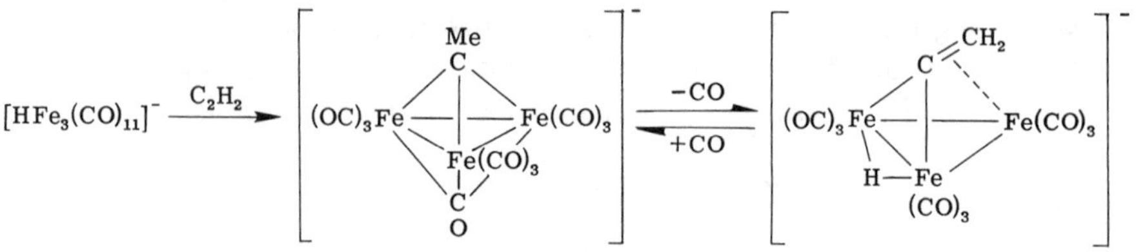

(77)   (78)

The hydrido-μ_2-η^1,η^2-vinyl complex (**79**), which is obtained from reactions between $H_2Os_3(CO)_{10}$ and acetylene (*124*) or ethylene (*125*) under mild conditions, loses a molecule of CO on heating to give the pale yellow μ_3-vinylidene derivative (**80**).

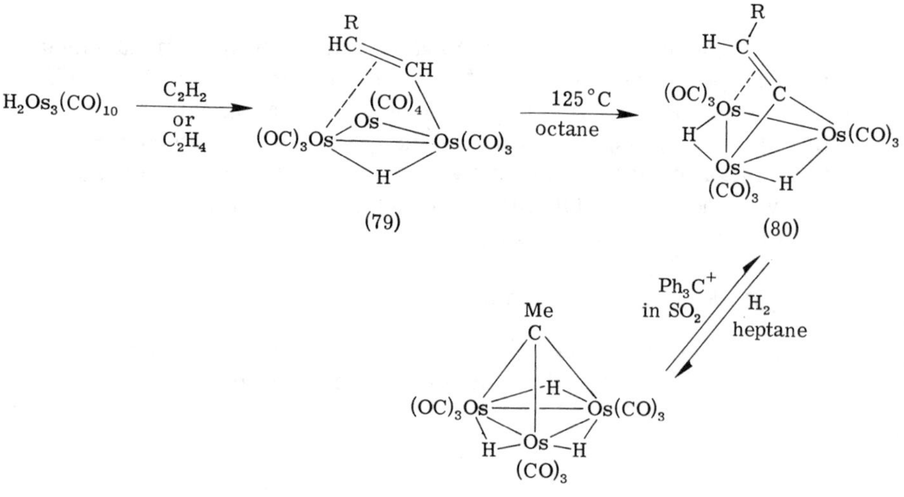

(79)   (80)

(81)

Complex **80** (R = H) is also formed directly from $Os_3(CO)_{12}$ and ethylene in octane at 125°C (*126*). It is fluxional, with hydrogen migration around the Os_3 core. Preliminary details of a crystal structure confirm the formulation shown. Propene reacts similarly with $Os_3(CO)_{12}$ to give an oil, which contains **80** (R = Me) as the major component; a similar mixture was obtained by pyrolysis of the μ_2-CH=CHMe complex formed from $H_2Os_3(CO)_{12}$ and propyne. The μ_2-styryl compound [from $H_2Os_3(CO)_{10}$ and HC_2Ph] afforded only white **80** (R = Ph) on heating. Hydrogenation of **80** (R = H) in refluxing heptane for 24 hr gives a nearly quantitative yield of $H_3Os_3(\mu_3$-CMe)(CO)$_9$ (**81**), a reaction that can be reversed by hydride abstraction from the ethylidyne complex with trityl tetrafluoroborate (127). The IR and Raman spectra of $H_2Os_3(\mu_3\text{-}C{=}CH_2)(CO)_9$ have been assigned; in the IR, a medium intensity band at 1467 cm^{-1} was assigned to a coupled $\nu(C{=}C)/\delta(CH_2)$ scissors mode, and absorptions at 1048m, 959ms, and 808m cm^{-1} to CH_2 rock, $\gamma(CH_2)$ wag, and $\gamma(CH)_2$ twist vibrations, respectively (128).

Thermal decomposition of $Os_3(CO)_{11}[Me_2As(CH{=}CH_2)]$ at 96°C gives a μ-vinyl complex $Os_3(\mu\text{-}AsMe_2)(\mu\text{-}CH{=}CH_2)(CO)_{10}$ (82), which at higher temperatures affords successively the white μ_3-vinylidene complex (83) and then the μ_3-alkyne derivative (84) (129). This sequence of reactions provides the first authenticated report of the interconversion of cluster-bound vinylidene to acetylene.

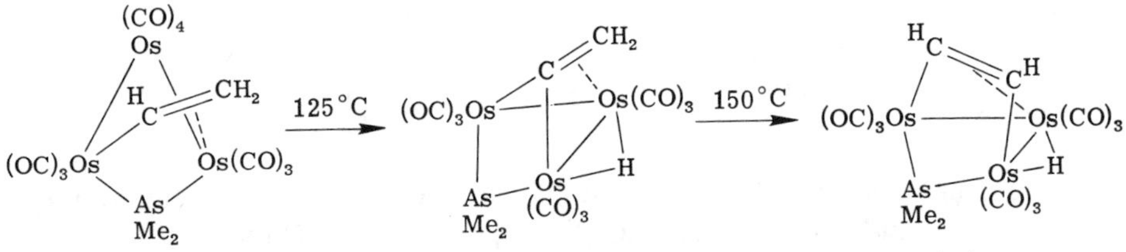

(82)   (83)   (84)

Although the ruthenium analogs of 80 (R = H or Me) were identified among the many products obtained from $Ru_3(CO)_{12}$ and $NaBH_4$ (111), the former was better obtained, together with the vinyl complex $HRu_3(\mu_2\text{-}CH{=}CH_2)(CO)_{10}$, from the reaction between $Ru_3(CO)_{12}$ and ethylene (100°C, 17 hr) (127, 130).

The mixed Ru_3Au_2 cluster (85) obtained from reactions between $[Ru_3(\mu_3\text{-}C_2Bu')(CO)_9]^-$ and $[O(AuPPh_3)_3]^+$ has been shown to contain a t-butylvinylidene ligand bridging the three ruthenium atoms (131). The two $Au(PPh_3)$ moieties, weakly interacting, are bonded to the opposite side of the Ru_3 triangle. Complex 85 is accompanied by smaller amounts of Ru_3Au and Ru_3Au_3 clusters, but it is not yet certain whether they also contain vinylidene ligands.

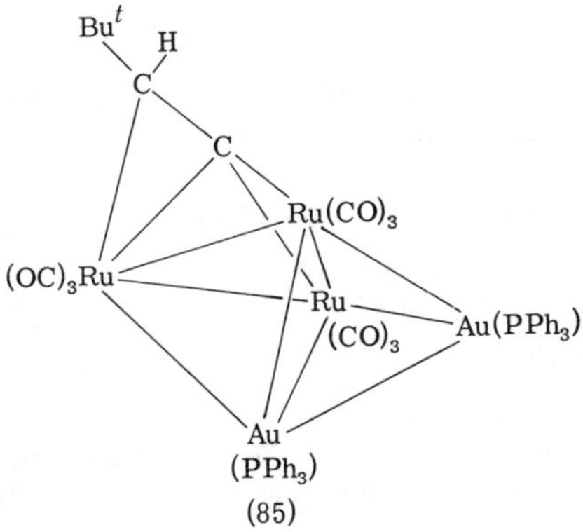

(85)

Yellow $H_2Os_3(\mu_3$-$C{=}C{=}O)(CO)_9$ (**86**) has been obtained by hydrogenation (in refluxing benzene) or pyrolysis (refluxing toluene) of the μ-methylene complex $Os_3(\mu$-$CH_2)(\mu$-$CO)(CO)_{12}$ (*132*). The mechanism of formation is not clear. Complex **86** undergoes rapid hydrogen equilibration; it is protonated by $HBF_4 \cdot OEt_2$ to give $[H_3Os_3(CCO)(CO)_9]^+$ [or $H_2Os_3(CCOH^+)(CO)_9$]. Either **86** or the protonated complex affords the ester derivative $H_3Os_3[\mu_3$-$C(CO_2Me)](CO)_9$ (**87**) in methanol.

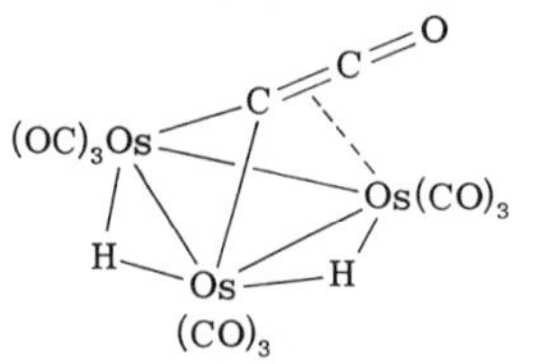

(86) 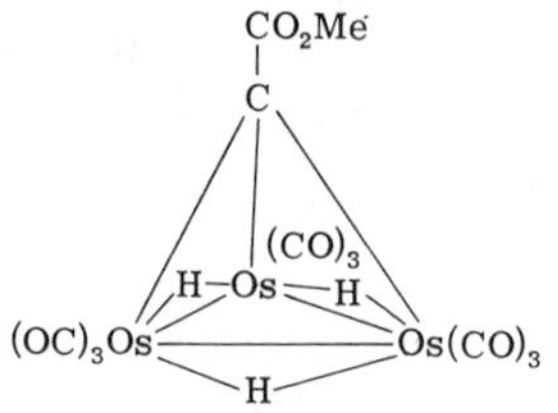   (87)

The iron analog of (**86**) has been obtained by oxidative degradation of the anionic carbido cluster $[Fe_4(\mu_4$-$C)(CO)_{12}]^{2-}$ with silver(I) ion (*133*).

C. Complexes Containing μ_4-Vinylidene Ligands

Addition of $Co_2(CO)_8$ to the μ_2-vinylidene complex (**40**) (Section IV,A,6) affords the heteronuclear cluster derivative (**88**) in 55% yield, in which the vinylidene group now bridges all four metal atoms:

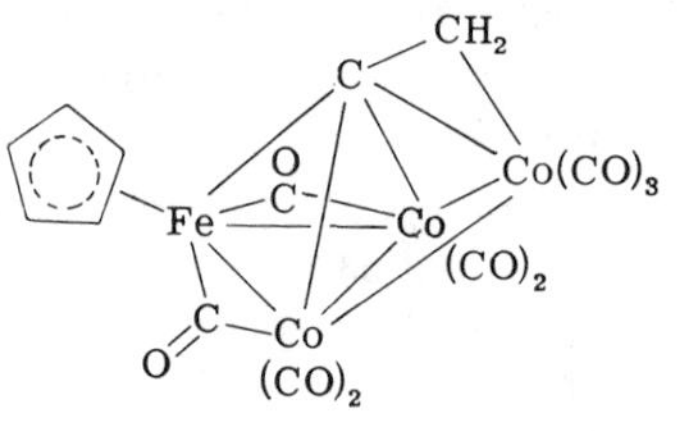

(88) 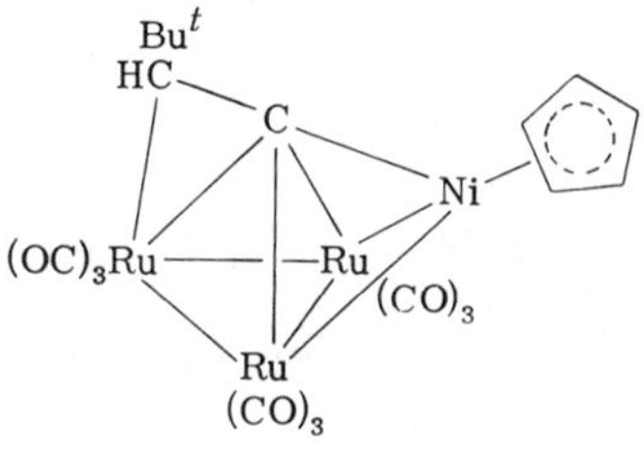   (89)

In solution, the two methylene protons are nonequivalent, although the solid state structure shows a plane of symmetry bisecting the CH_2 group (*70a*).

Conversion of the hydrido-acetylide complex $HRu_3(\mu_3$-$C_2Bu^t)(CO)_9$ to the black μ_4-vinylidene derivative (**89**) occurs on reaction with $[Ni(CO)(\eta$-$C_5H_5)]_2$ in refluxing heptane (*134*).* In contrast with the Ru_3Au_2 complex described in the previous section, the vinylidene ligand, presumably formed

* This complex has been shown to be the μ-hydrido derivative, $(\mu$-H$)Ru_3Ni(\mu_4$-η^2-$C{=}CHBu^t)(CO)_9(\eta$-$C_5H_5)$ by NMR and an X-ray structure of the isopropyl analog [A. J. Carty, N. J. Taylor, E. Sappa and A. Tiripicchio, *Inorg. Chem.* **22**, 1871 (1983)].

by migration of the metal-bonded proton to the β-carbon, now interacts with all four metal atoms, the α-carbon being approximately equivalently bonded to each of them.

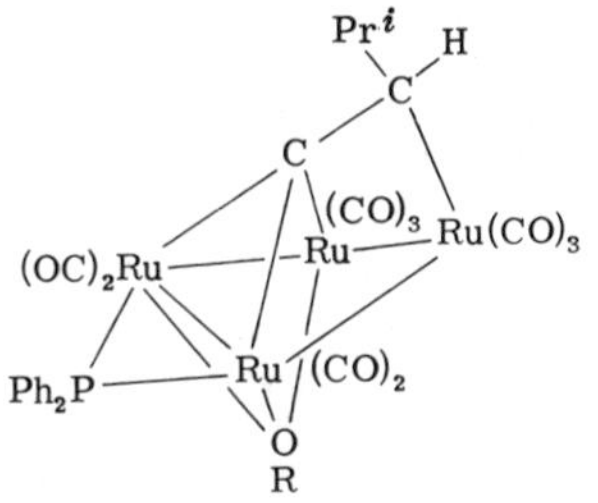

(90) 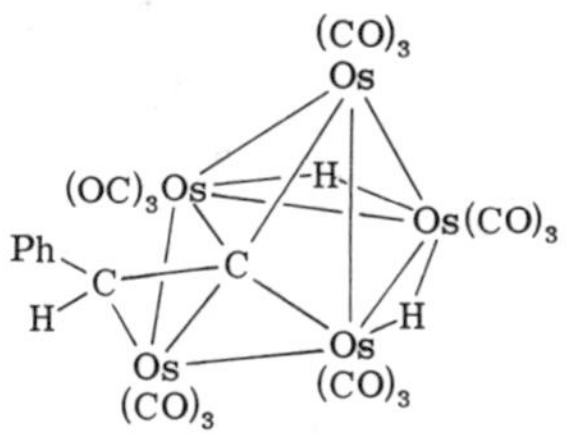   (91)

Treatment of $Ru_3(CO)_{11}(Ph_2PC\equiv CPr^i)$ in tetrahydrofuran with water or ethanol affords up to 30% of the red tetranuclear clusters (90; R = H or Et, respectively) (135). In these, the cleavage of the P–C bond is accompanied by addition of OH or OEt to the cluster, with transfer of hydrogen to the β-carbon. Unusually, the "hinge" metal atoms are separated by 3.455 Å, too far for a normal metal–metal interaction; this has the result of bringing the μ_3-O and μ_4-C atoms to within 2.65 Å of each other, suggesting the possibility of facile H transfer between the two atoms.

A reaction between $H_2Os_5(CO)_{15}$ and phenylacetylene afforded a yellow complex characterized as $H_2Os_5(CO)_{15}(CCPh)$; the carbon skeleton of the alkyne interacts with four of the five osmium atoms in a manner similar to that found in 89, and it is likely that this complex is another example of a cluster-bound vinylidene (91) (136).

D. Cluster-Bonded Propadienylidene Ligands

Addition of acid to the μ_3-acetylide complexes (92; M = Ru or Os) results in the formation of 93 (M = Ru or Os) by an acid-induced migration of the hydroxy group from carbon to metal (137). Only the orange osmium derivative could be isolated, and this was structurally characterized as the μ_3-diphenylpropadienylidene complex shown.

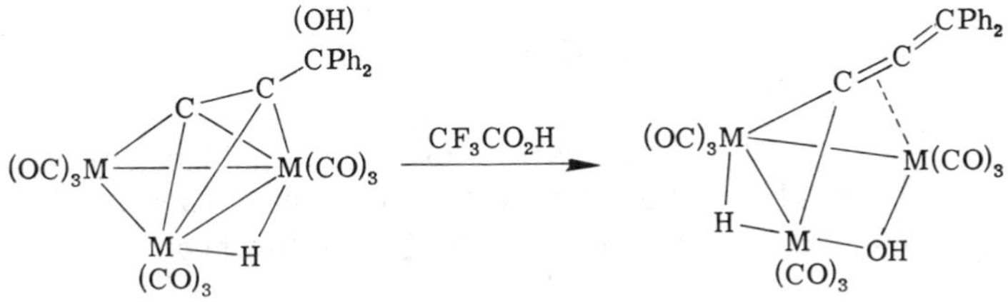

(92)   (93)

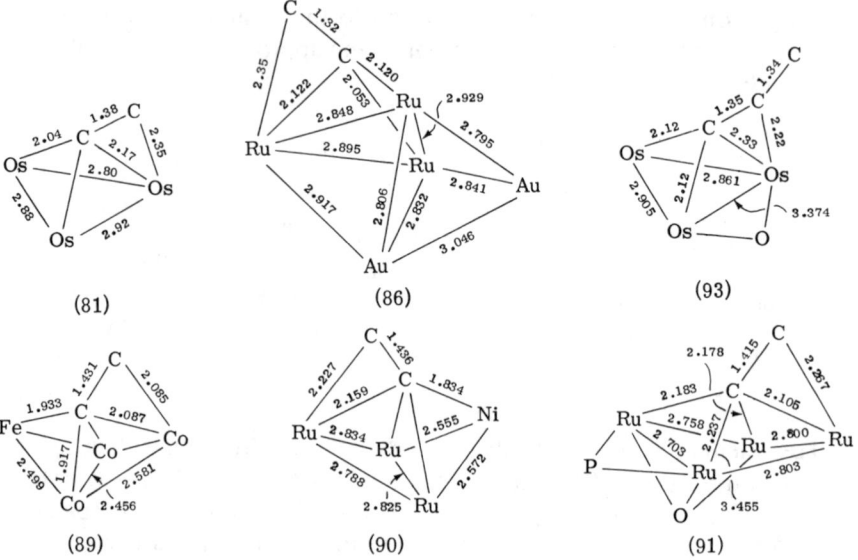

FIG. 2.   Metal-cluster–vinylidene skeletons for $H_2Os_3(\mu_3\text{-}C{=}CH_2)(CO)_9$ **(80)** (*148*), $Au_2Ru_3(\mu_3\text{-}C{=}CHBu^t)(CO)_9(PPh_3)_2$ **(85)** (*131*), $HOs_3(\mu_3\text{-}C{=}C{=}CPh_2)(\mu\text{-}OH)(CO)_9$ **(93)** (*137*), $Co_3Fe(\mu_4\text{-}C{=}CH_2)(CO)_9(\eta\text{-}C_5H_5)$ **(88)** (*70a*),$NiRu_3(\mu_4\text{-}C{=}CHBu^t)(CO)_9(\eta\text{-}C_5H_5)$ **(89)** (*134*) and $Ru_4(\mu_4\text{-}C{=}CHPr^t)(\mu_3\text{-}OH)(\mu\text{-}PPh_2)(CO)_{10}$ (*135*). Distances in angstroms.

The reaction is an intramolecular oxidative addition across one of the M–M bonds.

E. *Structural Studies*

Figure 2 summarizes the structural parameters of the vinylidene or propadienylidene ligands when bonded to various metal clusters. It can be seen that the C=C double bond is lengthened to ca. 1.4 Å as a result of the interaction with the cluster; the α-carbon bridges three or four metal atoms, with M–C distances consistent with normal single bonds.

XI

ADDENDUM[2]

1. The red complex originally reported as $M\bar{n}[C(\overset{+}{P}Ph_3){=}CH(CO_2Me)](CO)_2(\eta\text{-}C_5H_5)$ and obtained from the reaction between $Mn(CO)_2(\eta^2\text{-}HC_2CO_2Me)(\eta\text{-}C_5H_5)$ and PPh_3 (*48*), has been shown by an

[2] To December 1982.

X-ray study to be the isomeric $\overline{Mn}[C(CO_2Me){=}CH(\overset{+}{P}Ph_3)](CO)_2(\eta\text{-}C_5H_5)$; a similar derivative obtained in the reaction with dppe is formulated as $\overline{Mn}[C(CO_2Me){=}CH(\overset{+}{P}Ph_2CH_2CH_2PPh_2)](CO)_2(\eta\text{-}C_5H_5)$ [N. E. Kolobova, L. L. Ivanov, O. S. Zhvanko, I. N. Chechulina, A. S. Batsanov, and Y. T. Struchkov, *J. Organomet. Chem.* **238**, 223 (1982)].

2. Syntheses of the chiral complexes $[Re({=}C{=}CHR)(NO)(PPh_3)(\eta\text{-}C_5H_5)]^+$ (R = H, Me, Ph) from the corresponding acyls by reaction with triflic anhydride, KOBu^t, followed by further addition of triflic anhydride. The Me and Ph compounds each exist in two geometric isomers. Alkylation (with CD_3SO_3F) of $Re(C{\equiv}CMe)(NO)(PPh_3)(\eta\text{-}C_5H_5)$ afforded the $C{=}C(CH_3)(CD_3)$ complex initially as one isomer, which isomerized to an equilibrium mixture on warming; similar results were obtained with other acetylides and electrophiles. Thus, electrophilic attack of the acetylide initially affords only one (kinetic) isomer, which then equilibrates to the thermodynamic mixture. The isomers can be thermally or photochemically interconverted; for the C=CMePh complex, $\Delta H^{\#}_{rot}$ = 15.7 ± 1.7 kcal mol^{-1}. The important conclusion deriving from these results is that the rhenium chirality can be transferred through the C≡C bond (which has cylindrical symmetry); it is suggested that only one of the four p orbital lobes of the C≡C moiety is reactive toward electrophiles as a result of the presence of the bulky PPh$_3$ ligand [A. Wong and J. A. Gladysz, *J. Am. Chem. Soc.* **104**, 4948 (1982)].

3. Several reactions of $Rh({=}C{=}CHR')(PR_3)(\eta\text{-}C_5H_5)$ (PR_3 = PPr$_3^i$) have shown the way to a significant expansion of the potential of vinylidene complexes in synthesis. Thus HCl gives successively $RhCl(CH{=}CHR')(PR_3)(\eta\text{-}C_5H_5)$ and $RhCl(CHClCH_2R')(PR_3)(\eta\text{-}C_5H_5)$; CH_2N_2 affords $Rh(\eta^2\text{-}CH_2{=}C{=}CHR')(PR_3)(\eta\text{-}C_5H_5)$; chalcogens give $Rh(\eta^2\text{-}E{=}C{=}CHR')(PR_3)(\eta\text{-}C_5H_5)$ (E = S, Se, Te); and CuCl or RhCl(PR$_3$)$_2$ add to the Rh=C moiety. This extensive series of reactions involves the multiple bond between the metal and α-carbon (H. Werner, private communication).

4. The reaction between $Fe(CO)_5$ and $Ph_2C{=}C{=}O$, which afforded $Fe_2(\mu\text{-}C{=}CPh_2)(CO)_8$ (**32**) as the first-described vinylidene complex (**7**), has been reinvestigated; the major product is $Fe[\eta^3\text{-}PhC(Ph)C(O)](CO)_3$, which reacts further with $Fe_2(CO)_9$ to give $Fe_2(CO)_6[\mu\text{-}\eta^2\text{-}PhC(\eta^1,\eta^3\text{-}Ph)]$ [W. A. Herrmann, J. Gimeno, J. Weichmann, M. L. Ziegler, and B. Balbach, *J. Organomet. Chem.* **213**, C26 (1981)].

5. Reaction between $Mn({=}C{=}CHPh)(CO)_2(\eta\text{-}C_5H_5)$ and $Pt[P(OEt)_3]_4$ (benzene, 40°) affords a mixture of $(\eta\text{-}C_5H_5)(OC)\overline{Mn}(\mu\text{-}CO)(\mu\text{-}C{=}CHPh)Pt[P(OEt)_3]_2$ (20%) and $Mn(CO)_2[\eta^2\text{-}CHPh{=}CHP(O)(OEt)_2](\eta\text{-}C_5H_5)$ (30%) [A. B. Antonova, S. P. Gubin, and S. V. Kovalenko, *Izv. Akad. Nauk SSSR* p. 953 (1982)].

6. Decarboxylation of diketene in the reaction with $Mn(CO)_2(thf)(\eta\text{-}C_5H_5)$ affords $Mn(\eta^2\text{-}CH_2{=}C{=}CH_2)(CO)_2(\eta\text{-}C_5H_5)$, which readily complexes with a second $Mn(CO)_2(\eta\text{-}C_5H_5)$ fragment to give $Mn_2(\mu\text{-}C_3H_4)(CO)_4(\eta\text{-}C_5H_5)_2$ and complex **48** (Section IV,B) [W. A. Herrmann, J. Weichmann, M. L. Ziegler, and H. Pfisterer, *Angew. Chem.* **94**, 545 (1982); *Angew. Chem., Int. Ed. Engl.* **21**, 551 (1982); *Angew. Chem. Suppl.* 1223 (1982)].

7. The anionic cluster complex $[Fe_3(\mu_3,\eta^2\text{-}C{=}C{=}O)(CO)_9]^{2-}$ has been obtained by $Na[Ph_2CO]$-reduction of $[Fe_3[\mu\text{-}C(OR)](CO)_{10}]^-$ (R = Me or OMe); upon protonation or alkylation it affords $[Fe_3(\mu_3\text{-}CR)(\mu_3\text{-}CO)(CO)_9]^-$ (R = H or Me) [J. W. Kolis, E. M. Holt, M. Drezdzon, K. H. Whitmire, and D. F. Shriver, *J. Am. Chem. Soc.* **104**, 6134 (1982)].

REFERENCES

1. E. O. Fischer, *Pure Appl. Chem.* **24**, 407 (1970); **30**, 353 (1972); F. A. Cotton and C. M. Lukehart, *Prog. Inorg. Chem.* **16**, 487 (1972); D. J. Cardin, B. Cetinkaya, and M. F. Lappert, *Chem. Rev.* **72**, 545 (1972); E. O. Fischer, *Adv. Organomet. Chem.* **14**, 1 (1976); W. A. Herrmann, *Angew. Chem.* **90**, 855 (1978); *Angew. Chem., Int. Ed. Engl.* **17**, 800 (1978); R. R. Schrock, *Acc. Chem. Res.* **12**, 98 (1979).
2. W. A. Herrmann, *Adv. Organomet. Chem.* **20**, 159 (1982).
3. E. Band and E. L. Muetterties, *Chem. Rev.* **78**, 639 (1978); E. L. Muetterties, T. N. Rhodin, E. Band, C. F. Brucker, and W. R. Pretzer, *ibid.* **79**, 91 (1979); W. A. Herrmann, *Angew. Chem.* **94**, 118 (1982); *Angew. Chem., Int. Ed. Engl.* **21**, 117 (1982).
4. Y. Osamura, H. E. Schaeffer, S. K. Gray, and W. H. Miller, *J. Am. Chem. Soc.* **103**, 1904 (1981).
5. P. S. Skell, F. A. Fagone, and K. J. Klabunde, *J. Am. Chem. Soc.* **94**, 7862 (1972).
6. P. J. Stang, *Chem. Rev.* **78**, 383 (1978).
7. O. S. Mills and A. D. Redhouse, *Chem. Commun.* p. 444 (1966); *J. Chem. Soc. A* p. 1282 (1968).
8. R. B. King and M. S. Saran, *J. Am. Chem. Soc.* **94**, 1784 (1972).
9. R. B. King and M. S. Saran, *Chem. Commun.* p. 1053 (1972).
10. M. H. Chisholm and H. C. Clark, *Acc. Chem. Res.* **6**, 202 (1973).
11. A. N. Nesmeyanov, G. G. Aleksandrov, A. B. Antonova, K. N. Anisimov, N. E. Kolobova, and Y. T. Struchkov, *J. Organomet. Chem.* **110**, C36 (1976).
12. J. M. Bellerby and M. J. Mays, *J. Organomet. Chem.* **117**, C21 (1976).
13. A. Davison and J. P. Solar, *J. Organomet. Chem.* **155**, C8 (1978).
14. M. I. Bruce and R. C. Wallis, *J. Organomet. Chem.* **161**, C1 (1978).
15. D. Mansuy, *Pure Appl. Chem.* **52**, 681 (1980).
16. E. O. Fischer, H.-J. Kalder, A. Frank, F. H. Köhler, and G. Huttner, *Angew. Chem.* **88**, 683 (1976); *Angew. Chem., Int. Ed. Engl.* **15**, 623 (1976).
17. H. Berke, *Angew. Chem.* **88**, 684 (1976); *Angew. Chem., Int. Ed. Engl.* **15**, 624 (1976).
17a. J. Wolf, H. Werner, O. Serhadli, and M. L. Ziegler, *Angew. Chem.* **95**, 428 (1983); *Angew. Chem., Int. Ed. Engl.* **22**, 414 (1983).
18. A. B. Antonova, N. E. Kolobova, P. V. Petrovsky, B. V. Lokshin, and N. S. Obezyuk, *J. Organomet. Chem.* **137**, 55 (1977).
19. N. E. Kolobova, A. B. Antonova, O. M. Khitrova, M. Y. Antipin, and Y. T. Struchkov, *J. Organomet. Chem.* **137**, 69 (1977).

20. M. I. Bruce and R. C. Wallis, *Aust. J. Chem.* **32,** 1471 (1979).
21. N. E. Kolobova, L. L. Ivanov, and O. S. Zhvanko, *Izv. Akad. Nauk. SSSR, Ser. Khim.* pp. 478, 2646 (1980).
22. H. Berke, P. Härter, G. Huttner, and L. Zsolnai, *Z. Naturforsch., B: Anorg. Chem., Org. Chem.* **36B,** 929 (1981).
23. P. M. Treichel and D. A. Komar, *Inorg. Chim. Acta* **42,** 277 (1980).
24. A. Davison and J. P. Selegue, *J. Am. Chem. Soc.* **100,** 7763 (1978).
25. A. Davison and J. P. Selegue, *J. Am. Chem. Soc.* **102,** 2455 (1980).
26. H. Berke, *Z. Naturforsch., B: Anorg. Chem., Org. Chem.* **35B,** 86 (1980).
27. H. Berke, *Chem. Ber.* **113,** 1370 (1980).
28. H. Berke, P. Härter, G. Huttner, and J. von Seyerl, *J. Organomet. Chem.* **219,** 317 (1981).
29. J. P. Selegue, *J. Am. Chem. Soc.* **104,** 119 (1982).
30. R. B. King and M. S. Saran, *J. Am. Chem. Soc.* **98,** 1817 (1973).
31. R. A. Bell and M. H. Chisholm, *Inorg. Chem.* **16,** 687 (1977).
32. D. Mansuy, M. Lange, J.-C. Chottard, P. Guerin, P. Morliere, D. Brault, and M. Rougé, *J. Chem. Soc., Chem. Commun.* p. 648 (1977); D. Mansuy, M. Lange, J.-C. Chottard, J.-F. Bartoli, B. Chevoier, and R. Weiss, *Angew. Chem.* **90,** 828 (1978); *Angew. Chem., Int. Ed. Engl.* **17,** 781 (1978).
33. D. Mansuy, M. Lange, and J.-C. Chottard, *J. Am. Chem. Soc.* **100,** 3213 (1978).
34. P. K. Baker, G. K. Barker, M. Green, and A. J. Welch, *J. Am. Chem. Soc.* **102,** 7811 (1980).
35. D. S. Gill and M. Green, *J. Chem. Soc., Chem. Commun.* p. 1037 (1981).
36. W. R. Roper, J. M. Waters, L. J. Wright, and F. van Meurs, *J. Organomet. Chem.* **201,** C27 (1980).
37. B. E. Boland, S. A. Fam, and R. P. Hughes, *J. Organomet. Chem.* **172,** C29 (1979).
38. B. E. Boland-Lussier, M. R. Churchill, R. P. Hughes, and A. L. Rheingold, *Organometallics* **1,** 628 (1982).
39. R. D. Adams, A. Davison, and J. P. Selegue, *J. Am. Chem. Soc.* **101,** 7232 (1979).
40. K. G. Caulton, *J. Mol. Catal.* **13,** 71 (1981).
41. M. I. Bruce, A. G. Swincer, and R. C. Wallis, *J. Organomet. Chem.* **171,** C5 (1979).
42. M. I. Bruce and A. G. Swincer, *Aust. J. Chem.* **33,** 1471 (1980).
43. B. E. Boland-Lussier and R. P. Hughes, *Organometallics* **1,** 635 (1982).
44. C. A. Tolman, *Chem. Rev.* **77,** 313 (1977).
45. M. I. Bruce, A. G. Swincer, B. J. Thomson, and R. C. Wallis, *Aust. J. Chem.* **33,** 2605 (1980).
46. K. Oguro, M. Wada, and R. Okawara, *J. Organomet. Chem.* **159,** 417 (1978).
47. M. H. Chisholm and H. C. Clark, *J. Am. Chem. Soc.* **94,** 1532 (1972).
48. N. E. Kolobova, L. L. Ivanov, O. S. Zhvanko, V. F. Sizoi, and Y. S. Nekrasov, *Izv. Akad. Nauk SSSR, Ser. Khim.* p. 2393 (1979).
49. M. I. Bruce, F. S. Wong, B. W. Skelton, and A. H. White, *J. Chem. Soc., Dalton Trans.* p. 2203 (1982).
50. A. N. Nesmeyanov, N. E. Kolobova, A. B. Antonova, N. S. Obezyuk, and K. N. Anisimov, *Izv. Akad. Nauk SSSR, Ser. Khim.* p. 948 (1976).
51. N. E. Kolobova, V. V. Skripkin, G. G. Aleksandrov, and Y. T. Struchkov, *J. Organomet. Chem.* **169,** 293 (1979).
52. N. E. Kolobova, V. V. Skripkin, and T. V. Rozantseva, *Izv. Akad. Nauk SSSR, Ser. Khim.* p. 1665 (1979); Y. L. Slovokhotov, A. I. Yanovskii, V. G. Andrianov, and Y. T. Struchkov, *J. Organomet. Chem.* **184,** C57 (1980).
53. P. Hong, K. Sonogashira, and N. Hagihara, *J. Organomet. Chem.* **219,** 363 (1981).
54. A. N. Nesmeyanov, A. B. Antonova, N. E. Kolobova, and K. N. Anisimov, *Izv. Akad. Nauk SSSR, Ser. Khim.* p. 2873 (1974).

126   M. I. BRUCE and A. G. SWINCER

55. A. N. Nesmeyanov, N. E. Kolobova, A. B. Antonova, and K. N. Anisimov, *Dokl. Akad. Nauk SSSR* **220**, 105 (1975).
56. K. Folting, J. C. Huffman, L. N. Lewis, and K. G. Caulton, *Inorg. Chem.* **18**, 3483 (1979).
57. L. N. Lewis, J. C. Huffman, and K. G. Caulton, *J. Am. Chem. Soc.* **102**, 403 (1980).
58. N. E. Kolobova, T. V. Rozantseva, and P. V. Petrovskii, *Izv. Akad. Nauk SSSR, Ser. Khim.* p. 2063 (1979).
59. N. E. Kolobova, L. L. Ivanov, O. S. Zhvanko, and P. V. Petrovskii, *Izv. Akad. Nauk SSSR, Ser. Khim.* p. 432 (1981).
60. V. G. Andrianov, Y. T. Struchkov, N. E. Kolobova, A. B. Antonova, and N. S. Obezyuk, *J. Organomet. Chem.* **122**, C33 (1976).
61. Y. N. Al-Obaidi, M. Green, N. D. White, and G. E. Taylor, *J. Chem. Soc., Dalton Trans.* p. 319 (1982).
62. I. T. Horvath, G. Palyi, L. Marko, and G. Andreetti, *J. Chem. Soc., Chem. Commun.* p. 1054 (1979); *Inorg. Chem.* **22**, 1049 (1983).
63. R. B. King and M. S. Saran, *J. Am. Chem. Soc.* **95**, 1811 (1973).
64. D. F. Marten, E. V. Dehmlow, D. J. Hanlon, M. B. Hossain, and D. van der Helm, *J. Am. Chem. Soc.* **103**, 4940 (1981).
65. G. M. Dawkins, M. Green, J. C. Jeffery, and F. G. A. Stone, *J. Chem. Soc., Chem. Commun.* p. 1120 (1980); *J. Chem. Soc., Dalton Trans.* p. 499 (1983).
66. M. Cooke, D. L. Davies, J. E. Guerchais, S. A. R. Knox, K. A. Mead, J. Roué, and P. Woodward, *J. Chem. Soc., Chem. Commun.* p. 862 (1981).
67. M. Nitay, W. Priester, and M. Rosenblum, *J. Am. Chem. Soc.* **100**, 3620 (1978).
68. D. L. Davies, A. F. Dyke, A. Endesfelder, S. A. R. Knox, P. J. Naish, A. G. Orpen, D. Plaas, and G. E. Taylor, *J. Organomet. Chem.* **198**, C43 (1980).
69. R. Korswagen, R. Alt, D. Speth, and M. L. Ziegler, *Angew. Chem.* **93**, 1073 (1981); *Angew. Chem., Int. Ed. Engl.* **20**, 1049 (1981).
70. L. N. Lewis, J. C. Huffman, and K. G. Caulton, *Inorg. Chem.* **19**, 1246 (1980).
70a. P. Brun, G. M. Dawkins, M. Green, R. M. Mills, J.-Y. Salaün, F. G. A. Stone, and P. Woodward, *J. Chem. Soc., Chem. Commun.* p. 966 (1981).
71. J. P. Selegue, *Organometallics* **1**, 217 (1982).
72. J. P. Selegue, personal communication (1982).
73. H. Berke, P. Härter, G. Huttner, and L. Zsolnai, *Chem. Ber.* **115**, 695 (1982).
74. H. Berke, *J. Organomet. Chem.* **185**, 75 (1980).
75. H. Berke, G. Huttner, and J. von Seyerl, *Z. Naturforsch., B: Anorg. Chem., Org. Chem.* **36B**, 1277 (1981).
76. H. Berke, G. Huttner, and J. von Seyerl, *J. Organomet. Chem.* **218**, 193 (1981).
77. N. E. Kolobova, V. V. Skripkin, and T. V. Rozantseva, *Izv. Akad. Nauk SSSR, Ser. Khim.* p. 2667 (1980).
78. N. E. Kolobova, V. V. Skripkin, T. V. Rozantseva, Y. T. Struchkov, G. G. Aleksandrov, and V. G. Andrianov, *J. Organomet. Chem.* **218**, 351 (1981).
79. N. E. Kolobova, V. V. Skripkin, and T. V. Rozantseva, *Izv. Akad. Nauk SSSR, Ser. Khim.* p. 2393 (1979).
80. S. Raghu and M. Rosenblum, *J. Am. Chem. Soc.* **95**, 3060 (1973).
81. P. W. Jolly and R. Pettit, *J. Organomet. Chem.* **12**, 49 (1968).
82. O. M. Abu Salah and M. I. Bruce, *J. Chem. Soc., Dalton Trans.* p. 2302 (1974).
83. S.-B. Samuels, S. R. Berryhill, and M. Rosenblum, *J. Organomet. Chem.* **166**, C9 (1979); D. J. Bates, M. Rosenblum, and S.-B. Samuels, *ibid.* **209**, C55 (1981).
84. D. F. Marten, *J. Chem. Soc., Chem. Commun.* p. 341 (1980).
85. D. A. Stotter, R. D. Thomas, and M. T. Wilson, *Bioinorg. Chem.* **7**, 87 (1977).
86. D. Mansuy, M. Lange, and J. C. Chottard, *J. Am. Chem. Soc.* **101**, 6437 (1979).
87. L. Latos-Grazynski, R.-J. Cheng, G. N. La Mar, and A. L. Balch, *J. Am. Chem. Soc.* **103**,

4270 (1981); M. M. Olmstead, R.-J. Cheng, and A. L. Balch, *Inorg. Chem.* **21,** 4143 (1982).
87a. B. Chevrier, R. Weiss, M. Lange, J. C. Chottard, and D. Mansuy, *J. Am. Chem. Soc.* **103,** 2899 (1981).
88. H. M. Goff and M. A. Phillippi, *Inorg. Nucl. Chem. Lett.* **17,** 239 (1981).
89. T. J. Wisnieff, A. Gold, and S. A. Evans, *J. Am. Chem. Soc.* **103,** 5616 (1981).
90. D. Mansuy, I. Morgenstern-Badarau, M. Lange, and P. Gans, *Inorg. Chem.* **21,** 1427 (1982).
91. D. Lexa, J.-M. Savéant, J.-P. Battioni, M. Lange, and D. Mansuy, *Angew. Chem.* **93,** 585 (1981); *Angew. Chem., Int. Ed. Engl.* **20,** 578 (1981).
92. K. Oguro, M. Wada, and R. Okawara, *J. Chem. Soc., Chem. Commun.* p. 899 (1975).
93. M. Wada, K. Oguro, and Y. Kawasaki, *J. Organomet. Chem.* **178,** 261 (1979).
94. P. Carusi and A. Furlani, *Gazz. Chim. Ital.* **110,** 7 (1980).
95. R. Nast and V. Pank, *J. Organomet. Chem.* **129,** 265 (1977).
96. H. D. Empsall, B. L. Shaw, and A. J. Stringer, *J. Organomet. Chem.* **94,** 131 (1975).
97. R. Nast and A. Beyer, *J. Organomet. Chem.* **194,** 125, 379 (1980).
98. R. F. Walter and B. F. G. Johnson, *J. Chem. Soc., Dalton Trans.* p. 381 (1978).
99. R. Nast and A. Beyer, *J. Organomet. Chem.* **204,** 267 (1981).
100. R. Nast and A. Beyer, *Z. Naturforsch., B: Anorg. Chem., Org. Chem.* **35B,** 924 (1980).
101. M. V. Russo and A. Furlani, *J. Organomet. Chem.* **165,** 101 (1979).
102. R. A. Bell, M. H. Chisholm, D. A. Couch, and L. A. Rankel, *Inorg. Chem.* **16,** 677 (1977).
103. M. Wada and Y. Koyama, *J. Organomet. Chem.* **201,** 477 (1980).
104. G. K. Anderson, R. J. Cross, L. Manojlovic-Muir, K. Muir, and R. A. Wales, *Inorg. Chim. Acta* **29,** L193 (1978); *J. Chem. Soc., Dalton Trans.* p. 684 (1979).
105. V. B. Pukhnarevich, Y. T. Struchkov, G. G. Aleksandrov, S. P. Sushchinskaya, E. O. Tsetlina, and M. G. Voronkov, *Koord. Khim.* **5,** 1535 (1979); Y. T. Struchkov, G. G. Aleksandrov, V. B. Pukhnarevich, S. P. Sushchinskaya, and M. G. Voronkov, *J. Organomet. Chem.* **172,** 269 (1979).
106. M. H. Chisholm and R. K. Potkul, *Synth. React. Inorg. Met.-Org. Chem.* **8,** 65 (1978).
107. K. Weiss, E. O. Fischer, and J. Müller, *Chem. Ber.* **107,** 3548 (1974).
108. J. Levisalles, H. Rudler, Y. Jeannin, and F. Dahan, *J. Organomet. Chem.* **178,** C8 (1979); **187,** 233 (1980).
109. B. E. R. Schilling, R. Hoffmann, and D. L. Lichtenberger, *J. Am. Chem. Soc.* **101,** 585 (1979).
110. N. M. Kostic and R. F. Fenske, *Organometallics* **1,** 974 (1982).
111. R. M. Kirchner, J. A. Ibers, M. S. Saran, and R. B. King, *J. Am. Chem. Soc.* **95,** 5775 (1973).
111a. V. F. Sizoi, Y. S. Nekrasov, Y. N. Sukharev, N. E. Kolobova, O. M. Khitrova, N. S. Obezyuk, and A. B. Antonova, *J. Organomet. Chem.* **162,** 171 (1978).
112. R. B. King and M. S. Saran, *J. Chem. Soc., Chem. Commun.* p. 851 (1974).
113. H. Yamazaki, *J. Chem. Soc., Chem. Commun.* p. 841 (1976).
114. E. L. Muetterties, T. N. Rhodin, A. Band, C. F. Brucker, and W. R. Pretzer, *Chem. Rev.* **79,** 91 (1979); E. L. Muetterties, *Isr. J. Chem.* **20,** 84 (1980).
115. Fe: C. Brucker and T. Rhodin, *J. Catal.* **47,** 214 (1977). Ni: J. C. Hemminger, E. L. Muetterties, and G. A. Somorjai, *J. Am. Chem. Soc.* **101,** 62 (1979). Pt: J. E. Demuth, *Surf. Sci.* **80,** 367 (1979); H. Ibach, H. Hopster, and B. Sexton, *Appl. Surf. Sci.* **1,** 1 (1977); *Appl. Phys.* **14,** 21 (1977).
116. W. C. Swope and H. F. Schaefer, *Mol. Phys.* **34,** 1037 (1977).
117. C. R. Eady, B. F. G. Johnson, and J. Lewis *J. Chem. Soc., Dalton Trans.* p. 477 (1977).
118. D. Seyferth, G. H. Williams, and J. E. Hallgren, *J. Am. Chem. Soc.* **95,** 266 (1973); D. Seyferth, G. H. Williams, and D. D. Traficante, *ibid.* **96,** 604 (1974).

119. D. Seyferth, C. S. Eschbach, G. H. Williams, P. L. K. Hung, and Y. M. Cheng, *J. Organomet. Chem.* **78**, C13 (1974).
120. D. Seyferth, *Adv. Organomet. Chem.* **14**, 97 (1976).
121. D. Seyferth, C. S. Eschbach, and M. O. Nestle, *J. Organomet. Chem.* **97**, C11 (1975).
122. B. E. R. Schilling and R. Hoffmann, *J. Am. Chem. Soc.* **100**, 6274 (1978); **101**, 3456 (1979).
123. M. Lourdichi and R. Mathieu, *Nouv. J. Chim.* **6**, 231 (1982).
124. A. J. Deeming, S. Hasso, and M. Underhill, *J. Chem. Soc., Dalton Trans.* p. 1614 (1975).
125. J. B. Keister and J. R. Shapley, *J. Organomet. Chem.* **85**, C29 (1975).
126. A. J. Deeming and M. Underhill, *J. Organomet. Chem.* **42**, C60 (1972); *J. Chem. Soc., Dalton Trans.* p. 1415 (1974).
127. A. J. Deeming and M. Underhill, *J. Chem. Soc., Chem. Commun.* p. 277 (1973); A. J. Deeming, S. Hasso, M. Underhill, A. J. Canty, B. F. G. Johnson, W. G. Jackson, J. Lewis, and T. W. Matheson, *ibid.* p. 807 (1974).
128. J. R. Andrews, S. F. A. Kettle, D. B. Powell, and N. Sheppard, *Inorg. Chem.* **21**, 2874 (1982).
129. C. J. Cooksey, A. J. Deeming, and I. P. Rothwell, *J. Chem. Soc., Dalton Trans.* p. 1718 (1981).
130. J. Evans and G. S. McNulty, *J. Chem. Soc., Dalton Trans.* p. 2017 (1981).
131. M. I. Bruce, E. Horn, O. bin Shawkataly, and M. R. Snow, unpublished results.
132. A. C. Sievert, D. S. Strickland, J. R. Shapley, G. R. Steinmetz, and G. L. Geoffroy, *Organometallics* **1**, 214 (1982).
133. J. H. Davis, M. A. Beno, J. M. Williams, J. Zimmie, M. Tachikawa, and E. L. Muetterties, *Proc. Natl. Acad. Sci. U.S.A.* **78**, 668 (1981).
134. E. Sappa, A. Tiripicchio, and M. Tiripicchio-Camellini, *J. Chem. Soc., Chem. Commun.* p. 254 (1979); *Inorg. Chim. Acta* **41**, 11 (1980).
135. A. J. Carty, S. A. MacLaughlin, and N. J. Taylor, *J. Chem. Soc., Chem. Commun.* p. 476 (1981).
136. D. H. Farrar, G. R. John, B. F. G. Johnson, J. Lewis, P. R. Raithby, and M. J. Rosales, *J. Chem. Soc., Chem. Commun.* p. 886 (1981).
137. S. Aime, A. J. Deeming, M. B. Hursthouse, and J. D. J. Backer-Dirks, *J. Chem. Soc., Dalton Trans.* p. 1625 (1982).
138. S. G. Davies and F. Scott, *J. Organomet. Chem.* **188**, C41 (1980).
139. N. E. Kolobova, L. L. Ivanov, O. S. Zhvanko, G. G. Aleksandrov, and Y. T. Struchkov, *J. Organomet. Chem.* **228**, 265 (1982).
140. H. Berke and P. Härter, *Angew. Chem.* **92**, 224 (1980). *Angew. Chem., Int. Ed. Engl.* **19**, 225 (1980).
141. R. M. Kirchner, J. A. Ibers, M. S. Saran, and R. B. King, *J. Am. Chem. Soc.* **95**, 5775 (1973); R. M. Kirchner and J. A. Ibers, *Inorg. Chem.* **13**, 1667 (1974).
142. G. G. Aleksandrov, A. B. Antonova, N. E. Kolobova, and Y. T. Struchkov, *Koord. Khim.* **2**, 1684 (1976).
143. R. E. Davis, private communication (1978).
144. G. G. Aleksandrov, A. B. Antonova, N. E. Kolobova, and Y. T. Struchkov, *Koord. Khim.* **2**, 1561 (1976).
145. M. B. Hossain, D. J. Hanlon, D. F. Marten, D. van der Helm, and E. V. Dehmlow, *Acta Crystallogr., Sect. B* **B38**, 1457 (1982).
146. R. M. Kirchner and J. A. Ibers, *J. Organomet. Chem.* **82**, 243 (1974).
147. S. C. Kao, P. P. Y. Liu, and R. Pettit, *Organometallics,* **1**, 911 (1982).
148. R. Baker, unpublished results, cited in ref. *126*.
149. N. E. Kolobova, A. B. Antonova, and O. M. Khitrova, *J. Organomet. Chem.* **146**, C17 (1978).

ADVANCES IN ORGANOMETALLIC CHEMISTRY, VOL. 22

The Activation of Carbon Dioxide by Metal Complexes

DONALD J. DARENSBOURG
and
REBECCA A. KUDAROSKI

Department of Chemistry
Texas A&M University
College Station, Texas

I
INTRODUCTION

The chemistry of one-carbon-atom molecules (C_1 chemistry) is an important area of research for the organometallic chemist. The motivation for these efforts stems from the belief that the raw material base for commercial organic chemicals will shift from oil to coal, owing to both economic reasons and declining petroleum reserves. The major raw material for a C_1-based industry is carbon monoxide, and a great deal of the current research efforts are designed to investigate catalysis of the reduction of CO with hydrogen (*1*).

Carbon dioxide can also serve as a convenient reagent in selected reduction processes, and interest in its use as an inexpensive source of chemical carbon has greatly intensified over the past few years (*2, 3*). Carbon dioxide is a very stable molecule in as much as it is the thermodynamic end product of many energy producing processes [e.g., the combustion of hydrocarbons, Eq. (1)]. Hence, its use will require a major input of energy. The reactivity of

129

CO_2, however, can be greatly enhanced by the judicious use of metal catalysts. An additional important consideration in the utilization of carbon

$$C_2H_6 + \tfrac{7}{2}O_2 \rightarrow 2CO_2 + 3H_2O(l), \quad \Delta H_{298°C} = -1560.1 \text{ kJ} \qquad (1)$$

dioxide as a source of chemical carbon in fuel production obviously is the need for hydrogen derived from nonpetrochemical origins. This dilemma has been amply discussed by Eisenberg and Hendriksen (4). The present alternative commercial route to hydrogen production, the water–gas shift reaction [Eq. (2)], may be of importance, particularly in instances where CO_2 is a better C_1 feedstock than is carbon monoxide. Indeed this has been demonstrated to be the case in methanation reactions employing nickel-based catalysts where CO_2 proceeds more rapidly than CO and at lower temperatures (5). In addition, reactions of CO_2/H_2 on supported palladium

$$CO + H_2O \rightarrow CO_2 + H_2 \qquad (2)$$

catalysts can be made to be highly selective towards either CH_4 or CH_3OH production, depending on the nature of the support (6). It should be noted as well in this regard that major efforts are being conducted to commercially produce hydrogen from the water-splitting reaction based on energy from nuclear reactors (7).

In general the energy demands for the carbon dioxide reduction process decrease with the extent of retention of oxygen in the product molecules, e.g., reactions (3) and (4) are much less energy exhausting than is reaction (5). It is thus anticipated that in the short run the industrial utilization of

$$CO_2 \;+\; \text{/}O_2 \longrightarrow \qquad\qquad\qquad (3)$$

$$CO_2 \;+\; RC\equiv CH \longrightarrow \qquad\qquad\qquad (4)$$

$$CO_2 \;+\; H_2 \longrightarrow C_nH_{2n} + H_2O \qquad\qquad (5)$$

CO_2 as a feedstock for organic substances will continue to be in the production of urea, carboxylic acids, and organic carbonates. On the other hand, many laboratory syntheses of organic substances derived from CO_2 have been published, ranging from hydrocarbons to cyclic lactones. These have recently been surveyed by Denise and Sneeden (8).

The reduction of carbon dioxide by means of what might *formally* be considered insertion reactions [Eqs. (6) and (7)] represents important preliminary processes in the production of reduced carbon containing molecules derived from carbon dioxide. This is most likely the case whether MH

$$M-H + CO_2 \longrightarrow M-O-\overset{\displaystyle O}{\overset{\|}{C}}-H \qquad (6)$$

$$M-R + CO_2 \longrightarrow M-O-\overset{\displaystyle O}{\overset{\|}{C}}-R \qquad (7)$$

and MR symbolize homogeneous catalysts or surface metal species. It is not the intention of this article to focus on CO_2 reduction processes involving heterogeneous catalysts, but we wish to concentrate on homogeneous systems in hopes of providing some understanding of the basic mechanistic aspects of this important class of organometallic reactions, i.e., CO_2 insertion into metal–hydride and metal–carbon bonds.

II

COORDINATION CHEMISTRY OF CARBON DIOXIDE

Prior to a discussion of CO_2 insertion reactions into M–H and M–C bonds it is useful to review some of the known coordination chemistry of carbon dioxide, since activation of CO_2 by metal centers is assumed to be of significance in most of these processes. Carbon dioxide can interact with metal centers by three functionalities. These include the Lewis acid site at carbon (1), the Lewis base sites of the terminal oxygen atoms (2), and the η^2 C=O bond (3). It is possible as well that a combination

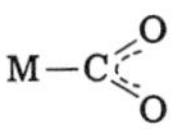

$$M-C\overset{\nearrow O}{\underset{\searrow O}{}} \qquad M-O{=}C{=}O \qquad M-\overset{\displaystyle C\overset{\nearrow O}{}}{\underset{\displaystyle O}{\|}}$$

| 1 | 2 | 3 |

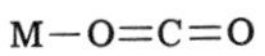

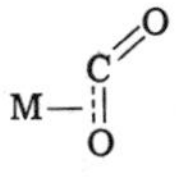

of these functionalities might be present in a given metal–CO_2 complex.

Infrared spectral identification of adduct formation involving carbon dioxide and a transition metal complex has often been in error because of subsequent reactions of CO_2 with concomitant production of carbonato–, hydrogen-carbonato–, or carboxylato–metal complexes. Indeed Mason and Ibers (9) have suggested that the only acceptable structural characterization for judging the authenticity of a class of transition metal–CO_2 complexes should be diffraction methods. X-ray structural studies have verified at least six CO_2 adducts which display all three types of bonding modes of

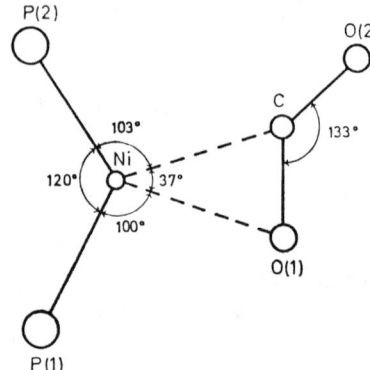

FIG. 1. Mode of bonding of CO_2 in $[Ni(CO_2)(PCy_3)_2]$,0.75-toluene. Ni–C, 1.84 Å; Ni–O(1), 1.99 Å; C–O(1), 1.22 Å; C–O(2), 1.17 Å, estimated SD 0.02 Å. Reproduced with permission from Ref. *10*. Copyright 1975 American Chemical Society.

CO_2 metals depicted above. These include the d^{10} and d^2 complexes with CO_2 η^2(C,O) bonded of $Ni[PCy_3](\eta^2\text{-}CO_2)$ (*10*) and $Nb(\eta^5\text{-}C_5H_4Me)_2(CH_2SiMe_3)(\eta^2\text{-}CO_2)$ (*11*), respectively, the structures of which are shown in Figs. 1 and 2. Herskovitz and co-workers have recently reported on an η^1-C bonded CO_2 complex of rhodium, $Rh(\eta^1\text{-}CO_2)(Cl)(diars)_2$, where CO_2 and chloride ligands are mutually trans and no metal–oxygen interactions are present (*12*). In the d^8 complex, $[Co(pr\text{-}salen)K(CO_2)THF]_n$, the CO_2 is bonded to the cobalt atom via the C functionality and O bonded to two different K^+ ions as seen in Fig. 3 (*13*). A similar bifunctional coordination of CO_2 exists in the metal cluster derivative, $[HOs_3(CO)_{10}(CO_2)\text{-}Os_6(CO)_{17}][Ph_3P)_2N]$, where carbon dioxide bridges two discrete Os_6 and

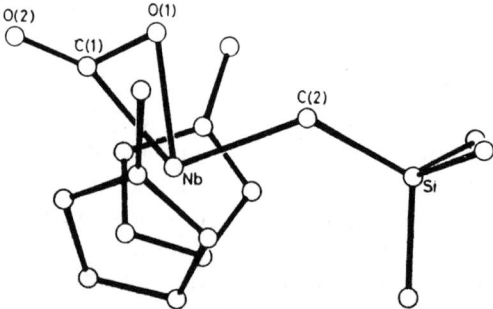

FIG. 2. The molecular structure of $[Nb(\eta\text{-}C_5H_4Me)_2(CH_2SiMe_3)(\eta^2\text{-}CO_2)]$ (**1**). Relevant dimensions are Nb–C(1), 2.144(7) Å; Nb–O(1), 2.173(4) Å; C(1)–O(1), 1.283(8) Å; and C(1)–O(2), 1.216(8) Å; O(1)–C(1)–O(2), 132.4(7)°. Reproduced with permission from Ref. *11*. Copyright 1981 American Chemical Society.

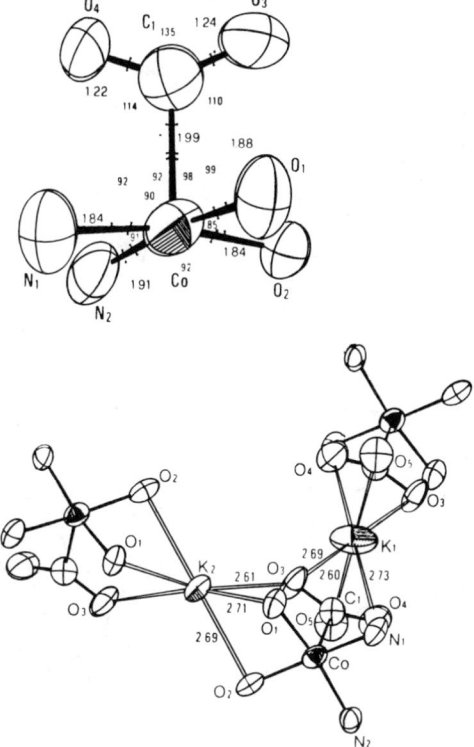

FIG. 3. An ORTEP view showing the repetitive unit of the polymeric [Co(pr-salen)KCO₂THF]ₙ and the coordination sphere around cobalt. Reproduced with permission from Ref. *13*. Copyright 1978 American Chemical Society.

Os$_3$ units (see Fig. 4) (*14, 15*). An analogous μ_3-CO$_2$ bridged tetranuclear complex of rhenium, [(CO)$_5$Re(CO$_2$)Re(CO)$_4$]$_2$, has recently been reported by Beck and co-workers (*16*) (Fig. 5). In addition, structural models for CO$_2$ η^2-C,O bonded have been crystallographically defined in the complexes (η^5-C$_5$H$_5$)$_2$V(diphenylketene) and [(η^5C$_5$H$_5$)$_2$Ti(diphenylketene)]$_2$ (*17, 18*), and η^1-O bonded in the complex [(η^5-C$_5$H$_5$)$_2$V(acetone)]BPh$_4$] (*17*) by Floriani and co-workers.

In these definitively characterized CO$_2$ adducts the O–C–O bond angle ranges from 132° to 135°. This deformation of the O–C–O bond from 180° is understandable upon examining a Walsh diagram of CO$_2$ in the ground state (*19*). The partial Walsh diagram for CO$_2$ illustrated in Fig. 6 shows that the energy of the unoccupied doubly degenerate $2\pi_u$ molecular orbitals decrease in going from the linear to the bent conformation, hence enhancing their effectiveness in backbonding with the metal center. This

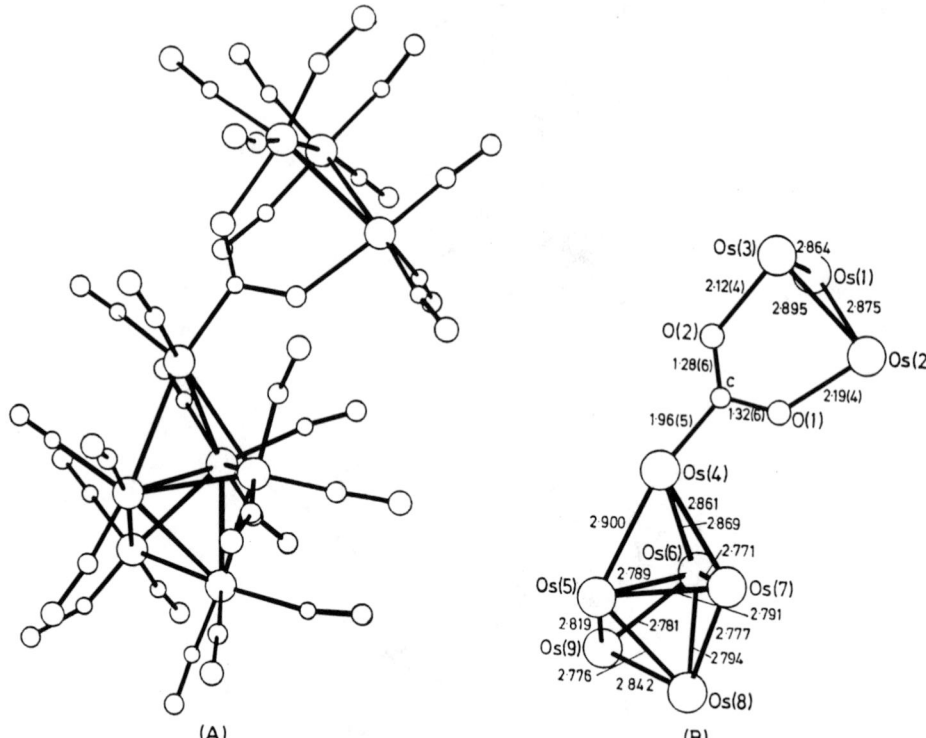

FIG. 4. (A) Structure of $[HOs_3(CO)_{10} \cdot O_2C \cdot Os_6(CO)_{17}]^-$; (B) significant bond lengths (in Å) within the cluster, with CO groups omitted for clarity. Reproduced with permission from Ref. *14*. Copyright 1976 American Chemical Society.

effect has been quantitatively interpreted in a recent ab initio molecular orbital study of CO_2 binding to transition metal centers by Sakaki and co-workers (*20*). The energy required to distort the CO_2 from its equilibrium structure is more than compensated for by the interaction energy, primarily backbonding from the metal to the $\pi_\parallel^*$ orbital of CO_2, with a minimum binding energy reached at an $O-C-O$ angle of $\sim 138°$.

These researchers (*20*) further concluded from their calculations, which critically contrasted the three coordination modes of CO_2, that the side-on coordination (η^2-C,O) is the most favored when a large back-donative stabilization is accomplished. This π-backbonding component decreases in the coordination modes η^2-C,O $\gg \eta^1$-C $> \eta^1$-O. When $M^{\delta+}$ and $C^{\delta+}$ repulsive interaction is great, for example in species containing positively charged metal centers, the end-on mode (η^1-O) is stabilized over the other two. Indeed, this type of interaction is seen between K^+ ion and the oxygen atoms

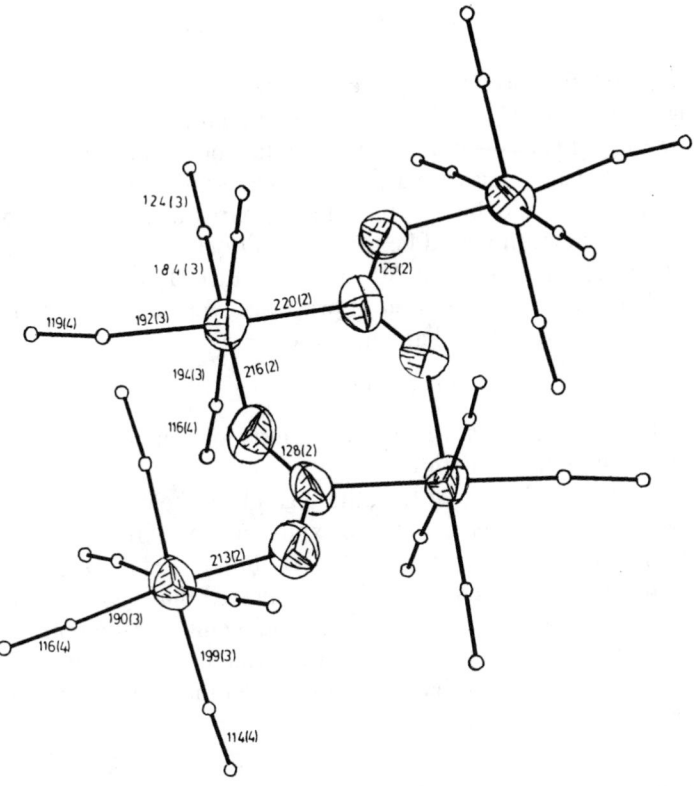

FIG. 5. A μ_3-CO_2 bridged tetranuclear complex of rhenium, $[(CO)_5Re(CO_2)Re(CO)_4]_2$. Reproduced with permission from Ref. 16. Copyright 1982 Verlag Chemie GmbH.

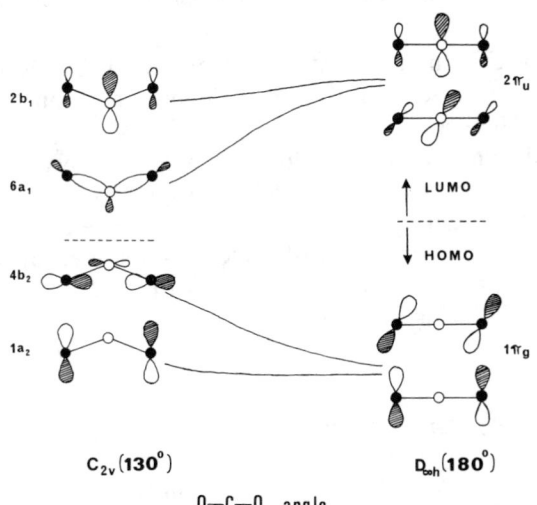

O—C—O angle

FIG. 6. Partial Walsh diagram for CO_2.

of CO_2 in the bifunctional complexes depicted in Fig. 3. On the other hand, binding through the C functionality (η^1-C) might be anticipated in species where the metal is electron rich and simultaneously the side-on coordination is not allowed for some reason such as coordination number. Since in the η^1-C binding mode the CO_2 ligand is primarily behaving as an acceptor group, this interaction would be expected in organometallic species which satisfy the 18-electron rule and form adducts with CO_2 without concomitant loss of other ligands. This is seen in processes typified by Eq. (8) (21–23). Additionally it is apparent that when η^1-C

$$W(CO)_5^{2-} + CO_2 \rightarrow W(CO)_5CO_2^{2-} \tag{8}$$

and η^1-O binding modes are simultaneously present in a complex, this represents a favorable situation.

The electronic influence of the other ligands in the metal's coordination sphere is a major consideration when exploring the coordination chemistry of carbon dioxide. Since the metal–CO_2 bond is stabilized mainly by back-donative interactions, it would be expected that good donor ligands (e.g., the ubiquitous phosphine ligands) would enhance the binding ability of CO_2 to the metal center. This has been verified both experimentally (10) and theoretically (20), and has as well been demonstrated to be of importance in CO_2 insertion processes (see below).

III

INSERTION REACTIONS OF CARBON DIOXIDE

Examples of CO_2 insertion into M–H and M–C bonds are numerous. Less common are instances of carbon dioxide insertion into M–O and M–N bonds. There are excellent reviews which encompass these areas by Eisenberg and Hendriksen (4), Volpin and Kolomnikov (24), Kolomnikov and Grigoryan (25), and Sneeden (26). Hence we will emphasize developments since the time of these reviews (i.e., from 1979 to present) with our perspective being primarily that of understanding mechanistic aspects of these insertion processes. Some overlap with these earlier reviews will necessarily occur during such efforts.

A. Insertion into Metal–Hydrogen Bonds

In principle the insertion of CO_2 into a transition metal hydrogen bond can result in either M–O or M–C bond formation, i.e., production of metalloformate (4) or metallocarboxylic acid (5) derivatives. Thus far, definitive identification of a metallocarboxylic acid derivative synthesized

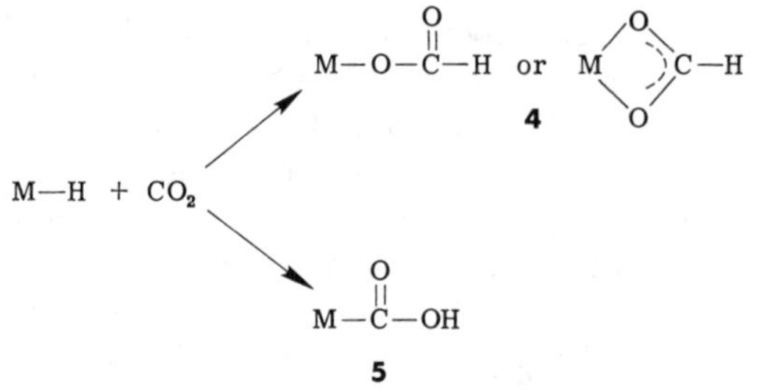

$$
M-H + CO_2 \longrightarrow
\begin{array}{c}
M-O-\overset{\overset{\displaystyle O}{\|}}{C}-H \text{ or } M\overset{O}{\underset{O}{\diagup}}C-H \\
\mathbf{4}
\end{array}
\qquad (9)
$$

$$
M-\overset{\overset{\displaystyle O}{\|}}{C}-OH
$$

5

by CO_2 insertion into a $M-H$ bond is lacking. Indirect evidence for its formation in the reaction of $[Ph_3P]_3Co(N_2)H$ with CO_2 in benzene has been presented on the basis of the production of methyl acetate from the sequence of reactions depicted in Eq. (10) (24, 27). In addition, methyl formate was found as well from the intermediacy of a cobalt formate complex.

$$
[Ph_3P]_3Co(N_2)H + CO_2 \xrightarrow{C_6H_6} \{[Ph_3P]_3Co(N_2)COOH\} \xrightarrow[-H_2O]{MeI \quad MeOH/BF_3} MeCOOMe \quad (10)
$$

Despite their equivocal presence in CO_2 insertion processes, the existence of metallocarboxylic acid species in organometallic chemistry and homogeneous catalysis is presently well documented. These species have been synthesized from the addition of OH^- to metal carbonyls (28–32), CO insertion into the $M-OH$ bond (33), and protonation of anionic metal–carbon dioxide adducts (22). Additionally, we have recently shown by isotopic labeling experiments that the group 6B metal derivatives, $M(CO)_5COOH^-$ and $M(CO)_5OC(O)H^-$, do not interconvert intramolecularly (34). These rather interesting metallocarboxylic acid species will be the subject of an upcoming review (35).

In general, the insertion reaction of carbon dioxide into metal hydrogen bonds is formally much akin to the analogous process involving olefins (Scheme 1). This analogy is particularly appropriate since the binding of

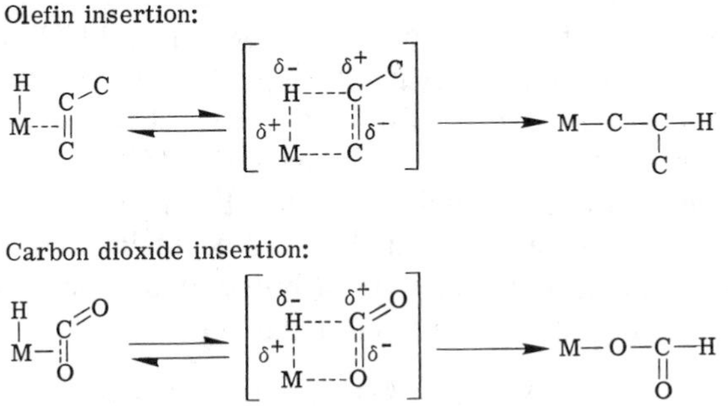

Olefin insertion:

Carbon dioxide insertion:

SCHEME 1

CO_2 to the metal center directly parallels that of metal–olefin binding (*36*). For the olefin insertion reaction, activation parameters support a concerted reaction process as depicted above with a relatively low enthalpy of activation, i.e., simultaneous bond breaking and bond making, and a negative activation entropy (*37*). Thus far, no studies determining the corresponding activation parameters for the CO_2 insertion process have been reported.

Since the earlier reviews in this area (*4, 24–26*), several papers have been published on the insertion of CO_2 into metal–hydrogen bonds, or its retrogradative counterpart, CO_2 extrusion from metalloformate species. Irradiation of $ReH_3(diphos)_2$ [diphos = bis(1,2-diphenylphosphino)ethane] in benzene results in elimination of dihydrogen to provide the $[ReH(diphos)_2]$ intermediate (*38*). $[ReH(diphos)_2]$ readily forms adducts with N_2, CO, and C_2H_4, whereas it reacts with carbon dioxide to afford the complex $Re(diphos)_2O_2CH$. The structure of $Re(diphos)_2O_2CH$ is assigned as having a bidentate formato ligand based on spectral data, and by analogy with the $Re(S_2CH)(CO)_2(PPh_3)_2$ derivative, which has been characterized by X-ray analysis (*39, 40*). Presumably, reaction of the unsaturated species, $[ReH(diphos)_2]$, with CO_2 occurs via adduct formation as well followed by rapid metal–hydrogen bond cleavage with concomitant C–H bond formation. This is further supported by the observation that $ReH_3(diphos)_2$ does not react with CO_2 in the absence of photoinduced H_2 reductive elimination. A parallel study has previously been reported of the photoinduced reaction of $H_4Fe(PPh_2Et)_3$ or $H_2Fe(N_2)(PPh_2Et)_3$ with CO_2 to yield $(HCO_2)_2Fe(PPhEt_2)_2$ (*41*).

Because of the analogy which can be drawn between olefin and carbon dioxide insertion processes (Scheme 1), it is of interest to note that although CO_2 insertion into the Re–H bond of $[Re(diphos)_2H]$ is very facile, similar insertion of ethylene to provide a rhenium alkyl species is not observed. This significant difference in reactivity for the C–H bond-forming process entailing C_2H_4 or CO_2 with metal hydrides is also seen in our studies involving anionic species (*34, 42–45*). The group 6B metal pentacarbonyl hydrides, $HM(CO)_5^-$, act as sponges toward carbon dioxide at pressures less than atmospheric to quantitatively yield formato derivatives [Eq. (11)]. Similar reactivities are noted for these metal hydrides toward COS and CS_2 with formation of thioformate derivatives (Fig. 7). On the other hand, no $C_2H_5M(CO)_5^-$ is afforded from the reaction of ethylene with $HM(CO)_5^-$ under fairly rigorous conditions (51.7 bars and 25°C) (*46*). The reaction of these inexpensive anionic metal hydrides with olefins to provide anionic metal alkyls is a pivotal step which must be realized in order for the formation of metallocarboxylates via carbon dioxide insertion into metal–carbon bonds to become utilitarian (see below).

$$HM(CO)_5^- + CO_2 \rightleftharpoons HCO_2M(CO)_5^- \qquad (11)$$

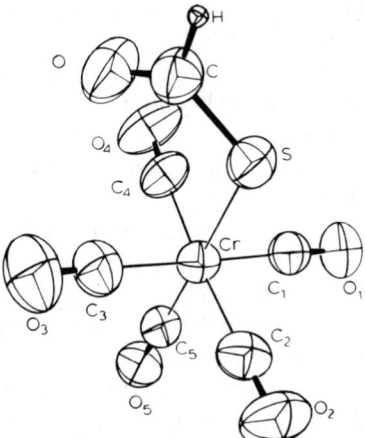

Fig. 7. Perspective drawing of the $Cr(CO)_5SC(O)H^-$ anion. Some bond lengths are as follows: Cr–S, 2.447(1) Å; Cr–C(eq)$_{av}$, 1.894(4) Å; Cr–C(ax), 1.837(4) Å; S–C, 1.725(5) Å; O–C, 1.206(6) Å; C–H, 1.06(4) Å. Reproduced with permission from Ref. 44. Copyright 1982 American Chemical Society.

In coordinatively saturated metal hydrides, such as the $HM(CO)_5^-$ (M = Cr, Mo, W) derivatives, formation of the four-centered transition state for CO_2 insertion (Scheme 1) may proceed with or without CO loss and concomitant coordination of CO_2 at the metal center. That is, CO_2 insertion may occur by means of dissociative (D) or dissociative interchange (I_d) processes, or an associative interchange (I_a) process (47, 48). In either instance an acid–base interaction between the anionic hydride ligand and the electrophilic carbon center of carbon dioxide as represented in 6 may occur prior to formation of the four-centered transition state depicted in Scheme 1. An interaction of this type has been observed for these $HM(CO)_5^-$ derivatives with Lewis acids such as BH_3 (49).

$$M-H\cdots\overset{\delta^-}{\underset{\underset{O}{\|}}{\overset{\overset{O}{\|}}{C}}}{}^{\delta^+}$$

6

This variance in intimate mechanisms is likely seen in the reactions of $HCr(CO)_5^-$ and $HW(CO)_5^-$ with CO_2 to provide $HCO_2M(CO)_5^-$ derivatives (45). For example, carbon monoxide dissociation in the chromium anion, as evinced by ^{13}CO exchange studies, occurs at a rate indistinguishable by conventional techniques from that of CO_2 insertion. Consistent with this observation, the rate of decarboxylation of $HCO_2Cr(CO)_5^-$ is retarded in an atmosphere of carbon monoxide. Similar behavior was noted in decarboxylation studies of the $(\eta^5\text{-}C_5H_5)Fe(CO)_2O_2CH$ derivative (50). On the other hand, insertion of CO_2 into the $HW(CO)_5^-$ species proceeds at a rate faster than CO ligand substitution. Hence, in this instance either $W-H\cdots CO_2$ (6) interaction enhances CO lability or CO_2 insertion takes place by a concerted (I_a) process. Definitive evidence for this mechanistic pathway has been obtained for carbon dioxide insertion into $CH_3W(CO)_5^-$ to afford $CH_3CO_2W(CO)_5^-$ (see below).

An elegant study of the decarboxylation of the *optically active* rhenium formate $(\eta^5\text{-}C_5H_5)Re(NO)(PPh_3)O_2CH$ has been recently reported by Merrifield and Gladysz (51). The hydride $(\eta^5\text{-}C_5H_5)Re(NO)(PPh_3)(H)$ formed in $>95\%$ yield at $70°-130°C$. Both ORD and CD spectra indicated that the decarboxylation proceeded with retention at rhenium. Rate studies and crossover experiments indicated that no PPh_3 dissociation occurred. Other data obtained include k_H/k_D ($112°C$) $= 1.55 \pm 0.08$, $\Delta H^{\neq} = 26.8 \pm 0.6$ kcal mol^{-1} and $\Delta S^{\neq} = -6.3 \pm 1.3$. Thus, if this decarboxylation requires, like $HCr(CO)_5^-$, a vacant coordination site on the metal, preequilibrium C_5H_5 slippage, NO bending, or tight ion pair formation seem to be the only possibilities. Unfortunately, the microscopic reverse of this decarboxylation could not be effected.

It is worthwhile to note that in the reaction of the group 6B metal hexacarbonyls with the azide anion to provide isocyanatometallates, a concerted mechanism is proposed based on kinetic parameters which involves a three-centered transition state (7), and that the activation enthalpy is some 22.6 kJ lower for tungsten than for chromium (52). This reactivity sequence correlates with an increase in M–CO bond distances going from chromium to tungsten hexacarbonyl (53–55).

$$\left[(CO)_5W \cdots \begin{array}{c} C=O \\ N \\ N_2 \end{array} \right]^-$$

7

Several reports of catalysis of the decomposition of formic acid involving homogeneous transition metal complexes and proceeding by means of metalloformate intermediates have recently appeared in the literature. For example, $Rh(C_6H_4PPh_2)(PPh_3)_2$ (8) catalyzes the decomposition of formic acid to CO_2 and H_2 via the intermediacy of the product of oxidative-addition of HCOOH, $Rh(HCO_2)(PPh_3)_3$ (56). β-Elimination of the hydride from the

$$\begin{array}{c} Ph_3P \quad\quad PPh_2 \\ Rh \\ Ph_3P \end{array}$$

8

monodentate formate ligand to the metal center occurs in a slower step to produce CO_2 and $Rh(H)(PPh_3)_3$ [Eq. (12)]. Reaction of $Rh(H)(PPh_3)_3$ with an additional mole of HCOOH yields H_2 and reforms $Rh(HCO_2)(PPh_3)_3$.

$$P_3RhO\diagdown_{H}C=O \longrightarrow \left[P_3Rh^{\text{\tiny|||||||}}\diagdown_{H}^{O}\diagup C=O \right] \longrightarrow P_3RhH + CO_2 \quad (12)$$

Alkaline solutions of $Ru(CO)_{12}$ (KOH in aqueous ethoxyethanol) have also been found to catalytically decompose formic acid (57, 58). Presumably this occurs by way of anionic ruthenium hydride derivatives [e.g., $HRu_3(CO)_{11}^-$] reacting with HCOOH to provide a ruthenium formate derivative and H_2. Subsequent β-elimination of hydride from the ruthenium formate led to regenerating the anionic ruthenium hydride species and carbon dioxide. We have recently synthesized and fully characterized a possible ruthenium formato intermediate for this process, $Ru_3(CO)_{10}$-$(O_2CH)^-$ (9) (59). Indeed this species in part extrudes CO_2 in the presence of CO with concomitant production of $Ru_3(CO)_{11}H^-$.

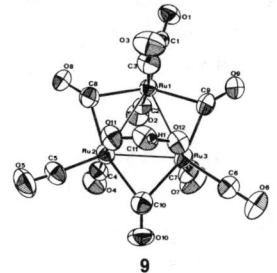

9

The complex *trans*-$PtH(O_2CH)[PEt_3]_2$ catalyzes the decomposition of formic acid in the presence of sodium formate. A mechanism based on the equilibria described in Scheme 2 has been proposed by Paonessa and Trogler (60). The role of formate ion is to promote catalysis by reaction with the platinum dimer (10) or the solvated complex [*trans*-$PtH(S)L_2$]$^+$, where S = acetone, to reform the catalytically active monomeric species 11 and 12.

A phosphine bridged molybdenum hydride dimer, μ-$H[Mo_2(CO)_8(\widehat{P\,P})]^-$

$$[Pt_2H_3L_4]^+ + [HCO_2]^- \underset{}{\overset{}{\rightleftharpoons}} \textit{trans}\text{-}PtH(O_2CH)L_2 + PtH_2L_2$$

$$\mathbf{10} \qquad\qquad\qquad\qquad \mathbf{11} \qquad\qquad \mathbf{12}$$

$$[\textit{trans}\text{-}PtH(S)L_2]^+ + [HCO_2]^-$$

SCHEME 2

($\overset{\frown}{\text{P P}}$ = bisdiphenylphosphinomethane), has been demonstrated to be a catalyst for the reaction of HCOOH to hydrogen and CO_2 (61). Infrared spectral evidence for the presence of the formato-bridged $Mo_2(CO)_8(\overset{\frown}{\text{P P}})(O_2CH)^-$ anion (13) has been obtained. Because of the very stable bridging phosphino ligand the intervention of monomer catalytically active species can be excluded.

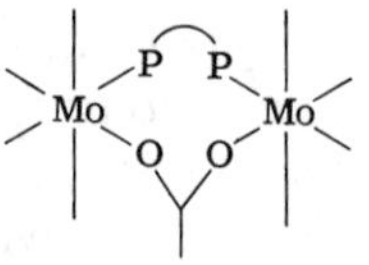

13

In a related process, a rhodium A-frame hydride complex (14), containing the bridging bisdiphenylphosphinomethane ligand, has been shown to undergo reaction with carbon dioxide at ambient pressure to yield the bridging formate derivative (15) (62). Further reaction of 15 with CO occurs with concomitant explusion of CO_2. This latter process is thought to transpire via a change in formate coordination from bridging to terminal followed by decarboxylation through a β-elimination step [Eq. (13)].

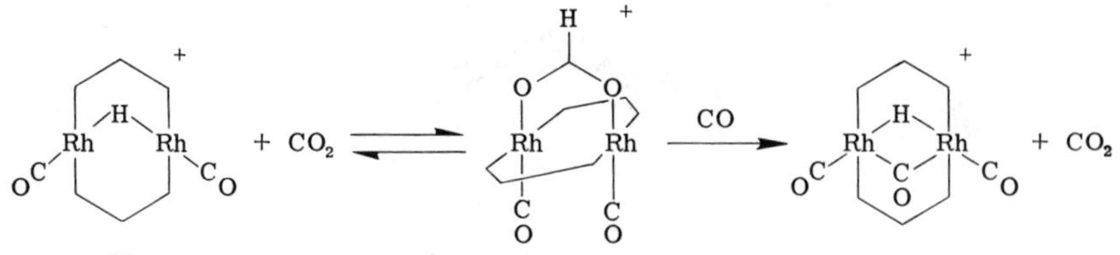

14   **15**

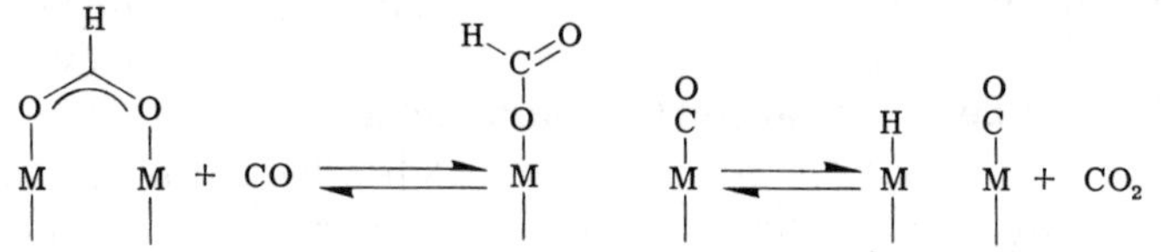

(13)

Scheme 3 forms a catalytic cycle for the water–gas shift reaction (63) employing $[Rh_2(\mu\text{-CO})(CO)_2(dpm)_2]$ in the presence of acid as a catalyst (62). It should be reiterated that alternative cycles might be written which do not involve formate intermediates. For example, a possible mechanism for catalysis of the water–gas shift reaction involving the binuclear metal species, $[Pt_2H_2(\mu\text{-H})(dpm)_2]^+$, is outlined below (Scheme 4) (64). We have critically discussed the role of formate versus carboxylic acid intermediates in homogeneous catalysis of the water–gas shift reaction by mononuclear metal catalysts elsewhere (34).

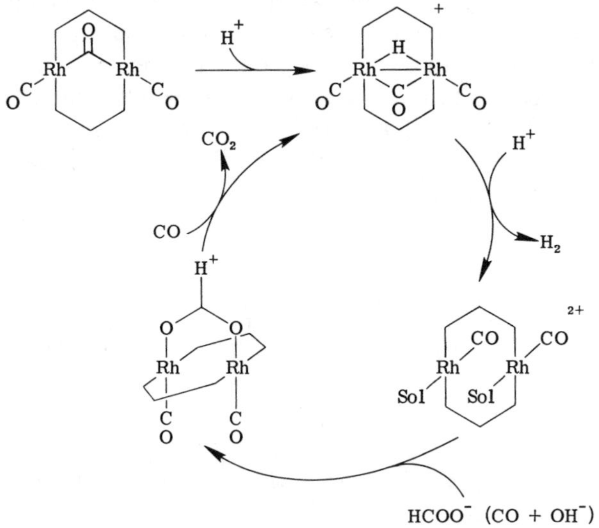

SCHEME 3

In all of these instances involving bridging formato ligands it has not been defined as to whether the metal–hydride bond forming process occurs at the same metal center containing the bound formate ligand [as is depicted in Eq. (13)] or at an adjacent metal center. This intimate mechanistic detail must await further experimental observations.

Beguin et $al.$ (65) have observed that the hexameric copper hydride, [HCuPPh$_3$]$_6$, reacts with CO_2 in benzene solution at room temperature to afford the formate $(Ph_3P)_2CuO_2CH$. In the presence of phosphine formation of the formate derivative is quantitative [Eq. (14)]. Production of **16** from

$$[HCuPPh_3]_6 + 6PPh_3 \xrightarrow{CO_2} \overset{\overset{O}{\parallel}}{HCOCu(PPh_3)_2}$$ (14)

16

[HCuPPh$_3$]$_6$ is also effected by carbon monoxide along with trace quantities of water. A plausible suggestion for formate formation in this instance is that

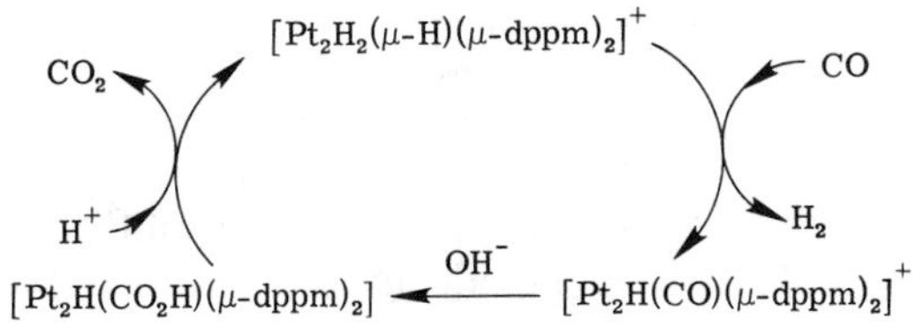

SCHEME 4

the copper complex catalyzes the water–gas shift reaction with the thus produced CO_2 reacting with hydride to yield **16**. Although production of the metallocarboxylic acid intermediate is proposed to result from CO insertion into a CuOH bond [Eq. (15)], a more attractive alternative mechanism would involve OH^- attack at a cationic copper carbonyl species [Eq. (16)]. Consistent with either proposal is the fact that formate production from CO/H_2O is inhibited by hydrogen.

$$L_n CuH \xrightarrow{H_2O} L_n CuOH \xrightarrow{CO} L_n Cu\overset{\overset{\displaystyle O}{\|}}{C}-OH \longrightarrow$$

$$L_n CuH + CO_2 \longrightarrow L_n CuO_2CH \quad (15)$$

$$L_n CuH \xrightarrow{H_2O} L_n CuH_2^+ + {}^-OH \xrightarrow{CO}$$

$$L_n CuCO^+ + H_2 \xrightarrow{{}^-OH} L_n Cu\overset{\overset{\displaystyle O}{\|}}{C}-OH \quad (16)$$

The results summarized in Scheme 5 indicate that in the presence of hydrogen, protic solvent, or hydride donors or acceptors, the formato ligand in $(Ph_3P)_2CuO_2CH$ is neither reduced nor transformed into organic formate (see below).

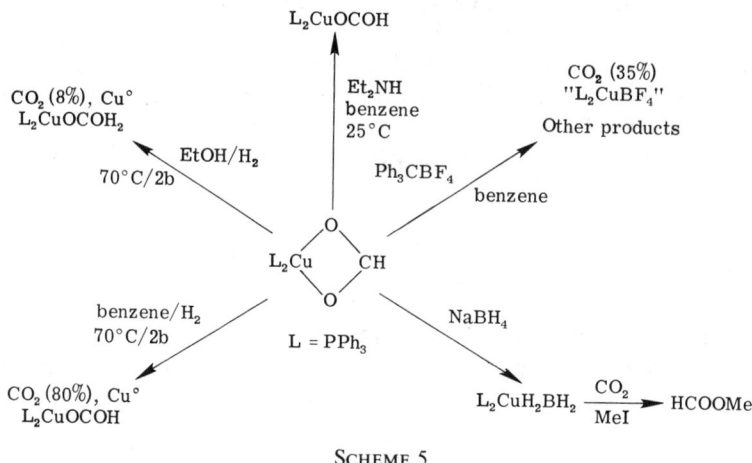

SCHEME 5

Other decarboxylation reactions of transition metal formato-derivatives appearing in the recent literature include the following investigations. Roper and Wright (66) have shown that the $MR(\eta^2\text{-}O_2CH)(CO)(PPh_3)_2$ complexes

(M = Ru or Os, R = o-tolyl) undergo decarboxylation to afford the cis-hydrido–aryl complex in the case M = Os(II) [Eq. (17)], with reductive-elimination of RH occurring additionally for the case M = Ru(II) [Eq. (18)]. In the latter instance ortho-metallation of triphenylphosphine occur concomitantly providing species **17**. A similar ortho-metallated product is noted during prolonged refluxing of OsR(H)(CO)(PPh$_3$)$_3$ in benzene. An intriguing reaction of Ru(CO)(PPh$_3$)$_3$ with formaldehyde resulted in production of the Ru(II) methyl derivative (**18**).

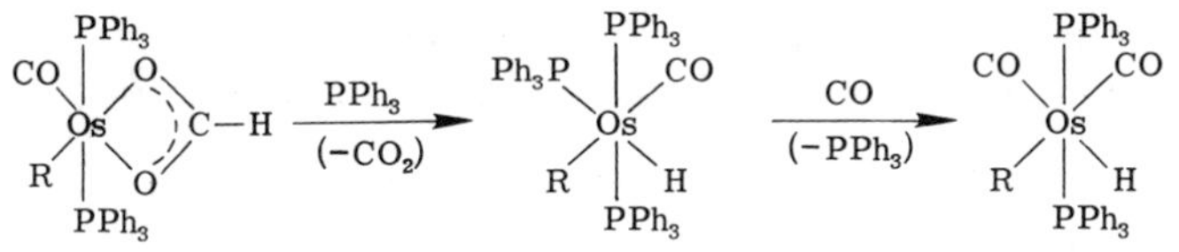

$$\text{(17)}$$

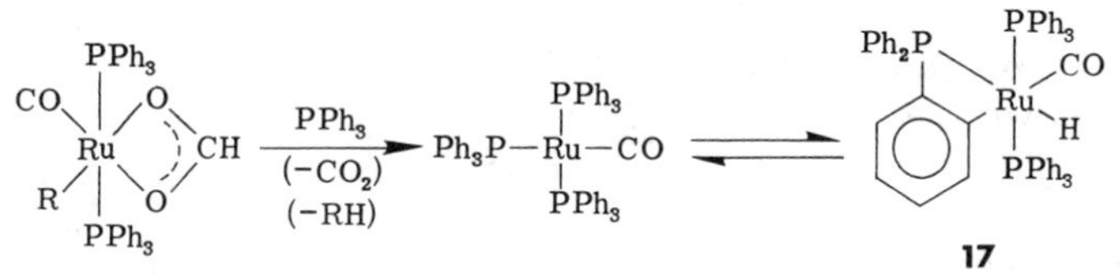

$$\text{(18)}$$

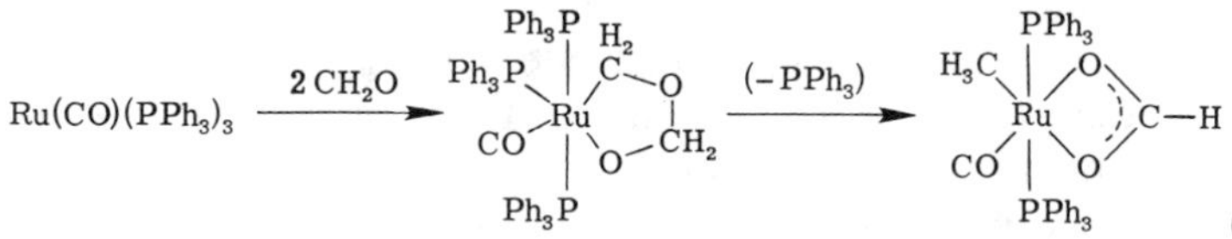

$$\text{(19)}$$

Treatment of the hydroxo(methyl)complexes, Pt(OH)Me(diphos) and Pt(OH)Me(dppp) (dppp = Ph$_2$PCH$_2$CH$_2$CH$_2$PPh$_2$), with formic acid has provided a synthesis of the corresponding O-bonded formato complexes, e.g., Pt(O$_2$CH)Me(diphos) (67). These complexes were shown to slowly lose CO$_2$ both in solution and the solid state to give the hydrido(methyl)complexes.

B. *Insertion into Metal–Carbon Bonds*

The pivotal step in carbon homologation processes using carbon dioxide is the insertion of carbon dioxide into metal–carbon bonds. Hence, this insertion reaction is of paramount importance in the utilization of carbon dioxide as a source of chemical carbon. The generalized reaction is schematically represented in Eq. (20), where a metal–alkyl or –aryl group is transformed into a coordinated carboxylate.

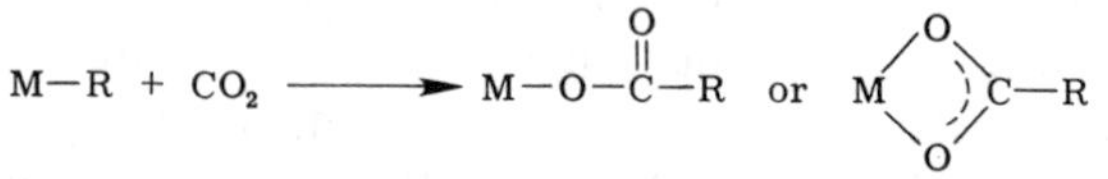

$$M-R + CO_2 \longrightarrow M-O-\overset{\overset{O}{\|}}{C}-R \ \text{ or } \ M\overset{\diagup O}{\underset{\diagdown O}{\diagdown}}C-R \tag{20}$$

The origin of the alkyl or aryl group (R) in Eq. (20) is as well an important issue in the development of C_1 chemistry involving CO_2. We will not be concerned with this question at present other than to mention that the carbanion in Eq. (20) is usually derived from lithium or Grignard reagents, or alkyl- or arylhalides [Eqs. (21) and (22)]. An alternate source, e.g., olefins,

$$M-X + RLi \rightarrow M-R + LiX \tag{21}$$

$$M^- + RX \rightarrow M-R + X^- \tag{22}$$

must be developed for large scale carbon–carbon bond-forming processes.

The known CO_2 insertion reactions involving metal–carbon bonds have all resulted in carbon–carbon bond formation with possibly one exception. Infrared spectral and chemical evidence has been presented for the formation of the metallocarboxylate ester $Co(CO_3)_n(COOEt)(PPh_3)$, $n = 0.5–1.0$ from the reaction of $Co(CO)(C_2H_5)(PPh_3)_2$ with carbon dioxide from Volpin's laboratory (68). Although these studies are not conclusive for "abnormal" CO_2 insertion, metallocarboxylate esters are well-known compounds which result from the nucleophilic addition of alkoxides on the carbon center in metal carbonyls (69).

Some very interesting work on the reversible insertion of carbon dioxide into palladium–carbon bonds has been reported recently by Braunstein and co-workers (70). Palladium(II) complexes of ethyl(diphenylphosphino)acetate (19 and 20) were found to react with CO_2 in tetrahydrofuran under ambient conditions [Eq. (23)] with C–C coupling to the nucleophilic α-phosphino carbon of the ligand taking place (complexes 21 and 22). This report represents the first example where the reversible CO_2 insertion in a molecular complex has been fully characterized by X-ray diffraction as resulting from a C–C bond forming process. Figure 8 depicts ORTEP drawings of complexes 19 and 22.

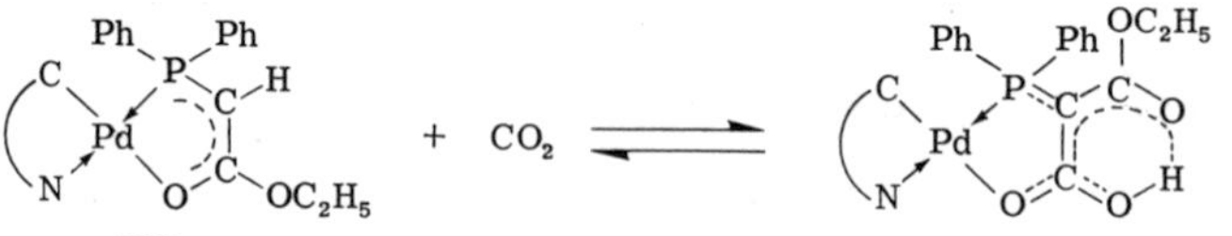

19 (C N) = dmba

20 (C N) = 8-mq

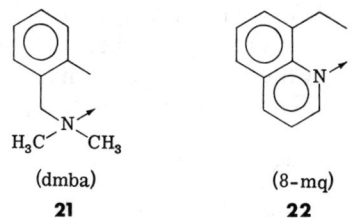

(dmba)

21

(8-mq)

22

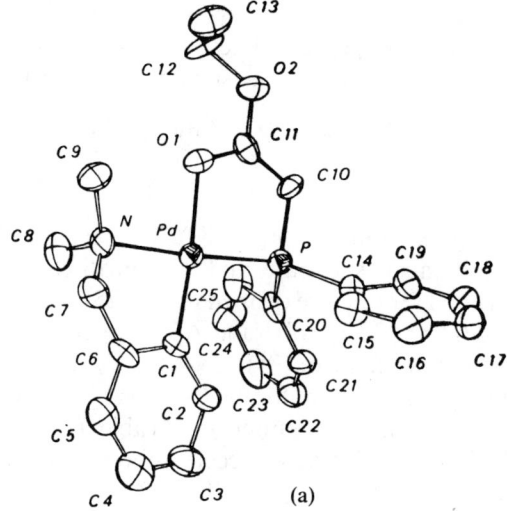

(a)

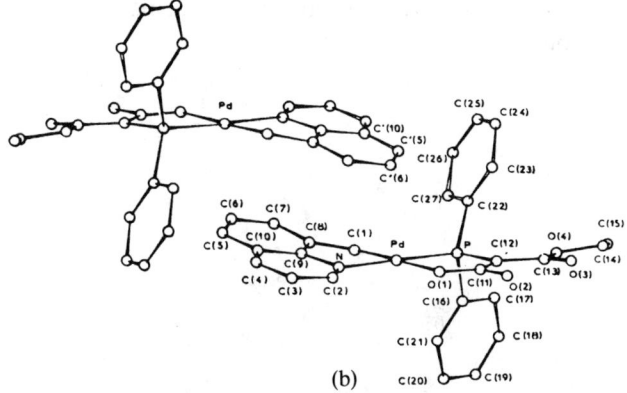

(b)

FIG. 8. (a) The molecular structure of **19**. (b) A view of the two molecules of **22** constituting a centrosymmetric "dimer." Reproduced with permission from Ref. *70*. Copyright 1981 American Chemical Society.

The chemistry of another reversible CO_2 carrier has been described by Saegusa and co-workers (71). These researchers reported on a Cu(I)–bicarbonato complex that reversibly decarboxylates in both organic and aqueous solvents [Eq. (24)]. This copper(I) complex has further been shown to act as

$$HOCO_2Cu(PEt_3)_3 \rightleftharpoons HOCu(PEt_3)_3 + CO_2 \qquad (24)$$

a carbon dioxide carrier to carboxylate cyclohexanone and propylene oxide, and this aspect will be examined in a later section.

We have recently observed that anionic alkyl and aryl derivatives of the group 6B metal carbonyls undergo insertion reactions with carbon dioxide to afford the corresponding carboxylates, e.g., as depicted in Eq. (25) (44, 72). An ORTEP representation of one such product of CO_2

$$CH_3W(CO)_5^- + CO_2 \rightarrow CH_3CO_2W(CO)_5^- \qquad (25)$$

23

insertion (**23**) is provided in Fig. 9 (73, 74). The reaction was found to be first-order in both metal substrate and carbon dioxide. The activation parameters measured for the case defined in Eq. (25) were found in tetrahydrofuran to be $\Delta H^* = 39$ kJ mol^{-1} and $\Delta S^* = -167$ J mol^{-1}, consistent with a concerted (I_a) mechanism (75).

A significant rate enhancement for the CO_2 insertion process was noted in the presence of alkali metal counterions (Table I), even in the highly coordinating THF solvent. This rate acceleration was not, however, catalytic in alkali metal counterion, since the once formed carboxylate was observed to form a tight ion pair (76, 77) via its uncoordinated oxygen atom with the alkali metal ion, as evinced by infrared spectroscopy in the $\nu(CO_2)$ region. That is, the counterion was consumed during the carbon dioxide insertion reaction.

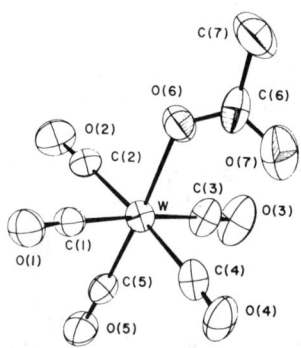

FIG. 9. ORTEP plot of the $W(CO)_5O_2CCH_3^-$ anion in **1** (40% thermal ellipsoids). Reproduced with permission from Ref. 74. Copyright 1982 American Chemical Society.

TABLE I
REACTION CONDITIONS OF CO_2 INSERTION INTO $CH_3W(CO)_5^{-a}$

No.	Additive	Time (hr)	Temp. (°C)	Extent of the reaction[b] (%)
1	none	24	ambient	~30
2	none	46	ambient	~46
3	none	24	52	100
4	Li salts[c]	20	ambient	100
5	LiCl[d]	6	ambient	100
6	Na[+e]	<25	ambient	100[f]

[a] $CH_3W(CO)_5^-$ PNP$^+$, 0.01 M solution in THF, CO_2 pressure 100–300 kPa. Reprinted with permission from Ref. 44. Copyright 1982 American Chemical Society.
[b] By IR spectroscopy of carbonyl region.
[c] LiCl, LiBr, LiO$_2$CCH$_3$ as by-products of the $CH_3Li \cdot LiBr + ClW(CO)_5^- \rightarrow CH_3W(CO)_5^-$ reaction pressurized *in situ* with CO_2.
[d] 1.1 molar excess with respect to $CH_3W(CO)_5^-$.
[e] NaBPh$_4$ fourfold molar excess with respect to $CH_3W(CO)_5^-$.
[f] Determined as $W(CO)_6$.

No infrared spectral evidence was obtained for the interaction of the alkali metal counterions with either the anionic metal substrate or free CO_2 in THF. Hence, we envisage the transition state for processes typified by Eq. (25) to result from a concerted attack of the nucleophilic $RM(CO)_5^-$ substrate on the electrophilic carbon or carbon–oxygen center of CO_2, with the metal counterion serving to reduce the buildup of negative charge on the oxygen atom of the incipient carboxylate ligand (Fig. 10). It must be emphasized that conclusive verification of this suggestion must await supplementary stereochemical studies (see below). This mode of CO_2 activation is akin to the bifunctional model complexes active in carbon dioxide fixation widely discussed by Floriani *et al.* (*13, 78*). Additionally, the enhancement of the reactivity of CO_2 attendant with alkali metal ions is analogous to the rate acceleration observed for the SO_2 insertion reaction with $(\eta^5\text{-}C_5H_5)$W(CO)$_3$CH$_3$ in the presence of Lewis acids (*79*).

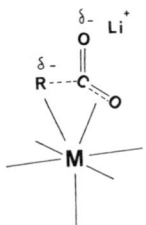

FIG. 10. Proposed intermediate for CO_2 insertion into metal alkyl bond in the presence of lithium counterion.

Carbon dioxide insertion into the W–C bond of $CH_3W(CO)_5^-$ was not retarded by excess carbon monoxide. In other words, as Fig. 10 illustrates, the CO_2 insertion process does not involve a coordinatively unsaturated intermediate. This observation could only be made when an alkali metal counterion was present, since the rate of CO_2 insertion was much faster than that of CO insertion under this condition. On the other hand, [PNP][$CH_3W(CO)_5$] undergoes CO insertion (80) at a much faster rate than carbon dioxide insertion. Both processes exhibited similar metal (W > Cr) and R(CH_3 > C_6H_5) dependences.

A major unanswered question in the carbon dioxide insertion reaction involving metal–carbon σ bonds is that of stereochemistry about the carbon center. Migratory insertion reactions involving carbon monoxide into M–C bonds occurs with retention of configuration at carbon (81, 82). In contrast, the stereochemistry at the α carbon for SO_2 insertion into $(\eta^5\text{-}C_5H_5)Fe(CO)_2R$ has been shown to take place with greater than 95% inversion of configuration at carbon (83). In addition, the chiral iron complex, $(\eta^5\text{-}1\text{-Me-}3\text{-}C_6H_5C_5H_3)Fe(CO)(PPh_3)R$, has been demonstrated to undergo SO_2 insertion with retention of configuration about the iron atom (84, 85).

These stereochemistry observations for SO_2 insertion, taken together with first-order kinetics with respect to metal substrate, retardation by steric bulk in R, and rate enhancement with increasing electron density at R, have been interpreted in terms of an S_E2 (inversion) mechanism involving a configurationally stable ion pair (24) as depicted in Scheme 6 (86–88).

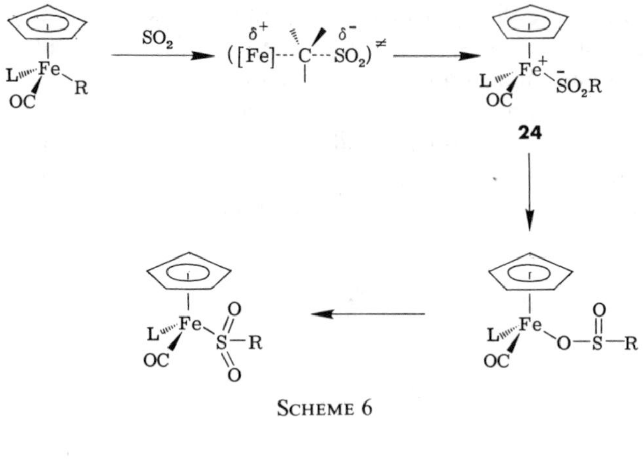

SCHEME 6

Whether CO_2 insertion processes into metal–carbon bonds are analogous to the corresponding processes involving SO_2 or CO must await stereochemical investigations employing *threo*-PhCHDCHDW(CO)$_5^-$ which are in progress in our laboratory. That is, our present observations on carbon dioxide insertion reactions will not allow us to distinguish between pathways involving the transition states in Fig. 10 (concerted) or **25** (dissociative/ionic). Because of the low reactivity of carbon dioxide we favor a process such as that shown in Fig. 10 where the metal center plays an activating role

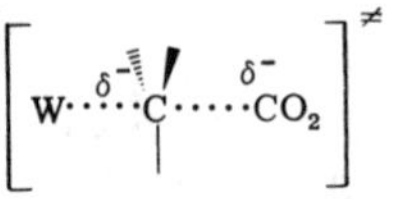

$$\left[W\cdots\overset{\delta-}{\underset{|}{C}}\cdots CO_2 \right]^{\neq}$$

25

in the insertion reaction. The increased reactivity of $CH_3W(CO)_5^-$ relative to $CH_3Cr(CO)_5^-$, as well as the greatly enhanced rate of COS insertion versus CO_2 insertion, are indeed consistent with this prejudice.

Other carbon dioxide insertion reactions into metal–carbon bonds reported since this area was last reviewed include the following (*89, 90*). Treatment of (η^5-C_5H_5)$_2$TiR complexes (R = Me, *t*-Bu, η^3-CHMeCHCH$_2$, and CMe=CHMe) with CO_2 have resulted in the formation of chelating carboxylate ligands bound to titanium. (*91–93*). The study involving the η^3-allyltitanium complex with a chiral cyclopentadienyl ligand provided the first demonstration of asymmetric carbon dioxide fixation (*92*). Similar products were obtained when vanadium complexes react with CO_2 in tetrahydrofuran [Eq. (26)] (*94*).

$$2R_3V \cdot nTHF + CO_2 \rightarrow V(OCOR)_2 + R_4V \cdot nTHF \qquad (26)$$

$$(R = C_6F_5, n = 2; R = CH_2SiMe_3, n = 0)$$

The paramagnetic chromium complex, Cr(*o*-CH$_2$C$_6$H$_4$NMe$_2$)$_3$, was found to insert one CO_2 molecule to afford a chelating carboxylate group (*95, 96*). (PhCH$_2$)$_2$Mn was also found to give a chelated carboxylate complex on reaction with carbon dioxide (*97*).

C. Insertion into M–O and M–N Bonds

The reactions of CO_2 with metal–hydroxides, –alkoxides, and –amides to provide metallobicarbonates, -alkyl carbonates, and -carbamates [Eqs. (27)–(29)] in general do not involve activation of carbon dioxide by prior coordination to the metal center. These processes generally entail either

$$M-OH + CO_2 \longrightarrow M-O-\overset{\overset{\displaystyle O}{\|}}{C}-OH \qquad (27)$$

$$M-OR + CO_2 \longrightarrow M-O-\overset{\overset{\displaystyle O}{\|}}{C}-OR \qquad (28)$$

$$M-NR_2 + CO_2 \longrightarrow M-O-\overset{\overset{\displaystyle O}{\|}}{C}-NR_2 \qquad (29)$$

reaction of a free ligand and CO_2 in solution followed by ligation of the new ligand, or direct attack of CO_2 on the coordinated ligand followed by a rearrangement to yield product.

The Cu(I) bicarbonate complex previously mentioned (71) was synthesized by the reactions summarized in Scheme 7, which includes CO_2 insertion into copper hydroxide and alkoxide species. The insertion reaction of CO_2 with metal hydroxides to form bicarbonates is believed to occur

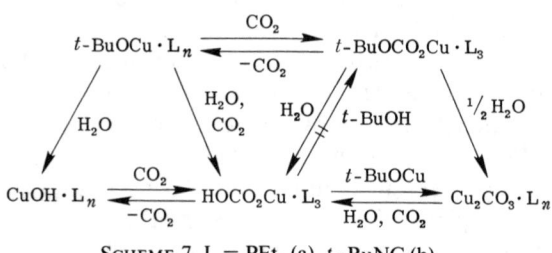

Scheme 7. L = PEt_3 (a), t-BuNC (b).

without rupture of the metal–oxygen bond [Eq. (30)]. Although transfer of an alkyl group is expected to be much less facile than a proton, CO_2 insertion

$$M-^*OH + CO_2 \rightleftharpoons M-^*O-\overset{\overset{\displaystyle O}{\|}}{C}-OH \qquad (30)$$

into metal–alkoxides to provide alkyl carbonates may proceed via a similar mechanism. Alternatively, this reaction can take place by a ligand displacement process as shown in Eqs. (31) and (32). Indeed this latter pathway has been proposed for the reaction of $Mo_2(OR)_6$ (R = Me_3Si, Me_3C, Me_2CH, and

$$ROH + CO_2 \rightleftharpoons ROCOOH \qquad (31)$$

$$MOR + ROCOOH \rightarrow MO_2COR + ROH \qquad (32)$$

Me_3CCH_2) with CO_2 in the presence of alcohol to afford the insertion products $Mo_2(OR)_4(O_2COR)_2$ (98).

An analogous process for the formation of carbamato derivatives containing the Mo≡Mo unit from the reactions of 1,2-$Mo_2R_2(NMe_2)_4$ compounds [R = CH_3 and $CH_2Si(CH_3)_3$] with carbon dioxide has been reported by Chisholm and co-workers (99, 100). That is, the insertion of CO_2 into

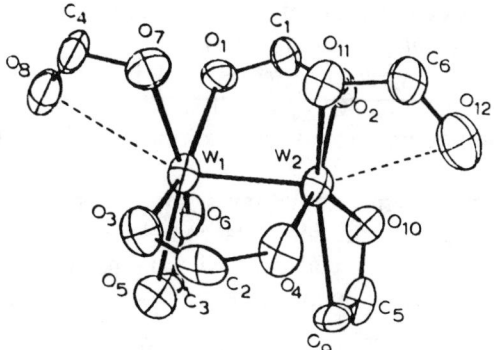

FIG. 11. Central skeleton of the $W_2(O_2CNMe_2)_6$ molecule; dotted lines indicate the long, quasi-axial W–O bonds. Reproduced with permission from Ref. *100*. Copyright 1982 American Chemical Society.

Mo–NMe$_2$ bonds proceeds via an amine-catalyzed sequence [Eqs. (33) and (34)]. In contrast to these results, when the 1,2-dialkyldimolybdenum com-

$$Me_2NH + CO_2 \rightleftharpoons Me_2NCOOH \qquad (33)$$

$$Mo—NMe_2 + Me_2NCOOH \rightarrow MoO_2CNMe_2 + Me_2NH \qquad (34)$$

pounds contain alkyl groups possessing β-hydrogen atoms, the reaction with CO$_2$ yielded Mo–Mo quadruply bonded compounds (e.g., as shown in Fig. 11) by reductive-elimination of alkenes and alkanes from carbamato intermediates [Eq. (35)].

$$Mo_2R_2(NMe_2)_4 + CO_2(excess) \rightarrow Mo_2(O_2CNMe)_4 + alkane + alkene \qquad (35)$$

The absorption of CO$_2$ by copper(I) amides [Eq. (36)] most likely also proceeds via a carbamic acid mechanism (*101*).

$$R_2NCu + CO_2 \xrightarrow{\text{t-BuNC}} R_2NCO_2Cu(\text{t-BuNC})_n \qquad (36)$$

IV

CARBON DIOXIDE REDUCTION AND/OR INCORPORATION

As we have previously discussed in Section I, the use of carbon dioxide as an industrial source of chemical carbon has been limited mainly to the production of organic carbonates, carboxylic acids, and ureas. In contrast, carbon dioxide has been utilized more extensively in the laboratory for the production of a greater variety of compounds including alkanes, alcohols, formates, oxalates, and lactones (8, 102).

This section will review some recent work in the areas of catalytic and stoichiometric reduction of CO_2, as well as incorporation of carbon dioxide into organic compounds, promoted by homogeneous transition metal complexes.

There have been several recent reports of the reduction of carbon dioxide to carbon monoxide by transition metal complexes. Maher and Cooper (21) have reported that several metal carbonyl dianions can effect the disproportionation of CO_2 to metal bound carbon monoxide [Eq. (37)] with Li_2CO_3

$$M_2[M'(CO)_n] + 2CO_2 \rightarrow M_2CO_3 + [M'(CO)_{n+1}] \tag{37}$$

$$n = 5, M = Na \text{ or } Li, M' = W$$

$$n = 4, M = Na, M' = Ru \text{ or } Fe$$

serving as the oxygen sink. A mechanism involving two intermediates has been proposed as seen in Scheme 8. The intermediacy of species A has recently been confirmed by infrared and ^{13}C NMR spectroscopy (22).

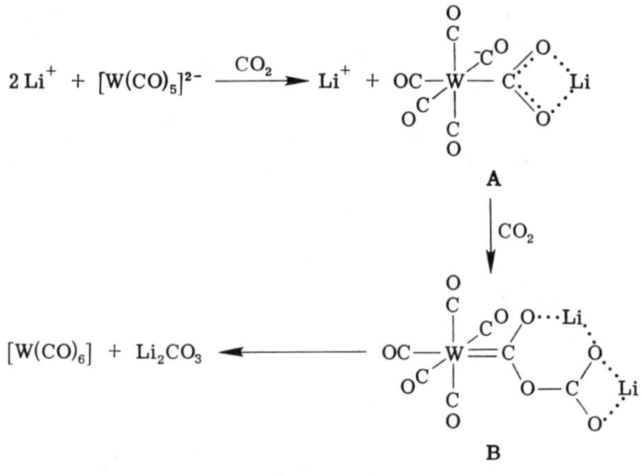

SCHEME 8

Although not catalytic, this system is cyclic in the sense that the $W(CO)_6$ may be converted back to the dianion by photolysis in the presence of NMe_3 followed by alkali metal reduction. Labeling studies have shown that the carbon of the sixth carbonyl is obtained from carbon dioxide and not from decomposition of the transition metal carbonyl.

$RhH(PR_3)_3$ and $Rh_2H_2(\mu\text{-}N_2)(PR_3)_4$ have also been effective at converting CO_2 to CO with hydrogen serving as the acceptor for the excess oxygen atom [Eq. (38)] (103). This process is essentially the reverse of the water–gas shift

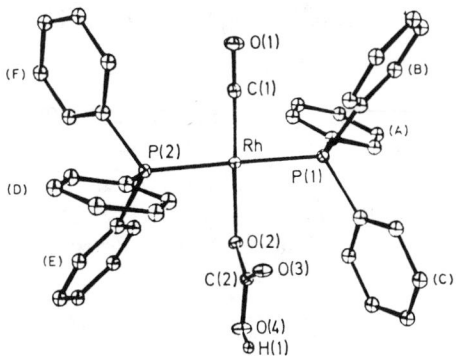

FIG. 12. A view of the structure of **1**. Principal bond lengths (Å): Rh–C(1), 1.798(4); Rh–P(1), 2.333(1); Rh–P(2), 2.332(1); Rh–O(2), 2.075(2); O(2)–C(2), 1.283(4); C(2)–O(3), 1.243(4); C(2)–O(4), 1.336(4). Principal bond angles (°): C(1)–Rh–P(1), 89.1(1); C(1)–Rh–P(2), 91.8(1); O(2)–Rh–P(1), 90.4(1); O(2)–Rh–P(2), 88.7(1); Rh–O(2)–C(2), 118.6(2); O(2)–C(2)–O(3), 124.8(3); O(2)–C(2)–O(4), 113.2(3); O(3)–C(2)–O(4), 122.0(3). Reproduced with permission from Ref. *104*. Copyright 1982 American Chemical Society.

$$\left.\begin{array}{c} RhH(PR_3)_3 \\ Rh_2H_2(\mu\text{-}N_2)(PR_3)_4 \end{array}\right\} \xrightarrow[\;H_2O\;]{CO_2} RhH_2(O_2COH)L_2 \xrightarrow[\;CO_2\;]{+\;dry} RhL_2(CO)(O_2COH) + H_2O$$

$$L = P(i\text{-}Pr)_3,\; P(c\text{-}C_6H_{11})_3$$

(38)

reaction [Eq. (2)], the driving force being the strength of the metal–carbon bond.

In a closely related study, complex **26**, where L = PPh$_3$, has been prepared in the absence of water by Nicholas and its structure has been determined by X-ray crystallography (Fig. 12) (*104*). Investigations employing a Rh–D complex, coupled with the lack of sensitivity of the reaction to H$_2$O, led these researchers to propose a route involving rhodium promoted reductive disproportionation of CO$_2$ followed by hydrogen transfer from Rh to the oxygen of the coordinated carbonate (Scheme 9) (*105*).

Phosphines may also serve as oxygen acceptors in CO$_2$ reduction. Both (Ph$_3$P)$_3$RhCl and [(cyclooctene)$_2$RhCl]$_2$ are capable of catalyzing the redox

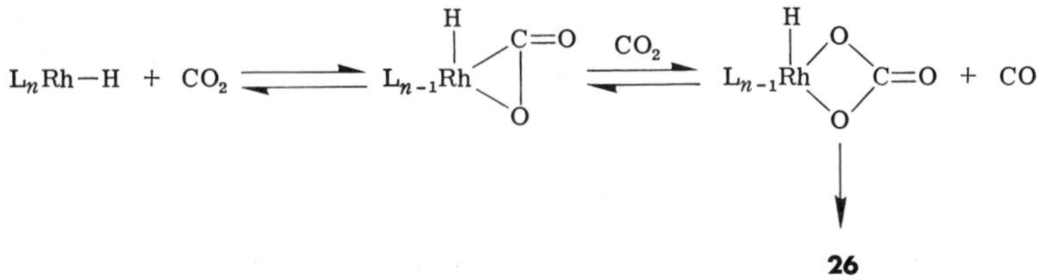

$$L_nRh\text{---}H + CO_2 \rightleftharpoons L_{n-1}Rh\overset{H}{\underset{O}{\diagdown}}C{=}O \overset{CO_2}{\rightleftharpoons} L_{n-1}Rh\overset{H}{\underset{O}{\diagdown}}\overset{O}{\underset{O}{\diagup}}C{=}O + CO$$

26

SCHEME 9

reaction of CO_2 with phosphine [Eq. (39)] (*106*). The reaction was conducted in

$$CO_2 + PR_3 \xrightarrow{\text{L}_3\text{RhCl}} OPR_3 + CO \tag{39}$$

refluxing decalin under an atmosphere of CO_2 and was found to be dependent upon the nature of the phosphine used. The author suggests a mechanism (Scheme 10); however, to date no detailed mechanistic study has been undertaken.

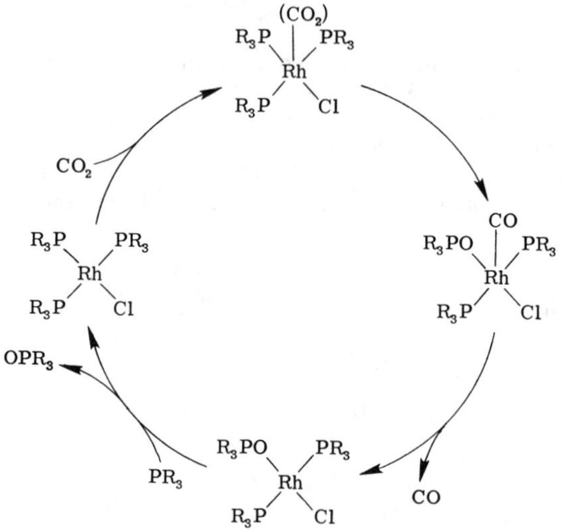

SCHEME 10

Fisher and Eisenberg (*107*) have reported on the electrocatalytic reduction of carbon dioxide using macrocycle complexes of nickel and cobalt (e.g., complex **27**). An indirect electrochemical reduction of CO_2 was ac-

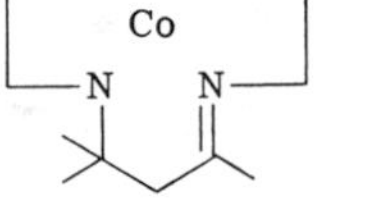

27

complished at potentials between -1.3 and -1.6 V versus SCE. The major reduction products were CO and H_2 in the presence of a source of protons.

The author suggests that reduction of CO_2 may proceed via direct attack of CO_2 on a metal hydride intermediate without prior coordination of CO_2 to the metal center (e.g., analogous to species 6).

Homogeneous reduction of carbon dioxide to ligated carbon monoxide has been achieved from the reaction of $[Na][(\eta^5\text{-}C_5H_5)Fe(CO)_2]$ with CO_2 and two equivalents of acid [Eq. (40)] (108). Evans (23) has previously presented infrared spectral evidence for the existence of 28. This reduction

$$(\eta^5\text{-}C_5H_5)Fe(CO)_2^- \xrightarrow{\ CO_2\ } (\eta^5\text{-}C_5H_5)Fe(CO)_2CO_2^- \xrightarrow[-H_2O]{\ 2H^+\ } (\eta^5\text{-}C_5H_5)Fe(CO)_3^+ \quad (40)$$

$$\qquad\qquad\qquad\qquad 28 \qquad\qquad\qquad\qquad\qquad\qquad 29$$

undoubtedly proceeds through intermediacy of the metallocarboxylic acid species, $(\eta^5\text{-}C_5H_5)Fe(CO)_2COOH$, earlier reported by Grice et al. (28). Cutler and co-workers (108) further reduced the carbon monoxide ligand in 29 to an alkoxymethyl ligand.

Several groups have been successful at the catalytic conversion of carbon dioxide, hydrogen, and alcohols into alkyl formate esters using neutral metal – phosphine complexes in conjunction with a Lewis acid or base (109). Denise and Sneeden (110) have recently investigated various copper and palladium systems for the product of ethyl formate and ethyl formamide. Their results are summarized in Table II. Of the mononuclear palladium complexes, the most active system for ethyl formate production was found to be the Pd(0) complex, $Pd(dpm)_2$, which generated 10 μmol HCOOEt per μmol metal complex per day. It was anticipated that complexes containing more than one metal center might aid in the formation of C_2 products; however, none of the multinuclear complexes produced substantial quantities of diethyl oxalate.

An obvious initial step in the reduction of CO_2 by homogeneous systems involves the insertion of CO_2 into the metal – hydrogen bond to give metal formates. However, subsequent work by Beguin et al. (65) has shed doubt on the intermediacy of the formato complex in their systems (see above). For example, these researchers were not successful in transforming a copper formate derivative into alkylformate [Eq. (41)]. On the other hand, they

$$6CO_2 + 6L + (HCuL)_6 \rightarrow L_2CuO_2CH \xrightarrow{\ H_2/EtOH\ } CO_2 + Cu(0) \quad (41)$$

were able to obtain methylformate by the hydrogenolysis of the methoxycarbonyl compound 30, suggesting the possible intermediacy of alkoxycarbonyl compound in these catalytic systems (102).

TABLE II

HYDROGENATION OF CO_2, $EtOH/Et_3N^a$

Catalyst	Products (μmol)		
(μmol metal)	CH_4	HCOOEt	$(COOEt)_2$
$(PdCldpm)_2^b$ (340)	250	5	0
$(PdCldpm)_2$ (520)	800	2000	1
$Pd(dpm)_2$ (440)	40	4000	0.4
$[Pddpm(EtOH)_x](BF_4)_2$ (420)	60	1000	1
$(Pddpm)_n$ (320)	200	500	0.2
$PdCl_2dpm$ (480)	300	250	1
$Pd_2Cu_2Cl_6dpm_3$ (460)	440	2000	0.6
$Pd(dpe)_2$ (400)	40	560	0
$PdCl_2dpe$ (710)	0	150	0
$Pd(PPh_3)_4$ (480)	20	200	0
$PdCl_2(PPh_3)_2$ (480)	0	350	0.3
$[(CuCl)_2dpm]_2$ (340)	0	30	0
$[Cu_3Cl_2dpm_3]Cl$ (410)	0	12	0
$(PdCldpm)_2^c$ (420)	40	$HCONEt_2$, 1080	$(CONEt)_2$, 0

^{a} Reaction conditions, CO_2 (15 bars), H_2 (15 bars), EtOH (50 ml), Et_3N (10 ml), 120°C, 24 hr. Reprinted with permission from Ref. *110*. Copyright 1981 Elsevier Sequoia S.A.
^{b} Solvent C_6H_6 (50 ml), EtOH (5 ml).
^{c} Solvent C_6H_6 (40 ml), Et_2NH (10 ml).

$$Cl_2PdL_2 \xrightarrow{\text{ClHgCOOCH}_3} L_2ClPdCOOCH_3 \xrightarrow{\text{H}_2/\text{MeOH}} HCOOCH_3 \ (25\%) \qquad (42)$$

30

Preliminary studies on the catalytic conversion of carbon dioxide, hydrogen, and alcohols into alkyl formates using tungsten and ruthenium catalysts are not, however, inconsistent with metalloformate intermediates (*111*). For example, we find comparable activity for the formation of methyl formate employing either μ-$H[W_2(CO)_{10}]^-$ or $W(CO)_5O_2CH^-$ as catalysts. Similarly $HRu_3(CO)_{11}^-$ and $Ru_3(CO)_{10}O_2CH^-$ (**9**) are equally active catalysts. It should nevertheless be clearly stated that the nature of the active catalyst in these cluster systems is not well understood at this stage in time. After 24 hr of catalysis (five to six turnovers at 125°C, 250 psi CO_2 and 250 psi H_2 loading pressures) the ruthenium species isolated in good yield is $H_3Ru_4(CO)_{12}^-$. This tetranuclear species was subsequently found to be a more effective catalyst than either $HRu_3(CO)_{11}^-$ or $HRu_3(CO)_{10}O_2CH^-$ (see Table III). Efforts are currently underway in our laboratory to better understand the mechanistic details of these important catalytic processes.

TABLE III

SYNTHESIS OF METHYLFORMATE BY CO_2 REDUCTION WITH MOLECULAR HYDROGEN IN
THE PRESENCE OF METHANOL USING RUTHENIUM CLUSTERS AS CATALYSTS

Run	Catalyst	[cat] ($M^{-1} \times 10^3$)	Turnovers per day
1	[PNP][HRu$_3$(CO)$_{11}$]	9.25	4.1
2	[PNP][Ru$_3$(CO)$_{10}$(μ-HCO$_2$)]	9.17	5.7
3	Ru$_3$(CO)$_{12}$	9.18	0.3
4	[PNP][H$_3$Ru$_4$(CO)$_{12}$][a]	9.20	7.3
5	[PNP][H$_3$Ru$_4$(CO)$_{12}$][b]	9.19	Traces

[a] Solid recovered from run No. 1.
[b] Solid recovered from runs 1 and 2. Reaction carried out in the absence of hydrogen.

In an earlier related study, Evans and Newell (*112*) demonstrated the anionic iron carbonyl hydrides [HFe$_3$(CO)$_{11}$][PNP] and [HFe(CO)$_4$][PNP] to be catalytically active for this reaction, generating yields no greater than 5.8 : 1 moles of formate per mole of catalyst precursor. The low yields were attributed to catalyst degradation caused by oxidation by carbon dioxide as evidenced by the detection of carbonates in the system.

2-Methyl-2-propenyl-3-methyl-3-butenoate and 2-methyl-2-propenyl-3-methyl-2-butenoate may be produced by the stoichiometric reaction of dichlorobis(2-methyl allyl)bis(diphenylphosphinoethane) dipalladium (**31**) with carbon dioxide in acetonitrile (*113*) [Eq. (43)], to give an overall 45% yield of the esters.

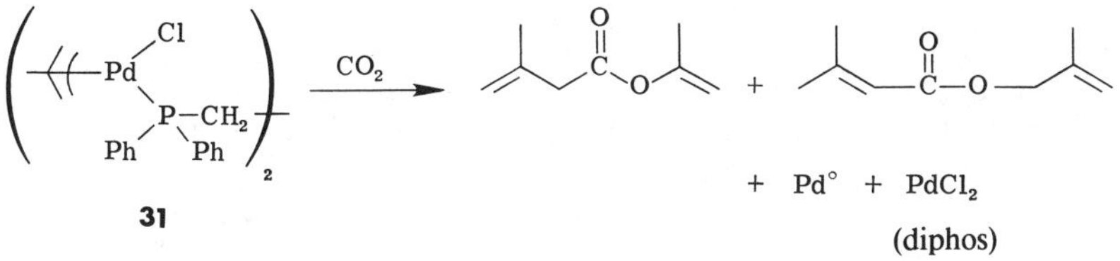

$$(43)$$

Methylation of the various RR'NCO$_2$Cu·(t-BuNC)$_n$ compounds, prepared by the addition of CO_2 and three equivalents of t-BuNC to a benzene solution of RR'NCu [Eq. (36)], has resulted in formation of the corresponding carbamates (see Table IV) (*101*). These researchers have as well found the copper(I) bicarbonate complexes, HOCO$_2$CuL$_3$ (**32a**, L = PEt$_3$ and **32b**, L = t-BuNC), to be effective at carboxylation of cyclohexanone in the presence of a stoichiometric quantity of water (*71*). The transcarboxylation of cyclohexanone [depicted in Eq. (44)] is proposed to proceed via a dicopper(I) complex (**33**).

TABLE IV

ANALYTICAL DATA AND PROPERTIES FOR $R^1R^2NCu^a$

R^1R^2NCu	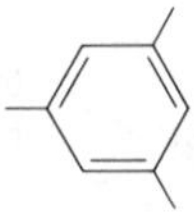 formation (%)	% Cu content		% R^1R^2-NCO$_2$Me	Dec. temp. (°C)
		Found	Calc.		
H$_2$NCu	100	80.0	79.9	17	ca. 90
n-BuNHCu	98	46.1	46.8	44	ca. 70
t-BuNHCu	87	46.2	46.8	68	> 140
Et$_2$NCu	92	47.0	46.8	71	ca. 90
n-Bu$_2$NCu	95	33.5	33.1	76	ca. 110
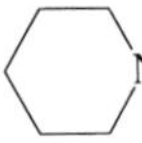 ⬡NCu	90	42.1	43.0	56	ca. 105
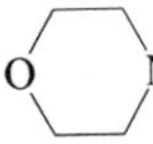 O⬡NCu	100	46.0	46.8	60	ca. 125
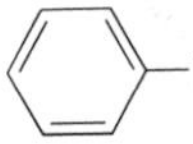 ⬡—NHCu	88	38.5	40.8	55	ca. 90

a Prepared by the reaction

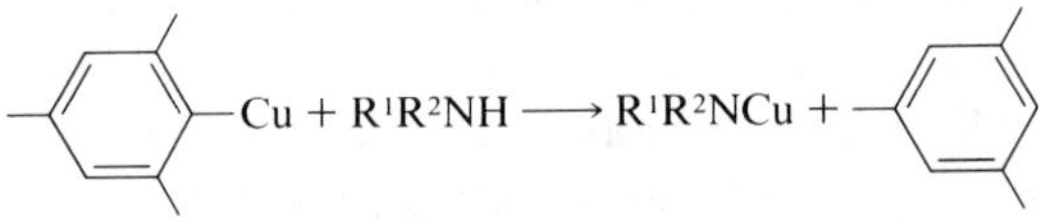

$$\text{(mesityl)}-Cu + R^1R^2NH \longrightarrow R^1R^2NCu + \text{(mesitylene)}$$

Reprinted with permission from Ref. *101*. Copyright 1981 American Chemical Society.

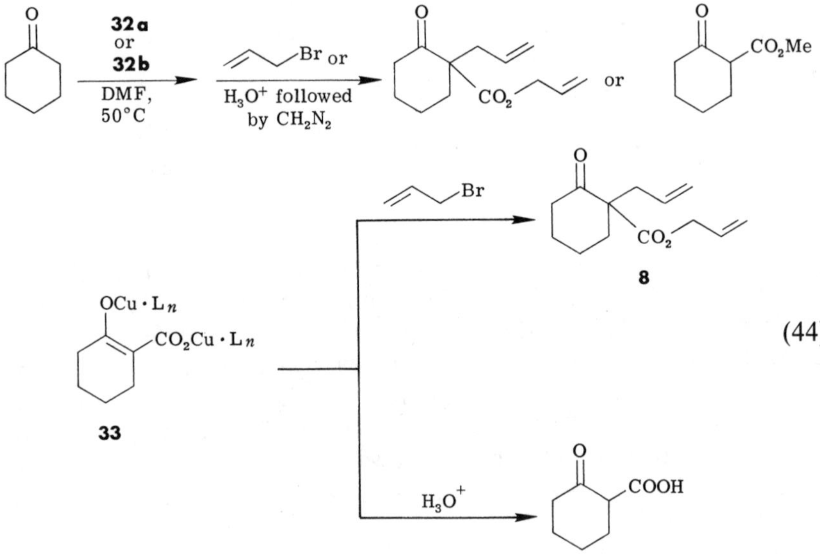

(44)

Transition metal complexes have proved very useful in both the catalytic and stoichiometric production of cyclic lactones. A series of palladium(0)–phosphine complexes have been shown to be effective for the conversion of three-membered ring systems to cyclic lactones [Eq. (45)] (*114*). When isopropylidenecyclopropane and [Pd(dba)$_2$]-PPh$_3$ (dba = dibenzylideneacetone)(4 : 1) in benzene were treated with 40 atm carbon dioxide at 126°C for 20 hr, 69% of the lactone (**34**) was formed. In contrast, when [Pd(diphos)$_2$] was used as the substrate under similar conditions 48% of **35** was produced with only trace amounts of **34**. None of the complexes appeared to be active for terminal alkenes such as **36** or **37**.

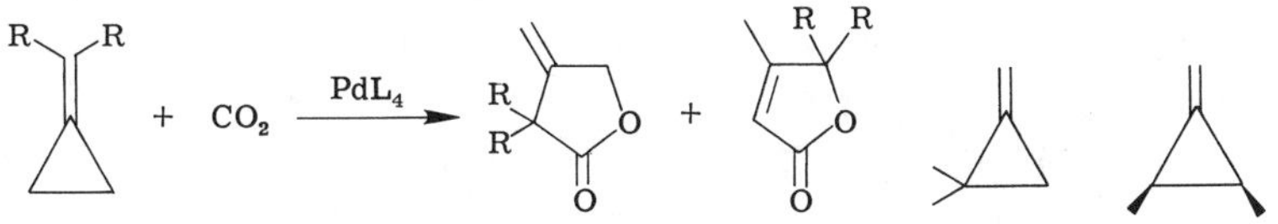

$$R = Me$$

$$(45)$$

Rh(diphos)(η-BPh$_4$) has been found to be a methylacetylene oligomerization catalyst which produces a variety of linear and branched dimers, as well as linear and cyclic trimers. When the reaction was carried out in the presence of carbon dioxide, a small quantity of 4,6-dimethyl-2-pyrone was also produced (*115*).

Lactones have also been prepared by the stoichiometric reaction of η^3-allylnickel complexes with carbon dioxide [Eq. (46)] (*116*). η^3-Allylnickelvinylacetate (**38**) has been suggested as an intermediate in this process.

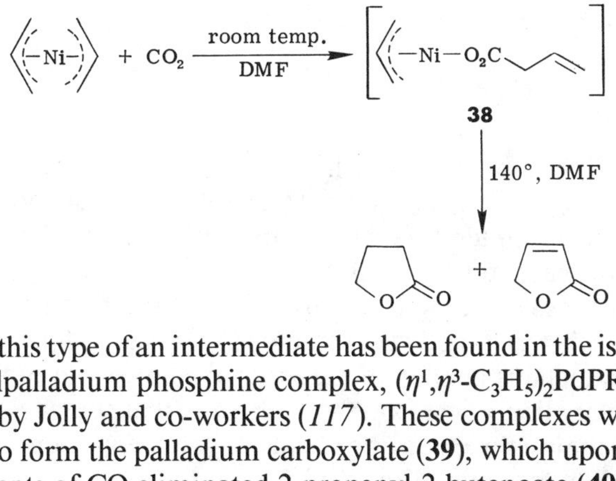

$$(46)$$

Support for this type of an intermediate has been found in the isolation of an η^1,η^3-bisallylpalladium phosphine complex, (η^1,η^3-C$_3$H$_5$)$_2$PdPR$_3$ (R = CH$_3$ and C$_6$H$_{11}$) by Jolly and co-workers (*117*). These complexes were found to insert CO$_2$ to form the palladium carboxylate (**39**), which upon addition of two equivalents of CO eliminated 2-propenyl-2-butenoate (**40**) [Eq. (47)].

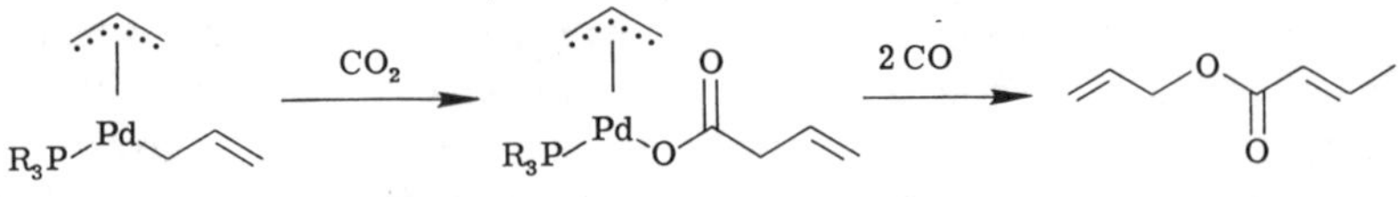

$$(47)$$

The oxanickelacyclopentenone (41), formed from 2-butyne, CO_2, and nickel(O), was found to undergo reaction with activated alkynes by insertion to provide oxanickelacycloheptadienones (42) [Eq. (48)] (*118*). These novel

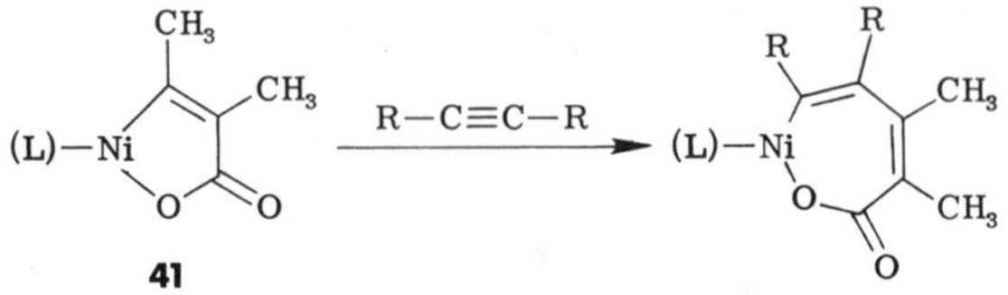

$$(48)$$

nickel complexes were suggested as intermediates in the nickel-catalyzed synthesis of 2-pyrones from alkynes and CO_2.

Dohring and Jolly (*119*) have reported on the catalytic cooligomerization of allene and carbon dioxide to give a mixture of esters, a lactone, and non-oxygen-containing oligomers. The transformation was accomplished using a catalyst mixture of $(\eta^3\text{-}C_3H_5)_2Pd$ and bisdicyclohexylphosphinoethane. The maximum yield of oxygen-containing compounds (43–45) was 40%, of which nearly 15% was the lactone.

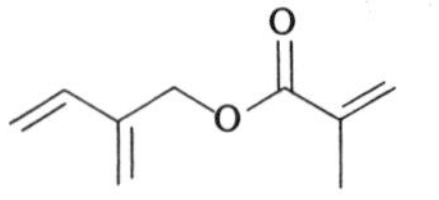

Another allyl compound which reacts stoichiometrically with carbon dioxide is $(\eta^5\text{-}C_5H_5)_2Ti(1\text{-methylallyl})$ (*120*). The titanium acetate complex which is formed is interesting in that the carbon dioxide carbon atom is attached to the substituted end of the allyl. It seems unlikely, then, that the product is the result of CO_2 insertion into the η^1-methylallyltitanium bond in view of the fact that methyl-substituted allyls tend to form η^1-complexes in which the metal is bonded to the least substituted end of the allyl. One possible explanation offered by the authors is that the allyl is bonded to titanium at the methylene carbon, but that rearrangement occurs subsequent to adduct formation [Eq. (49)].

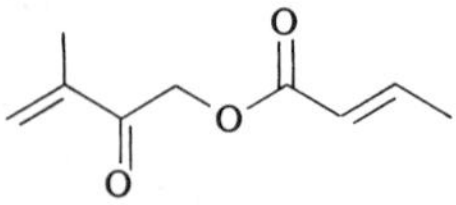

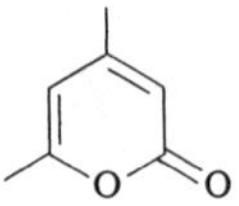

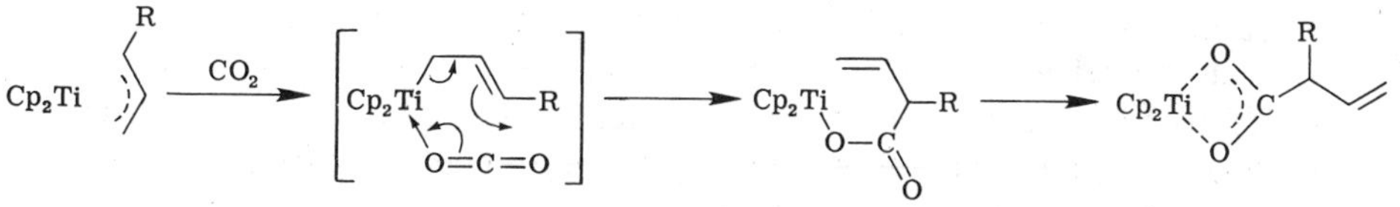

$$(49)$$

Several groups have been investigating the use of transition metal complexes for the catalytic conversion of hydrosilanes to silyl formates [Eq. (50)]. The anionic trinuclear cluster $[HRu_3(CO)_{11}]^{-1}$ has been effective for

$$CO_2 + HSiR_3 \rightarrow HCO_2SiR_3 \qquad (50)$$

this process and has the advantage of operating at moderate temperatures and pressures yielding high turnover numbers (121). $[HRu_3(CO)_{10}(SiEt_3)_2]^{-1}$ was isolated from the system and it is proposed that the initial step in the process is activation of the hydrosilane by the complex (Scheme 11).

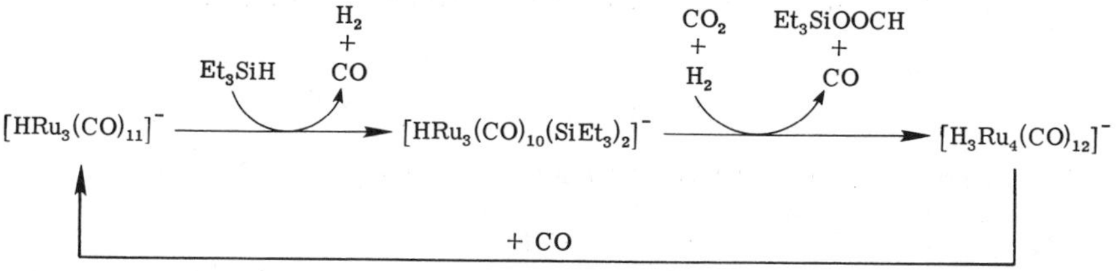

SCHEME 11

Under similar reaction conditions $[RuCl_2(PPh_3)_3]$ has also been found to catalyze the conversion of hydrosilanes to silyl formates (122).

V

CONCLUDING REMARKS

In this review we have attempted to cover the very recent literature relevant to the coordination chemistry of carbon dioxide and its use as a source of chemical carbon. We have omitted similar investigations involving the more reactive substrates, carbon disulfide and carbonyl sulfide. The reader is referred to a recent review by Ibers (123) which has contrasted the behavior of these sulfides to carbon dioxide. Likewise we have, as stated at the onset, elected to neglect heterogeneous processes involving the reduction of carbon dioxide. These can range from the use of organometallic compounds adsorbed on oxide supports [e.g., ruthenium carbonyl clusters (*124*)] to more traditional dispersed metal catalysts on inert supports. This omission is somewhat compensated for by our inclusion of Ref. *125–142,* which represent some of the important current efforts in this area.

It should be apparent from our coverage that we have just begun to understand the chemistry associated with the activation of carbon dioxide. However, efforts are presently being intensified in this area and it appears safe to predict that major advances in this field will occur during this decade. Clearly, one of the greatest challenges for the organometallic chemist will be to find potential catalysts which activate CO_2 and H_2 simultaneously.

ACKNOWLEDGMENTS

The authors are most grateful to the National Science Foundation, whose support has made possible their contributions to the research described herein. They are likewise appreciative to all their colleagues mentioned in the references, whose many original contributions have made this such an interesting and stimulating area of research to be involved in. Finally, we are thankful to Mrs. Melanie Gray for preparing this manuscript for publication.

REFERENCES

1. G. Henrici-Olive and S. Olive, *J. Mol. Catal.* **3**, 443 (1978); *Angew. Chem., Int. Ed. Engl.* **15**, 136 (1976); C. Masters, *Adv. Organomet. Chem.* **17**, 61 (1979).
2. J. Haggin, *Chem. Eng. News* Feb. 8, 13 (1982).
3. P. Dixneuf and R. D. Adams, Report on the International Seminor of the Activation of Carbon Dioxide and Related Heteroallenes on Metal Centers, Rennes, France, 1981.
4. R. Eisenberg and D. E. Hendriksen, *Adv. Catal.* **28**, 79 (1979).
5. R. Bardet, M. Perrin, M. Primet, and Y. Trambouze, *J. Chim. Phys.* **75**, 1079 (1978); R. Bardet and Y. Trambouze, *C. R. Hebd. Seances Acad. Sci., Ser C* **228**, 101 (1979); **290**, 153 (1980).
6. E. Ramaroson, R. Kieffer, and A. Kiennemann, *J. Chem. Soc., Chem. Commun.* p. 645 (1982).
7. J. Happin, *Chem. Eng. News* June 21, 7 (1982).
8. B. Denise and R. P. A. Sneeden, *CHEMTECH* **12**, 108 (1982).
9. M. G. Mason and J. A. Ibers, *J. Am. Chem. Soc.* **104**, 5153 (1982).
10. M. Aresta, C. F. Nobile, V. G. Albano, E. Forni, and M. Manassero, *J. Chem. Soc., Chem. Commun.* p. 636 (1975); M. Aresta and C. F. Nobile, *J. Chem. Soc., Dalton Trans.* p. 708 (1977).
11. G. S. Bristow, P. B. Hitchcock, and M. F. Lappert, *J. Chem. Soc., Chem. Commun.* p. 1145 (1981).
12. T. Herskovitz, J. C. Calabrese, and J. B. Kinney, *Abstr. Pap., 185th Nat. Meet. Am. Chem. Soc.* INOR 297 (1983).
13. G. Fachinetti, C. Floriani, and P. F. Zanazzi, *J. Am. Chem. Soc.* **100**, 7405 (1978); S. Gambarotta, F. Arena, C. Floriani, and P. F. Zanazzi, *ibid.* **104**, 5082 (1982).
14. C. R. Eady, J. J. Guy, B. F. G. Johnson, J. Lewis, M. C. Malatesta, and G. M. Sheldrick, *J. Chem. Soc., Chem. Commun.* p. 602 (1976).

15. J. J. Guy and G. M. Sheldrick, *Acta Crystallogr., Sect. B* **B34,** 1718 (1978).
16. W. Beck, K. Raab, U. Nagel, and M. Steimann, *Angew. Chem.* **94,** 526 (1982).
17. S. Gambarotta, M. Pasquali, C. Floriani, A. Chiesi-Villa, and C. Guastini, *Inorg. Chem.* **20,** 1173 (1981).
18. G. Fachinetti, C. Biran, C. Floriani, A. Chiesi-Villa, and C. Guastini, *Inorg. Chem.* **17,** 2995 (1978).
19. J. W. Rabalais, J. M. McDonald, V. Scherr, and S. P. McGlynn, *Chem. Rev.* **71,** 73 (1971).
20. S. Sakaki, K. Kitaura, and K. Morokuma, *Inorg. Chem.* **21,** 760 (1982).
21. J. M. Maher and N. J. Cooper, *J. Am. Chem. Soc.* **102,** 7606 (1980).
22. J. M. Maher, G. R. Lee, and N. J. Cooper, *J. Am. Chem. Soc.* **104,** 6796 (1982).
23. G. O. Evans, W. F. Walter, D. R. Mills, and C. A. Streit, *J. Organomet. Chem.* **144,** C34 (1978).
24. M. E. Volpin and I. S. Kolomnikov, *Organomet. React.* **5,** 313 (1975); *Pure Appl. Chem.* **33,** 567 (1973).
25. I. S. Kolomnikov and M. Kh. Grigoryan, *Russ. Chem. Rev. (Engl. Transl.)* **47,** 334 (1978).
26. R. P. A. Sneeden, *Actual. Chim.* **1,** 22 (1979).
27. I. S. Kolomnikov, G. Stepovska, S. Tyrlik, and M. E. Volpin, *Zh. Obshch. Khim.* **42,** 1652 (1972); *J. Gen. Chem. USSR (Engl. Transl.)* **42,** 1645 (1972).
28. N. Grice, S. C. Kao, and R. Pettit, *J. Am. Chem. Soc.* **101,** 1627 (1979).
29. A. J. Deeming and B. L. Shaw, *J. Chem. Soc. A* p. 443 (1969).
30. C. P. Casey, M. A. Andrews, and J. E. Rinz, *J. Am. Chem. Soc.* **101,** 741 (1979).
31. M. Catallani and J. Halpern, *Inorg. Chem.* **19,** 566 (1980).
32. J. R. Sweet and W. A. G. Graham, *Organometallics* **1,** 982 (1982).
33. T. G. Appleton and M. A. Bennett, *J. Organomet. Chem.* **55,** C89 (1973).
34. D. J. Darensbourg and A. Rokicki, *Organometallics* **1,** 1685 (1982).
35. D. J. Darensbourg, M. Y. Darensbourg, and S. C. Kao, in preparation. Earlier reviews in this area include D. J. Darensbourg, *Isr. J. Chem.* **15,** 247 (1977); J. Halpern, *Comments Inorg. Chem.* **1,** 3 (1981).
36. G. Henrici-Olivé and S. Olivé, "Coordination and Catalysis." Verlag Chemie, Weinheim, 1977.
37. R. Cramer, *J. Am. Chem. Soc.* **87,** 4714 (1965).
38. D. R. Roberts, G. L. Geoffroy, and M. G. Bradley, *J. Organomet. Chem.* **198,** C75 (1980).
39. M. Freni, D. Giusto, and P. Romiti, *J. Inorg. Nucl. Chem.* **33,** 4093 (1971).
40. V. G. Albano, P. L. Bellon, and G. Ciani, *J. Organomet. Chem.* **31,** 75 (1971).
41. V. D. Bianco, S. Doronzo, and M. Rossi, *J. Organomet. Chem.* **35,** 337 (1972).
42. D. J. Darensbourg, A. Rokicki, and M. Y. Darensbourg, *J. Am. Chem. Soc.* **103,** 3223 (1981).
43. D. J. Darensbourg and A. Rokicki, *ACS Symp. Ser.* **152,** 107 (1981).
44. D. J. Darensbourg and A. Rokicki, *J. Am. Chem. Soc.* **104,** 349 (1982).
45. S. G. Slater, R. Lusk, B. F. Schumann, and M. Y. Darensbourg, *Organometallics* **1,** 1662 (1982).
46. D. J. Darensbourg, M. Y. Darensbourg, and C. Ovalles, unpublished observations.
47. C. H. Langford and H. B. Gray, "Ligand Substitution Processes." Benjamin, New York, 1965.
48. D. J. Darensbourg, *Adv. Organomet. Chem.* **21,** 113 (1982).
49. M. Y. Darensbourg, R. Bau, M. Marks, R. R. Burch, Jr., J. C. Deaton, and S. Slater, *J. Am. Chem. Soc.* **104,** 6961 (1982).
50a. D. J. Darensbourg, M. B. Fischer, R. E. Schmidt, and B. J. Baldwin, *J. Am. Chem. Soc.* **103,** 1297 (1981).

50b. D. J. Darensbourg, C. S. Day, and M. B. Fischer, *Inorg. Chem.* **20**, 3577 (1981).
51. J. H. Merrifield and J. A. Gladysz, *Organometallics* **2**, 782 (1983).
52. H. Werner, W. Beck, and H. Engelmann, *Inorg. Chim. Acta* **3**, 331 (1969).
53. A. Whitaker and J. W. Jerrery, *Acta Crystallogr.* **23**, 977 (1967).
54. B. Rees and A. Mitschler, *J. Am. Chem. Soc.* **98**, 7918 (1976).
55. S. P. Arnesen and H. M. Seip, *Acta Chem. Scand.* **20**, 2711 (1966).
56. S. H. Strauss, K. H. Whitmire, and D. F. Shriver, *J. Organomet. Chem.* **174**, C59 (1979).
57. R. M. Laine, R. G. Rinber, and P. C. Ford, *J. Am. Chem. Soc.* **99**, 252 (1977).
58. C. Ungermann, V. Landis, S. A. Moya, H. Cohen, H. Walker, R. G. Pearson, R. G. Rinker, and P. C. Ford, *J. Am. Chem. Soc.* **101**, 5922 (1979).
59. D. J. Darensbourg and M. Pala, *Organometallics* **2**, in press (1983).
60. R. S. Paonessa and W. C. Trogler, *J. Am. Chem. Soc.* **104**, 3529 (1982).
61. M. Y. Darensbourg and R. El-Mehdawi, unpublished observations.
62. C. P. Kubiak, C. Woodcock, and R. Eisenberg, *Inorg. Chem.* **21**, 2119 (1982).
63. P. C. Ford, *Acc. Chem. Res.* **14**, 31 (1981).
64. J. R. Fisher, A. J. Mills, S. Sumner, M. P. Brown, M. A. Thompson, R. J. Puddephatt, A. A. Frew, L. Manojlović-Muir, and K. W. Muir, *Organometallics* **1**, 1421 (1982).
65. B. Beguin, B. Denise, and R. P. A. Sneeden, *J. Organomet. Chem.* **208**, C18 (1981).
66. W. R. Roper and L. J. Wright, *J. Organomet. Chem.* **234**, C5 (1982).
67. D. P. Arnold and M. A. Bennett, *J. Organomet. Chem.* **199**, C17 (1980).
68. I. S. Kolomnikov, G. Stepovska, S. Tyrlik, and M. E. Volpin, *J. Gen. Chem. USSR (Engl. Transl.)* **44**, 1710 (1974).
69. T. Kruck and M. Noack, *Chem. Ber.* **97**, 1693 (1964).
70. P. Braunstein, D. Matt, Y. Dusausoy, J. Fischer, A. Mitschler, and L. Ricard, *J. Am. Chem. Soc.* **103**, 5115 (1981).
71. T. Tsuda, Y. Chujo, and T. Saegusa, *J. Am. Chem. Soc.* **102**, 431 (1980).
72. D. J. Darensbourg and R. A. Kudaroski, unpublished observations.
73. F. A. Cotton, D. J. Darensbourg, and B. W. S. Kolthammer, *J. Am. Chem. Soc.* **103**, 398 (1981).
74. F. A. Cotton, D. J. Darensbourg, B. W. S. Kolthammer, and R. Kudaroski, *Inorg. Chem.* **21**, 1656 (1982).
75. These activation parameters were determined for CO_2 insertion into $[CH_3W(CO)_5][PNP]$, PNP = $[Ph_3PNPPh_3]^+$, in tetrahydrofuran solvent.
76. M. Y. Darensbourg and D. Burns, *Inorg. Chem.* **13**, 2970 (1974).
77. J. P. Collman, J. N. Cawse, and J. I. Brauman, *J. Am. Chem. Soc.* **94**, 5905 (1972).
78. G. Fachinetti, C. Floriani, P. F. Zanazzi, and A. R. Zanzari, *Inorg. Chem.* **18**, 3469 (1979).
79. R. G. Severson and A. Wojcicki, *J. Am. Chem. Soc.* **101**, 877 (1979).
80. C. P. Casey and S. W. Polichnowski, *J. Am. Chem. Soc.* **100**, 7565 (1978).
81. Flood has recently published an excellent contribution which includes a review of the stereochemistry of insertion reactions of CO and SO_2 into metal-carbon sigma bonds. *(82)*.
82. T. C. Flood, *Top. Inorg. Organomet. Stereochem.* **12**, 37 (1981).
83. P. L. Bock, D. J. Boschette, J. R. Rasmussen, J. P. Demers, and G. M. Whitesides, *J. Am. Chem. Soc.* **96**, 2814 (1974).
84. S. L. Miles, D. L. Milse, R. Bau, and T. C. Flood, *J. Am. Chem. Soc.* **100**, 7278 (1978).
85. T. G. Attig, R. G. Teller, S.-M. Wa, R. Bau, and A. Wojcicki, *J. Am. Chem. Soc.* **101**, 619 (1979).
86. A. Wojcicki, *Acc. Chem. Res.* **4**, 344 (1971).
87. S. E. Jacobson, P. Reich-Rohrwig, and A. Wojcicki, *Inorg. Chem.* **12**, 717 (1973).

88. S. E. Jacobson and A. Wojcicki, *J. Am. Chem. Soc.* **95**, 6962 (1973).
89. A general review of insertion reactions involving metal-carbon bonds, including those with CO_2, has recently been published by Alexander (*90*).
90. J. Alexander, *in* "Chemistry of the Metal-Carbon Bond Stage 2" (S. Patai and F. R. Hartley, eds.), in press. Wiley, New York, 1983.
91. E. Klei and J. H. Telgen, *J. Organomet. Chem.* **209**, 297 (1981).
92. F. Sato, S. Iijima, and M. Sato, *J. Chem. Soc. Chem. Commun.* p. 180 (1981).
93. E. Klei and J. H. Telgen, *J. Organomet. Chem.* **222**, 79 (1981).
94. G. A. Razuvaev, V. N. Latyaeva, L. I. Vyshinskaja, and V. V. Drobtinko, *J. Organomet. Chem.* **208**, 169 (1981).
95. L. E. Manzer, *J. Organomet. Chem.* **135**, C6 (1977).
96. L. E. Manzer, *J. Am. Chem. Soc.* **100**, 8068 (1978).
97. K. Jacob and K.-H. Thiele, *Z. Anorg. Allg. Chem.* **455**, 3 (1979).
98. M. H. Chisholm, F. A. Cotton, M. W. Extine, and W. W. Reichert, *J. Am. Chem. Soc.* **100**, 1727 (1978).
99. M. H. Chisholm and D. A. Haitko, *J. Am. Chem. Soc.* **101**, 6784 (1979).
100. M. J. Chetcuti, M. H. Chisholm, K. Folting, D. A. Haitko, and J. C. Huffman, *J. Am. Chem. Soc.* **104**, 2138 (1982).
101. T. Tsuda, K. Watanabe, K. Miyata, H. Yamamoto, and T. Saegusa, *Inorg. Chem.* **20**, 2728 (1981).
102. R. P. A. Sneeden, *J. Mol. Catal.* (in press); A. L. Lapidus and Y. Y. Ping, *Russ. Chem. Rev. (Engl. Transl.)* **50**, 63 (1981).
103. T. Yoshida, D. L. Thorn, T. Okano, J. A. Ibers, and S. Otsuka, *J. Am. Chem. Soc.* **101**, 4212 (1979).
104. S. F. Hossain, K. M. Nicholas, C. L. Teas, and R. E. Davis, *J. Chem. Soc., Chem. Commun.* p. 268 (1981).
105. T. Herskovitz, *J. Am. Chem. Soc.* **99**, 2391 (1977).
106. K. M. Nicholas, *J. Organomet. Chem.* **188**, C10 (1980).
107. B. Fisher and R. Eisenberg, *J. Am. Chem. Soc.* **102**, 7361 (1980).
108. T. Bodnar, E. Coman, K. Menard, and A. Cutler, *Inorg. Chem.* **21**, 1275 (1982).
109. These earlier catalytic investigations have been reviewed (*4*).
110. B. Denise and R. P. A. Sneeden, *J. Organomet. Chem.* **221**, 111 (1981).
111. D. J. Darensbourg, C. Ovalles, and M. Pala, unpublished observations.
112. G. O. Evans and C. J. Newell, *Inorg. Chim. Acta* **31**, L387 (1978).
113. R. Santi and M. Marchi, *J. Organomet. Chem.* **182**, 117 (1979).
114. Y. Inoue, T. Hibe, M. Satake, and H. Hashimoto, *J. Chem. Soc. Chem. Commun.* p. 982 (1979).
115. P. Albano and M. Aresta, *J. Organomet. Chem.* **190**, 243 (1980).
116. T. Tsuda, Y. Chiyo, and T. Saegusa, *Synth. Commun.* **9**, 427 (1979).
117. T. Hung, P. W. Jolly, and G. Wilke, *J. Organomet. Chem.* **190**, C5 (1980).
118. H. Hoberg and D. Schaefer, *J. Organomet. Chem.* **238**, 383 (1982).
119. A. Dohring and P. W. Jolly, *Tetrahedron Lett.* **21**, 3021 (1980).
120. E. Klei, J. H. Teugen, H. J. DeLiefe Meiger, E. J. Kwak, and A. P. Bruins, *J. Organomet. Chem.* **224**, 327 (1982).
121. G. Suss-Fink and J. Reiner, *J. Organomet. Chem.* **221**, C36 (1981).
122. H. Koinuma, F. Kawakami, H. Kato, and H. Hirai, *J. Chem. Soc., Chem. Commun.* p. 213 (1981).
123. J. A. Ibers, *Chem. Soc. Rev.* **11**, 57 (1982).
124. H. E. Ferkul, D. J. Stanton, J. D. McCowan, and M. C. Baird, *J. Chem. Soc., Chem. Commun.* p. 955 (1982).

125. B. W. Krupay and Y. Amenomiya, *J. Catal.* **67**, 362 (1981).
126. M. Saito and R. B. Anderson, *J. Catal.* **67**, 296 (1981).
127. G. D. Weatherbee and C. H. Bartholemew, *J. Catal.* **68**, 67 (1981).
128. F. Solymosi, A. Erdöhelyi, and T. Bánsági, *J. Catal.* **68**, 371 (1981).
129. E. Zagli and J. L. Falconer, *J. Catal.* **69**, 1 (1981).
130. E. Baumgarten and A. Zachos, *J. Catal.* **69**, 121 (1981).
131. J. E. Kubsh, Y. Chen, and J. A. Dumesic, *J. Catal.* **71**, 192 (1981).
132. N. M. Gupta, V. S. Kamble, and R. M. Iyer, *J. Catal.* **66**, 101 (1980).
133. J. L. Falconer and A. Ercument Zagli, *J. Catal.* **62**, 280 (1980).
134. R. Maatman and S. Hiemstra, *J. Catal.* **62**, 349 (1980).
135. M. I. Vass and P. Budrugeac, *J. Catal.* **64**, 68 (1980).
136. F. Solymosi, A. Erdohelyi, and M. Kocsis, *J. Catal.* **65**, 428 (1980).
137. N. M. Gupta, V. S. Kamble, K. A. Rao, and R. M. Iyer, *J. Catal.* **60**, 57 (1979).
138. M. Kobayashi and R. Futaya, *J. Catal.* **56**, 73 (1979).
139. K. Klier, V. Chatikavanij, R. G. Herman, and G. W. Simmons, *J. Catal.* **74**, 343 (1982).
140. W. L. Holstein and M. Boudart, *J. Catal.* **75**, 337 (1982).
141. E. Ramaroson, R. Kieffer, and A. Kiennemann, *Appl. Catal.* **4**, 281 (1982).
142. B. Denise, R. P. A. Sneeden, and C. Hamon, *J. Mol. Catal.* (in press).

ADVANCES IN ORGANOMETALLIC CHEMISTRY, VOL. 22

Basic Metal Cluster Reactions

H. VAHRENKAMP

Institut für Anorganische Chemie
Albert-Ludwigs-Universität
Freiburg, Federal Republic of Germany

I

INTRODUCTION

Cluster chemistry is textbook chemistry today (*1, 2*). A tremendous amount of work has been accumulated during the last decade, allowing one to state that the chaos of facts is gaining shape and character. There is a solid basis for the understanding of structure, bonding, and stability of organometallic cluster compounds. It rests on the work of many synthetic cluster chemists around the world, led by the late P. Chini and his successors in Italy, by the teams of J. Lewis and F. G. A. Stone in England, by L. F. Dahl, H. D. Kaesz, and many others in the United States, by L. Marko in Hungary, and by active research groups in Germany and France. Clusters with up to 55 metal atoms have been prepared (*3*), and a reasonable systematization of synthetic strategies and cluster compositions exists.

On this basis cluster reactivity, too, is beginning to be understood.

Reactions on the outside of clusters have been investigated in considerable detail, including reactivity modifications of organic ligands (4), unusual binding of organic substrates to several metal atoms (5, 6), reactions of cluster-bound carbon atoms (7) or methylene groups (8), and ligand fluxionality (9). The relation between clusters and surfaces (10, 11), and the cluster catalysis discussion (12, 13) are to be considered in this context. In all these cases the cluster can be called a substituent of the organic moiety. The cluster core modifies external properties, but hardly ever changes itself.

If one considers that, except for the heaviest transition metals, metal–metal bonding is not stronger than metal–ligand bonding (14), it becomes obvious that the cluster core itself should participate in many cluster reactions irrespective of the ligand types, sizes, or geometries. This way of looking at cluster chemistry was the reason for choosing the title for this review. Basic metal cluster reactions are understood to be those affecting the cluster core, its composition, shape, size, charge, oxidation state, or bonding situation. Basic cluster chemistry is the "metallic" part of organometallic cluster chemistry. It utilizes the properties of metal–metal bonds and is therefore an extension of metal–metal bond chemistry (15). Basic cluster reactions are not limited to certain metal or cluster types. One prerequisite for facile reactions, metal–metal bond lability, is exemplified by the light transition elements (15, 16). An equally important one, nondisruption of the remaining cluster framework, is better illustrated by the heavy transition elements (17, 18). And furthermore, metal–metal bond polarity is best provided by the large goup of mixed metal clusters (16, 19). The field of basic metal cluster reactions, although many have yet to be discovered, should be wide and varied.

For the purposes of this review a cluster is considered to contain at least three interconnected metal atoms. The chemistry considered is "core" chemistry as outlined above. A clear-cut separation between basic and external cluster chemistry is not possible. A few borderline cases will be treated here. Others, such as the complete breakdown of clusters into mono- or dinuclear units, the incorporation of interstitial atoms, metal–metal bond length variations due to ligand variations, cluster transformations with unsystematic changes in cluster size, reactions not starting from clusters, oxidative additions of ligand units to clusters, and cluster-assisted ligand transformations (exemplified by many hydrocarbons and organic nitrogen compounds), will not be considered. The literature of cluster chemistry has been surveyed up to and including 1982.

This is a selective review. It tries to be comprehensive only in naming all types of reactions and stressing the important ones. As compared to the voluminous part of the cluster literature, which justifies itself by repetitive reference to organic synthesis and catalysis, it attempts to present an inorganic chemist's view of the field: closer to basic research, emphasizing possibilities rather than applications, and once in a while allowing the aesthetic thrill to be its only motivation.

II

ELECTRON TRANSFER REACTIONS

Addition or removal of electrons, the simplest type of chemical reaction, is much more common for transition metal than for main group element compounds. And among the metal complexes the clusters are predestined for this process. Just as increasingly delocalized systems open up the field of organic electrochemistry, so accumulation of metal atoms brings variety into transition metal redox chemistry. Nature "knew" this in using iron-sulfur cluster proteins for multistep redox systems, which chemists are now trying to model in more and more detail (20, 21). And cluster-associated organic chemistry has shown that the cluster can act as an electron reservoir with charge donating as well as charge accepting properties (4, 6). Amazingly little use has been made of this in redox catalysis, and the number of cluster types subjected to redox investigations remains relatively small.

It cannot be stated with certainty yet which types of organometallic cluster are well suited as redox systems, or whether the number of accessible oxidation states increases with the size of the cluster. It seems that the binary metal carbonyls and simple derivatives thereof do not give stable radical ions (22). Cyclopentadienylmetal clusters, however, have a rich redox chemistry (23). The highest number of reversible electron transfers has been reported for clusters with μ_3 bridging ligands, culminating in the cubane-derived Fe_4S_4 systems. The bridged structures include species with $M_3E(E = S,$ PR, CR, etc.), $M_3E_2(E = S, PR, CO)$, and $M_4E_4(E = S, SR, NR, CO)$ cores. Bridging sulfur ligands especially favor the accessibility of more than one oxidation state, but not without the presence of other favorable ligands. Thus, the electrochemistry of the compound $S_2Fe_3(CO)_9$ is poor compared with that of the isoelectronic $S_2Co_3Cp_3$ (24), as is the electrochemistry of $[(CO)_3Mn-SR]_4$ or $[(CO)_3Fe-AsR]_4$ compared with that of similar cubane-like Fe_4S_4 systems (25). A number of stable paramagnetic ionic cluster compounds have been isolated, and the highest number of accessible redox states, namely, between -1 and $+5$, was reported for $S_2Ni_3Cp_3$ (24). These numbers cannot compare, however, with the number of stable anionic cluster compounds in which negative charges provide the correct electron count, and which were formed by hard base induced disproportionation reactions, or by chemical reductions accompanied by ligand removal.

The redox potentials of clusters in general follow the trends predicted from electron density arguments (24–28). Thus electron donating ligands like Cp or PR_3 groups make reductions more difficult, and provide stability for cations. In a series of M_3E clusters (26) with a variety of metal (Fe, Co, Ni, Mo, W) and μ_3 ligands (S, PR, CR, GeR), the first reduction occurs over a narrow potential range supporting the idea (see below) that the HOMO and LUMO have predominantly metal d orbital character. Ligand influences decrease in the order Cp/(CO)$_3$ > μ_3-PR/μ_3-S > μ_3-CMe/μ_3-CPh where the ligand listed lastly in each pair makes reduction easier. The effect of changing the metal atoms seems to be small, but it is difficult to assess this factor when it is accompanied by a change of ligands. The presence of one electron in excess of the noble gas configuration is not unusual for these systems, as is shown by the stability of the paramagnetic clusters $SCo_3(CO)_9$ and $RPCo_3(CO)_9$.

By means of their electron spin resonance (ESR) activity the cluster radical ions permit examination of their electronic structure. This was first demonstrated fifteen years ago by an elegant single crystal ESR study of paramagnetic $SCo_3(CO)_9$ diluted in the diamagnetic host $SFeCo_2(CO)_9$ (29), and later confirmed by a solution ESR study of $PhPCo_3(CO)_9$ and $PhPFeCo_2(CO)_9^-$ (30). The electron in the HOMO of $ECo_3(CO)_9$ or in the LUMO of $EFeCo_2(CO)_9$ was identified as residing in a metal–metal antibonding orbital with predominantly metal d character, with no contribution of the μ_3 ligand, and only small contributions of the terminal ligands. Detailed ESR investigations on the radical anions of $SFeCo_2(CO)_9$ (31), $RCCo_3(CO)_9$ (27, 28), and a series of other M_3E systems (26), supported by EHMO calculations (32), have quantified these findings and lead to the conclusion that in the mixed metal M_3E clusters the spin density is no longer equally distributed over all three metal atoms. For instance, in $SFeCo_2(CO)_9^-$ it is 30% at each cobalt and 15% at the iron atom. Common to all ESR observations is the metal–metal antibonding nature of the LUMO according to a weakening of the metal framework upon reduction.

These relationships between orbital occupancy and metal–metal bonding are the basis for the systematic structural variations between clusters of equal or similar composition but different electron count. This again was first noted for the prototype "couple" $SCo_3(CO)_9/SFeCo_2(CO)_9$ (29) and repeated for the "couple" $PhPCo_3(CO)_9/PhPFeCo_2(CO)_9$ (30), where owing to one antibonding electron the metal–metal bonds in the Co_3 clusters are 10 pm longer than in the $FeCo_2$ clusters. In larger cluster pairs such as $(CpNi)_6/(CpNi)_6^+$ (33) this effect is much smaller. If one adopts metal replacements as above (e.g., Fe for Co) so they are equivalent to one electron transfers, then this approach can be extended to larger series of clusters. Dahl and co-workers have investigated two such series of M_3E_2 (34–37) and

TABLE I

METAL–METAL BONDING IN M_3E_2 AND M_4E_4 CLUSTERS

M_3E_2 cluster	Z	M–M bond orders	M_4E_4 cluster	Z	M–M bond orders
$Cp_3Co_3(S)(CO)$	48	$3 \cdot 1$	$Cp_4Fe_4(CO)_4$	60	$6 \cdot 1$
$Cp_3CoNi_2(CO)_2$	48	$3 \cdot 1$	$[Cp_4Fe_4S_4]^{2+}$	66	$2 \cdot 0, 4 \cdot \frac{1}{4}$
$Cp_3Ni_3(CO)_2$	49	$3 \cdot \frac{5}{6}$	$[Cp_4Fe_4S_4]^+$	67	$4 \cdot \frac{1}{8}, 2 \cdot 1$
$[Cp_3Co_3S_2]^+$	49	$1 \cdot 1, 2 \cdot \frac{3}{4}$	$Cp_4Fe_4S_4$	68	$4 \cdot 0, 2 \cdot 1$
$[Cp_3Ni_3(CO)_2]^-$	50	$3 \cdot \frac{2}{4}$	$[Cp_4Co_4S_4]^+$	71	$2 \cdot 0, 4 \cdot \frac{1}{8}$
$Cp_3Co_3S_2$	50	$3 \cdot \frac{2}{3}$	$Cp_4Co_4S_4$	72	$6 \cdot 0$
$Cp_3Ni_3S_2$	53	$3 \cdot \frac{1}{6}$			

M_4E_4 (38–41) compounds where the total number of valence electrons, Z, is directly related to the degree of metal–metal bonding in the M_3 resp. M_4 frameworks. Table I summarizes their results.

Whereas in ligand bridged dinuclear complexes, removal or addition of two electrons makes or breaks one metal–metal bond (15) this does not seem to be the case for clusters, presumably because of their delocalized bonding. At least for one case, however, two-electron reduction can induce a significant change in cluster shape (18, 42): the 84-electron cluster $Os_6(CO)_{18}$ with framework 1 is easily reduced to the 86-electron anion $Os_6(CO)_{18}^{2-}$ with framework 2, in accordance with skeletal electron counting rules.

$Os_6(CO)_{18}$

1

$[Os_6(CO)_{18}]^{2-}$

2

III

REACTIONS INVOLVING UNIDENTATE REAGENTS

Possible unidentate reagents are atomic cations and other simple electrophiles as well as atomic anions and other simple nucleophiles. They can react with the cluster with or without change in the electron count as well as with or without change in cluster core shape. Not considered here are nucleophilic ligand substitutions or reactions in which the number of metal atoms in the cluster changes (cf. Section V).

A. *Acid–Base Reactions*

Many neutral clusters and all anionic clusters are Brønsted acids. The corresponding protonation reactions and the stability and fluxionality of their products have been reviewed in detail (*43*). Up to two protons can be added to electron-rich neutral clusters, and up to four negative charges can be accommodated on cluster anions. By nuclear magnetic resonance (NMR) spectroscopy and crystal structure analysis, it has been found that in most cases the hydrogen atoms in clusters occupy edge-bridging positions (*43–45*), in accord with theoretical calculations (*46*) and electron density arguments.

No other simple electrophile has been found to add to the metal atoms in clusters. Size effects seem to prevent this and lead to attachment at the point of highest electron density in the ligand sphere. This is the reason for the coordination of main group element electrophiles at the oxygen centers of bridging CO ligands (*47*), or of metal carbonyl fragments at μ_3 sulfur ligands (*48*). Notable examples of unusual reaction pathways resulting from protonation and methylation are **3** to **4** and **3** to **5** (*49*), and the protonation of the isostructural iron and osmium anions **6** to **7** and **6** to **8**, which in the iron case even results in an O–H bond (*50*). The fact that many cluster anions cannot be protonated to give stable neutral cluster hydrides may be due to inaccessibility of the cluster core to the proton.

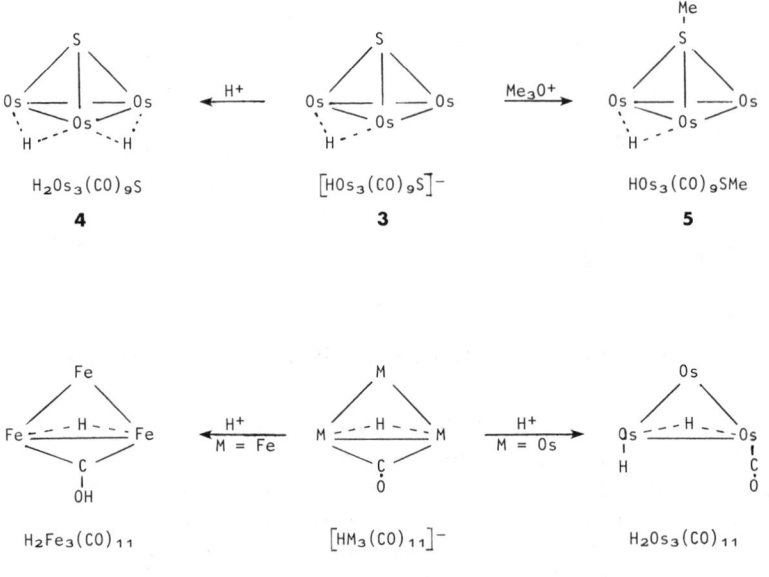

Some striking demonstrations of metal–metal bond lability are provided by cluster rearrangements due to protonation. This is the case for some anionic osmium clusters (cf. Section VI). It involves ligand activation for some tetrairon clusters (*51–53*). Thus, the clusters **9** and **11** open up upon protonation, and compensation for the lost iron–iron bonds in the products **10** and **12** comes from the bonding between one iron atom and a carbonyl oxygen. The relation of these unusual nucleophile–electrophile interactions to cluster-induced CO transformations is obvious.

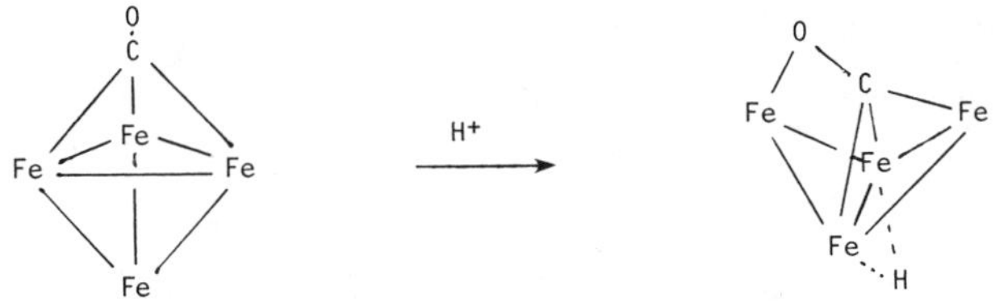

$[Fe_4(CO)_{13}]^{2-}$

9

$[HFe_4(CO)_{13}]^-$

10

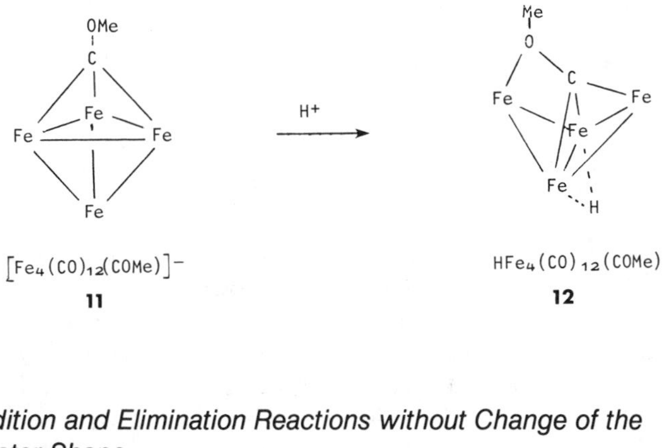

$[Fe_4(CO)_{12}(COMe)]^-$

11

$HFe_4(CO)_{12}(COMe)$

12

B. *Addition and Elimination Reactions without Change of the Cluster Shape*

The weakness of most metal–metal bonds compared with metal–ligand bonds makes cleavage of metal–metal bonds by nucleophiles a common process (*15*). In the case of metal–metal double bonds this corresponds to nucleophilic addition to the metal–metal bonded systems. Since unsaturated clusters exist which can be considered to contain metal–metal double bonds, this should be an important aspect of substrate activation by clusters.

Most of the clusters not obeying the 18-electron rule contain early or late transition metals where the d^4 situation or 16-electron configurations are not unusual (54–56). Correspondingly, these clusters do not clearly show unsaturation in their reactions. There are, so far, three cluster types for which localized metal–metal double bonds are documented. Of these **13** and **14** are isoelectronic and structurally related, whereas type **15** allows variations of M and R.

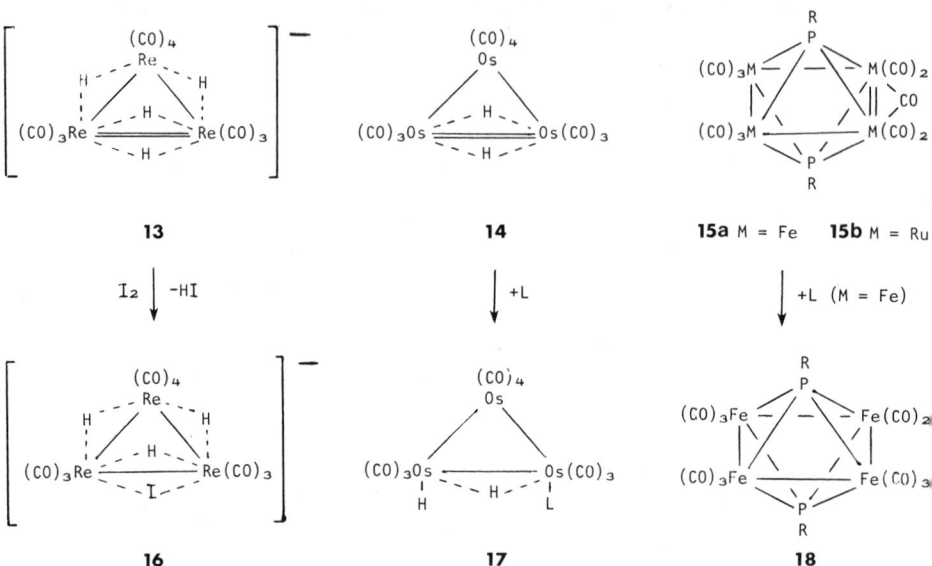

The cluster **13** does not react with simple nucleophiles. The electrophile iodine, however, introduces I$^+$ as a two electron donor leading to the iodo bridged cluster **16** (57). The cluster **14** adds several different donor ligands to form **17**. Some of these reactions are reversible with reformation of **14** or derivatives thereof (58, 59). The iron cluster **15a** reversibly adds CO. After the addition of phosphine ligands, CO can be eliminated, and several PR$_3$ groups can be introduced into **15a** by such addition–elimination cycles (60, 61). In each case the cluster shape does not change significantly upon addition, and the incoming ligand is accommodated in the ligand sphere. Of the cluster pairs **13/16**, **14/17**, and **15/18**, all three unsaturated clusters were discovered first. No known cluster has been found so far to undergo ligand elimination with formation of metal–metal double bonds. However, ligand substitutions including those with potential in catalysis may proceed through such intermediates, just as the addition/elimination cycles of **14** and **15a** correspond to CO substitutions.

C. Metal–Metal Bond Breaking or Making by Addition or Elimination of Nucleophiles

Degradation by nucleophiles is the most common cluster reaction (5, 6). Accordingly, clusters like $Fe_3(CO)_{12}$ or $Co_4(CO)_{12}$ are frequently used as sources of monometallic units. In order not to obtain total degradation strong metal–metal bonds must be present, or the cluster core must be stabilized by bridging ligands. In these cases, single metal–metal bonds may be broken by adding donor ligands with concomitant opening of the cluster framework. This reaction and its reversal make new cluster shapes and new modes of substrate activation accessible. The significance of this was only recently discovered (12, 15), making it an active field of current cluster chemistry (16, 17, 62).

A prototype reaction scheme is illustrated by the interconversions among **19, 20,** and **21** (63). Although two of the three metal–metal bonds in **19** are lost by addition of CO or phosphine ligands, the cluster does not lose its identity and is reformed by CO removal. The rigidity given to the cluster framework is sufficient even to maintain the optical activity of the clusters **22** after an analogous opening–closing cycle with CO (64). Several other μ_3-ligand bridged trinuclear clusters similar to **19** (65) or **22** (66) or like **23** (67) and **24** (68) show similar opening–closing patterns upon reaction with CO or phosphine ligands. One case of such reactions without the presence of a μ_3-ligand is known: the cyclic clusters $Os_3(CO)_{10}(\mu_2\text{-Hal})_2$ are reversibly transformed into the open chain compounds $Hal—Os_3(CO)_{12}—Hal$ by CO (69).

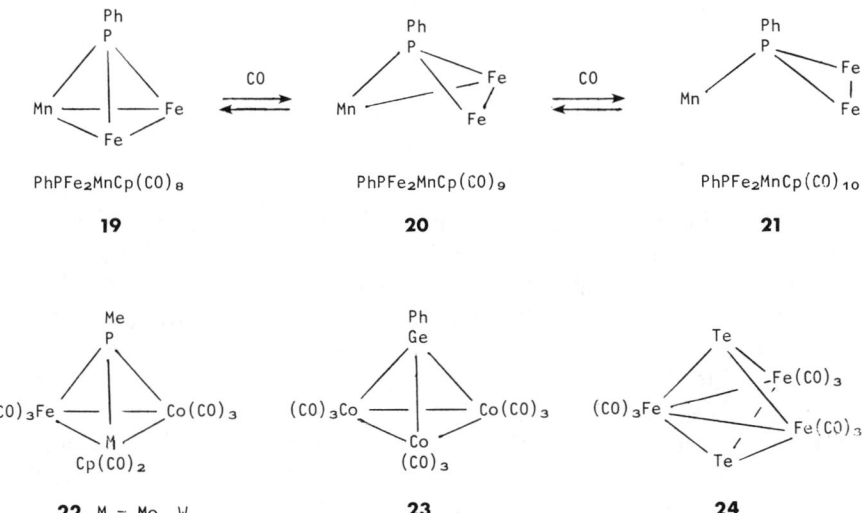

Increasing the nuclearity of the clusters increases the number of possible opening modes. Among the tetranuclear clusters the tetrahedron – butterfly interconversion is the simplest such mode, as observed for $H_2Os_3Pt(CO)_{10}PR_3$ with CO, PR_3, or AsR_3 (55). The ligand bridged systems 25 reversibly lose two metal – metal bonds with CO, unfolding to 26 (70), a reaction type applicable for an extensive range of metal and bridging atoms (71). Moreover, in 15b the bridging phosphorus ligands allow a structural rearrangement under CO which involves loss of the Ru – Ru double bond as well as one Ru – Ru single bond during the formation of 27 (72, 73).

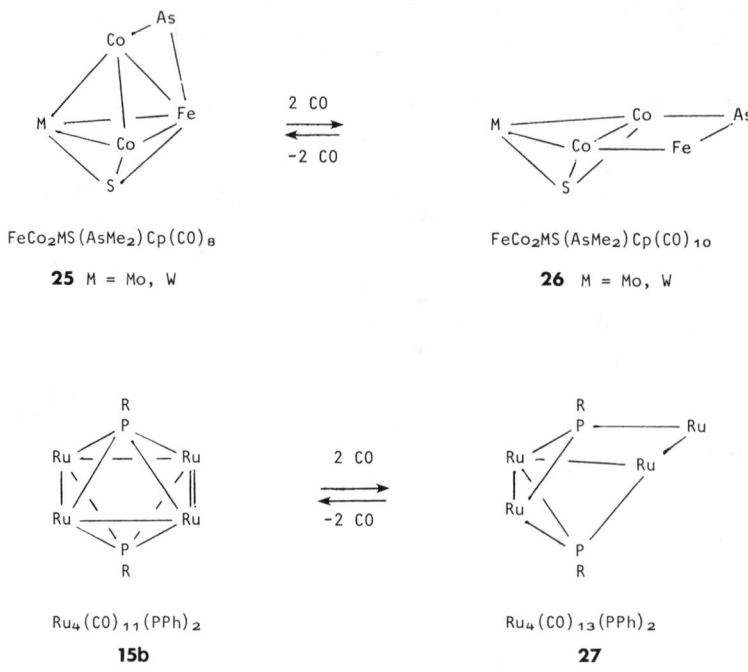

FeCo₂MS(AsMe₂)Cp(CO)₈

25 M = Mo, W

FeCo₂MS(AsMe₂)Cp(CO)₁₀

26 M = Mo, W

Ru₄(CO)₁₁(PPh)₂

15b

Ru₄(CO)₁₃(PPh)₂

27

The opening of a trigonal bipyramidal cluster skeleton can be envisaged as proceeding through 28 → 31. Likewise the bicapped tetrahedral framework 1 may be opened through the sequence 32 → 34. Each step involves addition of one donor ligand like CO. Both beautiful series have been partly realized for osmium carbonyls. Thus, $H_2Os_5(CO)_{15}$ adds nucleophiles to form $H_2Os_5(CO)_{15}L$ with framework 29 (74), whereas $Os_5(CO)_{16}$ adds CO to form $Os_5(CO)_{19}$ with structure 31 (75). The cluster $Os_6(CO)_{18}$ and CO yield $Os_6(CO)_{20}$ with the presumed structure 33. The latter reacts with $P(OMe)_3$ to give $Os_6(CO)_{17}[P(OMe)_3]_4$ with the planar framework 34 (76). This is the most complete cluster unfolding sequence discovered so far.

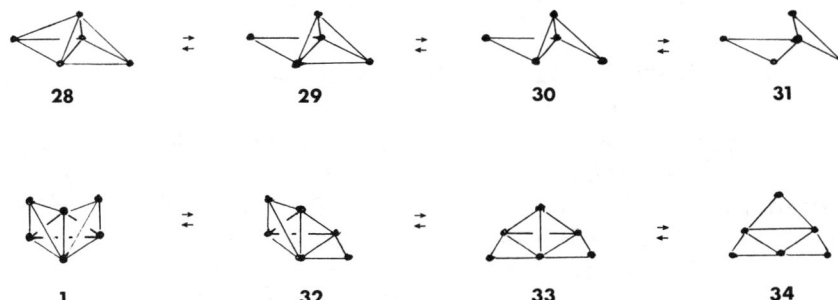

28 **29** **30** **31**

1 **32** **33** **34**

Two types of cluster unfolding involve initial attack of an electrophile. One is the above mentioned conversion of a two electron μ-CO into a four electron η-CO in Fe_4 clusters after protonation or methylation (51–53). The other occurs after attack of electrophiles with unshared electron pairs like I_2 or NO^+ at anionic clusters. The neutral clusters formed rearrange, such that a metal–metal bond is lost and the attacking electrophile becomes a 3e-μ_2 ligand. Examples are the NO^+ induced rearrangement of **35** to **36** (77), and the halogen induced conversion of **37** to **38** (78), which is reversible in one case. Even the decanuclear cluster $Os_{10}C(CO)_{24}^{2-}$ undergoes analogous reversible opening of two Os–Os edges with iodine (79). The number of possible cluster framework interconversions is greater than that of carbon containing frameworks, and a lively development is to be expected in this field.

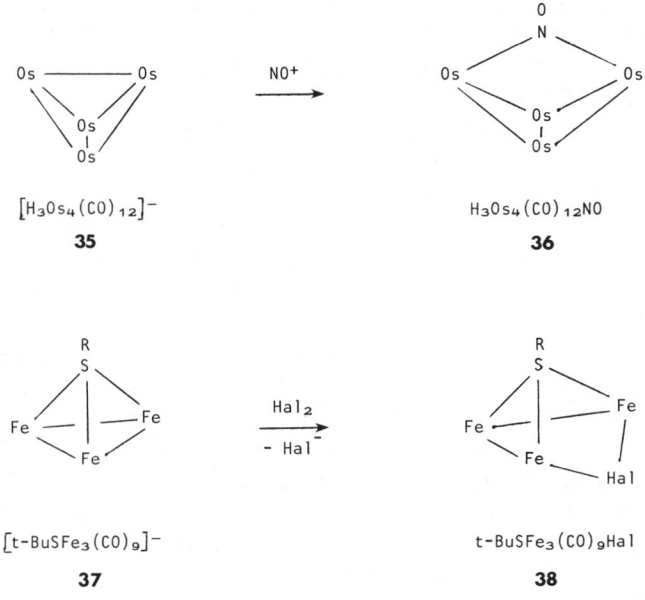

$[H_3Os_4(CO)_{12}]^-$

35

$H_3Os_4(CO)_{12}NO$

36

$[t\text{-BuSFe}_3(CO)_9]^-$

37

$t\text{-BuSFe}_3(CO)_9Hal$

38

IV

REACTIONS INVOLVING POLYFUNCTIONAL REAGENTS

Under this heading all systems having an AB_2 composition in the most general sense, all unsaturated compounds, and all reagents capable of oxidative addition are considered to be polyfunctional. In general their reaction with clusters will involve transformations of the cluster as well as of the reagent. Such reactions are therefore borderline cases in the context of basic metal cluster chemistry as defined above. Accordingly, most reactions of clusters with organic substrates, reactions where the reagent attacks the cluster and a ligand, or the ligand alone, and application-oriented phenomena will not be treated here, and reference is made to recent reviews (5,6, 8, 13, 43, 54, 62). The discussion is limited mostly to inorganic or organoelement reagents which add to or insert into one or more metal–metal bonds.

A. Two-Center Additions without Metal–Metal Bond Breaking

Reactions of substrates X–Y with clusters after which X and Y are bound to two metal atoms may be called two-center oxidative additions, irrespective of the mechanistic meaning of this term (80, 81). Depending upon the number of electrons introduced by X and Y, such reactions will involve different degrees of ligand substitution or metal–metal bond cleavage. If the cluster core geometry is to be maintained there has to be either a "cluster unsaturation" before the reaction, or a definite number of ligands, in general CO, has to be expelled.

Hydrogenation is the best-known reaction of this type, due to its ease and due to the versatility of hydrido metal clusters (43). Hydrogen reacts under mild conditions with polynuclear metal carbonyls making this as important for hydrido cluster synthesis as is the protonation of cluster anions. In simple cases just one CO ligand is replaced by two hydride ligands, but frequently the cluster nuclearity changes indicating metal–metal bond cleavage of an intermediate by hydrogen. Hydrogenation of metal carbonyls is the standard preparative procedure for synthesis of the important cluster hydrides of rhenium, ruthenium, and osmium (43). More recently, it has been applied to mixed metal clusters (19), two current examples being the conversion of $Ru_2Co_2(CO)_{13}$ to $H_2Ru_2Co_2(CO)_{12}$ (82), and of the mixed hydrides of the iron triad $H_2MM_3'(CO)_{13}$ to $H_4MM_3'(CO)_{12}$ (83).

The favorable bonding situation between hydrogen and cluster metal atoms also provides some driving force for the many observed HX oxidative additions to clusters ranging from C–H cleavage to cluster expansion with mononuclear organometallic hydrides (5, 6). Of main group element HX additions, X can be Cl, Br, I, O, S, N, P, As, Si, Sn, and B. If X has no unshared electron pairs, e.g., Si or Sn, the addition can be compared to H_2 addition, i.e., one HX expels one CO as in the formation of **39** from $Ru_3(CO)_{12}$ or $Os_3(CO)_{12}$ and $HSiCl_3$ (84). All other substrates HX that have unshared electron pairs can add in one of two ways, as depicted in **40** and **41** for ruthenium and osmium carbonyls. Of these, **40** corresponds to H_2 addition with H and X as one electron ligands, whereas in **41** X has become a three electron ligand so that a further CO ligand has been expelled.

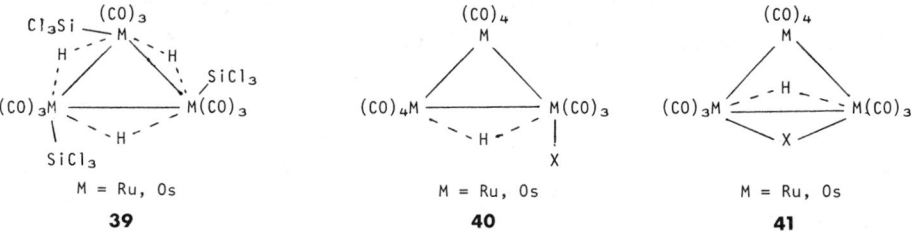

The composition **40** has been observed for hydrogen halide additions to $Os_3(CO)_{11}(NCMe)$ (85). The resulting compounds $HOs_3(CO)_{11}Hal$ are labile with respect to CO elimination and formation of $HOs_3(CO)_{10}Hal$ of type **41**. This class of compound is preferentially formed by the addition of HHal, H_2O, HSR, H_2NR, and H_2PR to ruthenium or osmium carbonyls (86–92). Some mechanistic insight is provided by the observations that the "double-bonded" cluster $H_4Re_3(CO)_{\overline{10}}$ (**13**) adds hydrogen halide easily without CO elimination (93), that $Os_3(CO)_{10}(NCMe)_2$, containing two labile acetonitrile ligands, adds several kinds of HX easily forming products of type **41** (86), and that the reaction can be performed stepwise starting with CO substitution by halide anions (94) or RPH_2 (92). All this means that oxidative addition to clusters has to be preceded by ligand elimination, i.e., development of cluster unsaturation.

B. *Capping of Metal Atom Triangles*

Almost all clusters contain metal atom triangles, which can be envisaged as defining incomplete tetrahedral structures. This makes it understandable that capping of these triangles, i.e., filling of such interstices, is a common cluster reaction. It occurs in cluster growth (see Section V,A), as well as during the multicenter activation of organic substrates. Especially, the triosmium cluster is able to bind to quite different triply bridging units X (organic, organoelement, and inorganic) which have in common the fact that they result from precursors H_2X (*5, 6, 95*).

As a ligand the capping unit is a four- or five-electron donor. Its introduction therefore requires the expulsion of other ligands. And if the capping unit comes from a precursor HX or H_2X, a H–X splitting reaction must also occur which may be an oxidative addition to the cluster. Nevertheless many capping reactions proceed easily, demonstrating the preferred $M_3(\mu_3X)$ bonding situation. The stepwise nature of the capping reactions is obvious in the transformation of the μ_2-iodo bridged ruthenium and osmium clusters 41 to the μ_3-iodo bridged clusters 42 (*85, 94*). The reversal of such a reaction happens in the opening of 43 to 44 by donor ligands like C_2H_4, CO, PR_3, and MeCN, where 43 can be re-formed for L = C_2H_4 (*96*).

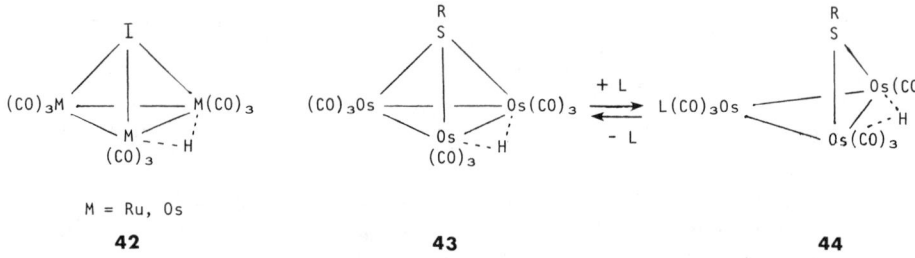

M = Ru, Os

42   **43**   **44**

Generally, no intermediates have been observed for capping reactions introducing sulfur containing μ_3-ligands as found in $HFe_3(CO)_9SR$ or $HRu_3(CO)_9SR$ from HSR, or in $H_2Ru_3(CO)_9S$ from H_2S (*97, 98*), or in the formation of $H_2Fe_3(CO)_9PR$ from $Fe_3(CO)_{12}$ and H_2PR (*99*). The inertness of Ru_3 and Os_3 clusters, however, allows the isolation of the intermediates $HM_3(CO)_{10}(\mu\text{-PHR})$ with structure 41 which are converted upon heating to $H_2M_3(CO)_9PR$ (*91, 92, 100*). Since the capping group comes in as a four- or five-electron ligand, cluster unsaturation should favor capping. This is the case for the formation of $[H_3Re_3(CO)_9(\mu_3\text{-ER})]^-$ from $[H_4Re_3(CO)_{10}]^-$ and HOR or HSR (*101, 102*). And for the same reason the unsaturated compound $H_2Os_3(CO)_{10}$ is the best precursor for the various $H_2Os_3(CO)_9(\mu_3\text{-E})$ clusters (*6*).

Some four-electron capping units enter as such. This is the case for the many reactions forming μ_3-sulfur ligands from elemental sulfur (*103*). It also holds for the triruthenium μ_3-nitrene cluster 45, formed from Me_3SiN_3 and $Ru_3(CO)_{12}$ (*104*). A versatile four-electron ligand is the acetylene moiety, which can add facially to M_3 units as a two-center capping group, as found in the clusters 46, which can be obtained from trinuclear carbonyls of iron, ruthenium, and osmium (*105, 106*).

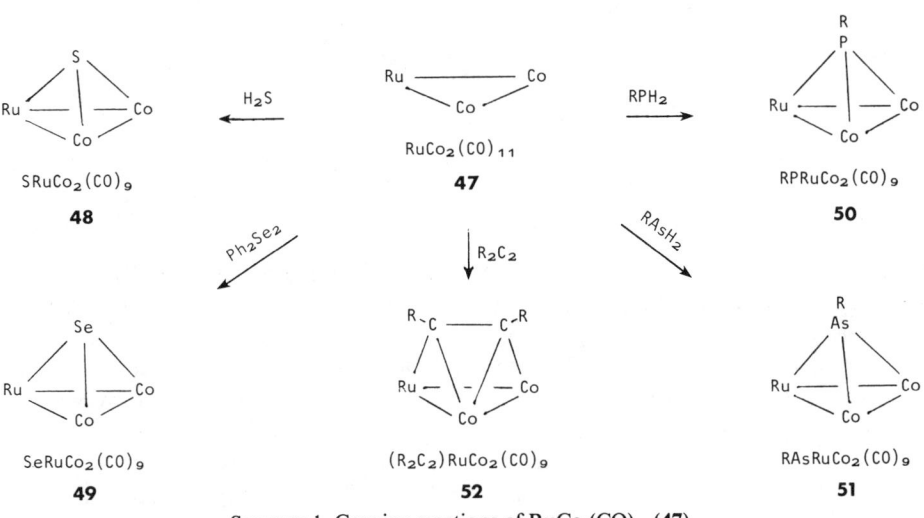

(Me₃SiN)Ru₃(CO)₁₀

45

M₃(CO)₉(RCCR),   M = Fe, Ru, Os

46

The cluster which has so far been found to be most susceptible to main group element capping is $RuCo_2(CO)_{11}$ (**47**). It reacts under mild conditions with all types of capping reagents according to Scheme 1, introducing capping units involving elements from the sixth, fifth, and fourth main groups (*107*). The driving force for these reactions comes both from the lability of the starting cluster **47** and from the fact that all its capping reactions produce the cluster type $M_3(CO)_9(\mu_3\text{-X})$ which is widely occurring due to its pronounced stability. The capping reaction, i.e., the μ_3-binding by

Scheme 1. Capping reactions of $RuCo_2(CO)_{11}$ (**47**).

substrates, achieves two things at the same time. It puts the capping atom into a geometrically (and mostly electronically) favorable situation, and it adds stability to the cluster framework due to the "clamp" effect of the μ_3-ligand. It is for this reason that more often than not clusters of the $M_3(\mu_3\text{-X})$ type are formed from the capping precursors and simple organometallic compounds which are not yet clusters (*5, 103*).

C. Addition Reactions with Cluster Unfolding

If a two-center oxidative addition is not accompanied by ligand elimination, it must involve breaking or making of a metal–metal bond (80). Consequently, some cluster-assisted activation reactions of organic substrates are accompanied by cluster opening (5, 62, 108). Their number is, however, small compared with the number of substrate activations maintaining the cluster shape. On the other hand, the number of simple oxidative additions to clusters leading to nothing but the breaking of one metal–metal bond is also small. There is only one known case of such a halogen addition, the conversion of cyclic $Os_3(CO)_{12}$ to linear $Os_3(CO)_{12}Hal_2$ (109). However, more HX additions of this type have to be found. The unusual Os_3 cluster 53 adds HCl to yield 54 with a more open structure (110). And one apical Ru–Ru bond of square pyramidal $Ru_5C(CO)_{15}$ is opened by HCl or HBr yielding 55 where the C atom is more encapsulated, although the cluster framework is more open (111).

$Os_3S(SCH_2)(CO)_8(PMe_2Ph)$	$HOs_3S(SCH_2)(CO)_8(PMe_2Ph)Cl$	$HRu_5C(CO)_{15}Hal$
53	**54**	**55**

Another reagent capable of metal–metal bond cleavage is $SnCl_4$. And again it requires the inertness of the trinuclear ruthenium and osmium carbonyls to achieve simple reactions. $Os_3(CO)_{12}$ and $SnCl_4$ afford linear $Cl\text{-}Os_3(CO)_{12}SnCl_3$ (112). The cluster $H_2Ru_3(CO)_9S$ is not only opened by $SnCl_4$, but the resulting cloride ligand also replaces one CO group moving into the bridging position of 56 (113). Ligand replacement and bridge formation also occurs upon oxidative addition of R_2S_2 or R_2Se_2 to $Os_3(CO)_{11}(NCMe)$ with formation of 57 (114). It is possible that a linear Os_3 compound is an intermediate in this reaction, since linear compounds $Os_3(CO)_{12}Hal_2$ are converted by heating to triangular $Os_3(CO)_{10}Hal_2$ with a structure analogous to that of 57 (115). Bridging and capping ligands stabilize polynuclear compounds, and hence they make possible cleavage reactions that otherwise might lead to complete rupture of a cluster. The formation of 58 instead of cluster destruction in the reaction of $RPFe_3(CO)_9^{2-}$ with the aggressive reagent SCl_2 (116) must therefore be favored because of the introduction of sulfur into a capping position.

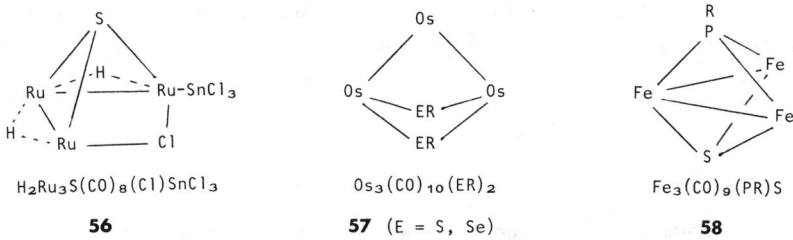

56   57 (E = S, Se)   58

$H_2Ru_3S(CO)_8(Cl)SnCl_3$ $Os_3(CO)_{10}(ER)_2$ $Fe_3(CO)_9(PR)S$

No tetrahedron–butterfly interconversion has been achieved so far by metal–metal bond opening via oxidative addition to tetranuclear clusters. Acetylenes, however, have a strong tendency to insert into tetrahedral clusters of iron, cobalt, and their congeners. Reactions of this type are among the oldest known in cluster chemistry, the prototype being the formation of **59** from $Co_4(CO)_{12}$ and $RC \equiv CR$ (*117*). During the reactions, ligands equivalent to four electrons are lost, one M–M bond is opened, and the C–C bond length approaches the single bond value. Cluster hydrides are good starting materials for such reactions, losing two hydrogens and one CO. Mixed metal clusters allow the formation of isomeric $M_4(R_2C_2)$ units as found in $FeRu_3(CO)_{12}(R_2C_2)$ (*118*), but can also show exclusive formation of one product, e.g., from the cleavage of the Co–Co bond in $Ru_2Co_2(CO)_{13}$ (*82*).

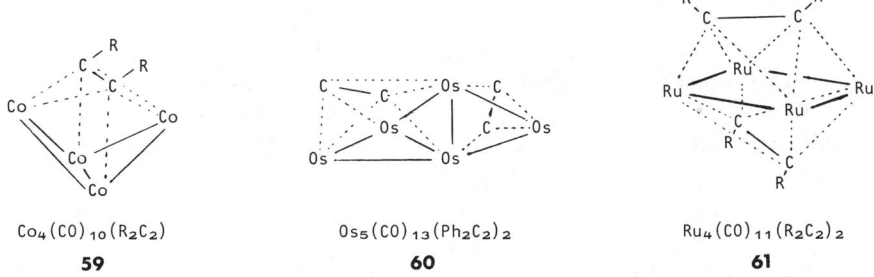

$Co_4(CO)_{10}(R_2C_2)$ $Os_5(CO)_{13}(Ph_2C_2)_2$ $Ru_4(CO)_{11}(R_2C_2)_2$

59   60   61

Acetylenes can also cleave more than one metal–metal bond in clusters. Thus $H_2Os_5(CO)_{15}$ is completely unfolded to **60** by two acetylene ligands (*119*). The monoacetylene butterfly species $Ru_4(CO)_{12}(R_2C_2)$ add a second acetylene molecule losing another Ru–Ru bond, to form the double-capped unsaturated clusters **61**, the structure of which corresponds to that of the analogous iron compound (*120, 121*). In all these cases the multicenter interactions between the acetylene and several metal atoms, which also make the alkyne a good capping unit for metal triangles, serve to stabilize these unusual tetranuclear clusters. The possibilities available for bridging and capping therefore are a major aspect of the uniqueness of cluster reactions.

V

REACTIONS CHANGING THE METAL ATOM COMPOSITION

All cluster syntheses involve growth or fragmentation of metal atom aggregates. But until very recently, the buildup of clusters or the use of clusters as fragment sources involved empirical approaches with little predictability. Only in a few cases, e.g., in the growth of rhodium clusters by redox condensations ($Rh_4 \rightleftarrows Rh_5 \rightleftarrows Rh_6 \rightleftarrows Rh_7$; $Rh_{13} \rightleftarrows Rh_{14} \rightleftarrows Rh_{15}$) (122), or in the CO induced breakdown of ruthenium carbonyls (123), was the stepwise addition or removal of monometallic units recognized. Today a growing number of designed changes of cluster metal atom compositions is emerging, and synthetic strategies based thereon are being developed (16, 62). Yet generalizations in this central area of basic cluster reactions are still difficult to make.

A. Cluster Expansion

Such reactions, starting from a cluster and ending with a cluster of higher nuclearity, must be mechanistically complex. Normally several ligands have to be removed and several metal–metal bonds formed. Complete mechanistic information is not yet available for any case. Therefore, the material will be organized according to reagent types.

1. Miscellaneous Reagents, Mechanistic Possibilities

The key step in a cluster expansion reaction is the attachment of the incoming metal unit. Once this has taken place, a sequence of metal–metal bond formations accompanied by ligand eliminations can occur which is the reversal of the cluster unfolding reactions described in Section III,C. In uncontrolled cluster expansions, the first step is the combination of coordinatively unsaturated cluster and monometallic units, and the reaction is unlikely to stop at this stage. Under mild conditions the attachment may result from a nucleophile/electrophile combination, the products of which have been isolable in a few cases (see below). More insight into possible reaction pathways is to be expected when the incoming metal unit has a reactivity pattern other than the ability to lose ligands. Thus, oxidative addition of a M–X compound to a cluster can provide the first connection which is then followed by aggregation. M–Hal as well as M–H oxidative addition to clusters have been reported. Thus, Ph_3P–AuCl adds to $Os_3(CO)_{12}$ to form **62** (*124*) with external gold; and $Ir(PPh_3)_2(N_2)Cl$ adds to $H_2Os_3(CO)_{10}$ to form **63** (*125*) where due to the lability of both the iridium and osmium reagents further ligand elimination and closing of the new cluster have occurred.

62   **63**

Both $HRe(CO)_5$ and $H_2Os(CO)_4$ can be oxidatively added to $Os_3(CO)_{11}(NCMe)$ (*126–128*). This leads to external attachment of the new metal carbonyl unit as in **64** (*127*), and a second $HRe(CO)_5$ molecule can be incorporated the same way (*126*). In both cases just one metal–metal bond has been formed in the first step. CO elimination from **64** introduces one more metal–metal bond, one possible result of which is rhomboidal **65** (*126*), whereas further CO elimination under H_2 leads to full aggregation to tetrahedral **66** (*127*). All three steps of a $M_3 + M'$ capping sequence have thereby been performed. A similar two-step sequence leads from $Os_6(CO)_{17}(NCMe)$ and $H_2Os(CO)_4$ via $H_2Os_7(CO)_{21}$ to $H_2Os_7(CO)_{20}$ (*128*).

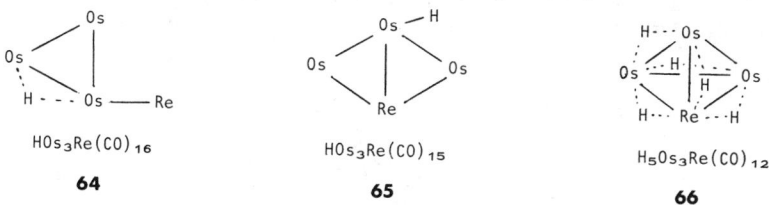

$HOs_3Re(CO)_{16}$ $HOs_3Re(CO)_{15}$ $H_5Os_3Re(CO)_{12}$

64   **65**   **66**

The attachment of the incoming metal unit is possible even without making a metal–metal bond in the first step. Organometal dimethylarsenides $Cp(CO)_3M$–$AsMe_2$(M = Mo, W) act as Lewis bases, and replace CO in $FeCo_2(CO)_9S$ to form **67**. Controlled CO elimination from **67** allows the external Mo or W atom to utilize its capping potential and leads to the tetrahedral $FeCo_2M$ clusters **25** (*70*). This aggregation reaction works for the starting clusters $FeCo_2(CO)_9S$, $RuCo_2(CO)_9S$, $FeCo_2(CO)_9PR$, and $Co_3(CO)_9CR$, in each case capping the metal triangle therein by a Mo or W atom (*71*). This sequence applied to the starting cluster $FeCoMoSCp(CO)_8$ leads to **68**, the first cluster with four different metal atoms (*129*). All these reactions, although not straightforward at first glance, demonstrate the stepwise nature of cluster expansions.

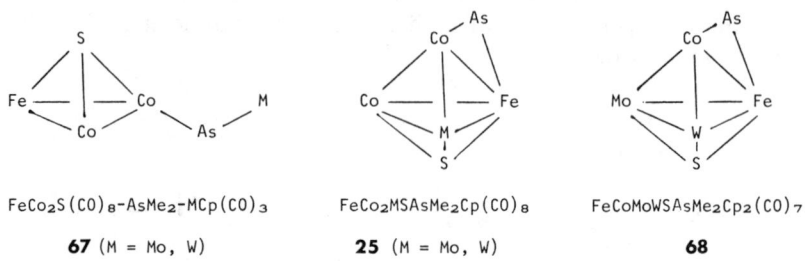

$FeCo_2S(CO)_8-AsMe_2-MCp(CO)_3$

67 (M = Mo, W)

$FeCo_2MSAsMe_2Cp(CO)_8$

25 (M = Mo, W)

$FeCoMoWSAsMe_2Cp_2(CO)_7$

68

2. Anionic Clusters as Reagents

Anionic clusters are good nucleophiles (see Section III,A) and are often easy to make. On the other hand, the electrophilic nature of most monometallic complexes is obvious from ligand substitutions. The combination of these properties makes a strategy for cluster expansion. This strategy was used for the first time by Hieber (*130*) in making $Fe_4(CO)_{13}^{2-}$ from $Fe_3(CO)_{11}^{2-}$ and $Fe(CO)_5$. It is probably active in many syntheses of large metal carbonyl clusters because the Re, Os, Rh, Ir, Ni, and Pt clusters involved are almost always anionic. However, simple stoichiometries can rarely be written for such reactions (*122*). This route makes mixed metal clusters accessible, e.g., **69** from $HRu_3(CO)_9PPh^-$ and $Rh(CO)_3(PEt_3)_2^+$ (*131*), **70** from $Rh_6(CO)_{15}^{2-}$ and $Ni(CO)_4$ (*132*), or **71** from $Ni_6(CO)_{12}^{2-}$ and $W(CO)_6$ (*133*). The potential of this strategy has not as yet been fully evaluated or exploited.

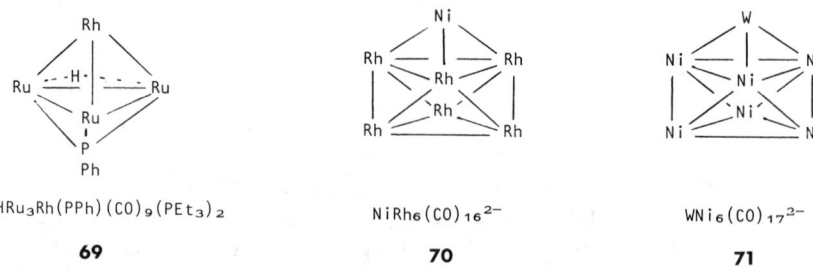

$HRu_3Rh(PPh)(CO)_9(PEt_3)_2$

69

$NiRh_6(CO)_{16}^{2-}$

70

$WNi_6(CO)_{17}^{2-}$

71

Only two kinds of reagent have been extensively used for this type of reaction so far. These are the anionic carbido carbonyl clusters of iron and the mononuclear phosphine gold compounds. Starting from $Fe_4C(CO)_{12}^{2-}$ and the appropriate reagents the range of clusters **72** was obtained with $ML_n = Cr(CO)_3^{2-}$, $W(CO)_3^{2-}$, $Rh(CO)_2^{-}$, $Rh(COD)^-$, $Ir(COD)^-$, $Pd(C_3H_5)^-$, and $Cu(NCMe)^-$. And starting from $Fe_5C(CO)_{14}^{2-}$ the even larger range of clusters **73** resulted with $ML_n = Fe(CO)_2^{2-}$, $Cr(CO)_3^{2-}$, $Mo(CO)_3^{2-}$, $W(CO)_3^{2-}$, $Rh(COD)^-$, $Rh(CO)_2^{-}$, $Ni(CO)_2$, $Pd(CO)_2$, and $Cu(NCMe)^-$ (*134*, *135*). In both cases the incoming metal unit has completed a previously incomplete metal atom polyhedron.

$CFe_4(CO)_{12}ML_n$	$CFe_5(CO)_{14}ML_n$
72	**73**

The attachment of $AuPR_3$ units to clusters was initiated with the aim of experimentally testing the isolobal analogy between the H atom and the AuL unit (*136*). It then turned out that almost all hydrido clusters or their anions react smoothly with R_3PAuX (X = Cl, Br, CH_3, PF_6) introducing $AuPR_3$ without ligand substitution; and within the last year more than a dozen aurations have been described (*136–146*). Prominent compounds are **74** as the original example (*136*), **75** (*140*), and **76** (*142*) as clusters with μ_2-Au units, **77** (*137*), and **78** (*145*) as bisaurated clusters with different core shape, and **79** (*140*) as trisaurated cluster. These compounds again teach something about cluster growth. Like the H atom the incoming metal unit finds the place of lowest steric crowding which means that $AuPR_3$ units can be μ_2 bridging and that several $AuPR_3$ units will be neighboring. The comparison of **77** and **78** shows two types of M_6 clusters with equal numbers of M–M bonds which may well be successive steps in a general growth pattern of naked clusters or metallic crystallites. Finally, auration can be the last step in the synthesis of another cluster with four different metal atoms, like in the sequence **48** → **80** → **81** which involves also metal exchange as described in Section V,C (*147*).

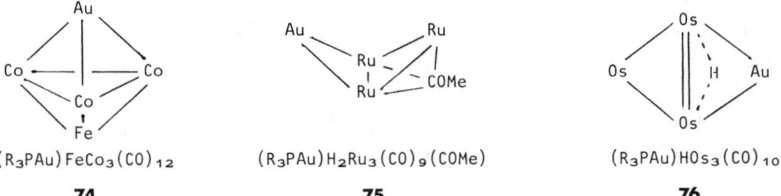

$(R_3PAu)FeCo_3(CO)_{12}$	$(R_3PAu)H_2Ru_3(CO)_9(COMe)$	$(R_3PAu)HOs_3(CO)_{10}$
74	**75**	**76**

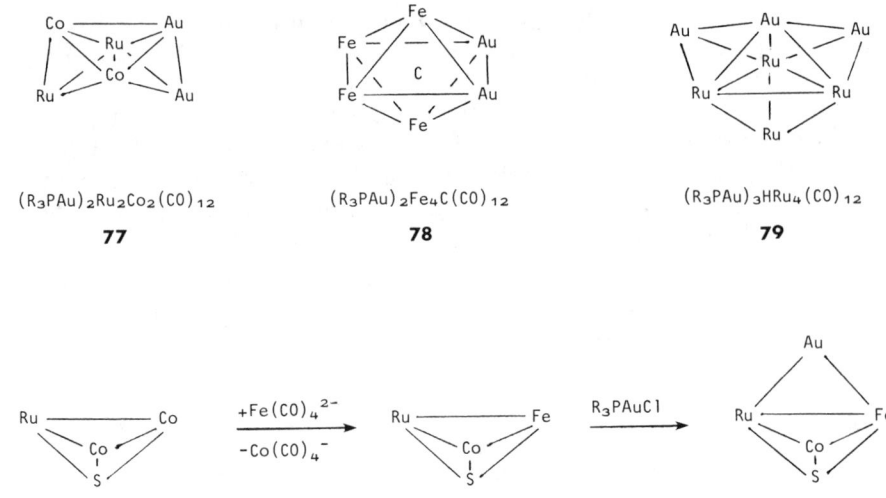

3. *Anionic Mononuclear Reagents*

The reaction of neutral clusters with anionic mononuclear compounds has not been nearly as fruitful as the method described previously. Considerable difficulties were found in reacting the $M_3(CO)_{12}$ clusters of Fe, Ru, and Os with $Mn(CO)_5^-$ and $Re(CO)_5^-$, due to formation of complicated product mixtures (*148*). In one case the simplest possible reaction product resulting from CO substitution by the metal nucleophile was obtained: $Ru_3(CO)_{12}$ and $Re(CO)_5^-$ yielded $ReRu_3(CO)_{16}^-$ which corresponds to **64**. Acidification of the product mixtures gave hydrido clusters like $HReOs_3(CO)_{15}$ (**65**) and $H_3MOs_3(CO)_{13}$ (M = Mn, Re). The latter should have a tetrahedral MOs_3 arrangement.

More successful were the reactions of $M_3(CO)_{12}$ (M = Fe, Ru, Os) with $Fe(CO)_4^{2-}$. They result in capping of the metal triangles to form the anionic clusters $FeM_3(CO)_{13}^{2-}$, or the corresponding hydrido clusters (*149, 150*). The reaction works for mixed-metal clusters also, and has yielded the tetrahedral Fe_2Ru_2, $FeRu_3$, $FeRu_2Os$, $FeRuOs_2$, and $FeOs_3$ carbonyls. Similarly $M_3(CO)_{12}$ (M = Ru, Os) reacts with $Co(CO)_4^-$ to yield $CoRu_3$, $CoRu_2Os$, $CoRuOs_2$, and $CoOs_3$ tetrahedral clusters (*151*). One of the cluster expansion reactions in rhodium chemistry has been found to proceed this way: $Rh_6(CO)_{16}$ adds $Rh(CO)_4^-$ to form $Rh_7(CO)_{16}^{3-}$ (*152*). It is not obvious why the nucleophile/electrophile combination between clusters and mononuclear compounds should work better one way, and it may have been only the greater number of experiments or available reagents which makes the use of anionic clusters appear more successful.

4. *Unsaturated or Labile Cluster Reagents*

Since cluster expansion normally requires removal of ligands this process should be facilitated by lack of ligands or ligand lability. Both assumptions are borne out by osmium cluster reactions. The unsaturated cluster $H_2Os_3(CO)_{10}$ reacts with several simple mononuclear organometallics (*56, 153–158*). In each case the mononuclear compound is the source of the capping fragment and a tetrahedral dihydrido cluster is formed. Initially assumed to work only for nucleophilic mononuclear compounds (*153*), the process has now been generalized for 14-electron fragments (*157*), and Os_3M clusters with M = Fe, Ru, Co, Rh, Ir, Ni, Pt, and Au have been obtained from $H_2Os_3(CO)_{10}$.

Further Os_3M clusters become available from labile $Os_3(CO)_{11}L$ or $Os_3(CO)_{10}L_2$ systems (L = Me_3N, MeCN, C_8H_{14}). The stepwise Os_3Re aggregation sequence **64** → **65** → **66** (*127*) proceeding this way has already been mentioned. Similarly, Os_3W capping occurs starting from $Os_3(CO)_{10}(NCMe)_2$ and $HWCp(CO)_3$ (*159*). The interesting carbyne reagent $Cp(CO)_2W\equiv CTol$ which is unsaturated itself adds to the Os_3 unit of $Os_3(CO)_{10}(C_8H_{14})_2$ to form **82**, the structure and electron count of which can be derived from a Os_3W tetrahedron or a Os_3WC trigonal bipyramid (*160*).

The cluster $RuCo_2(CO)_{11}$ (**47**) is quite labile, as was seen from its tendency for main group element capping leading to the clusters **48–52**. Its tendency for transition element capping is equally pronounced, which makes it react with $Co(CO)_4^-$ to form $RuCo_3(CO)_{12}^-$ and with $Ru(CO)_n$ fragments resulting from its own decomposition to form $Ru_2Co_2(CO)_{13}$ (*161*). It also adds $Cp(CO)_2W\equiv CTol$ with formation **83** similar to **82** (*107*). Cluster unsaturation, especially in the hidden form of cluster lability, here again proves to be a valuable source of basic cluster reactions.

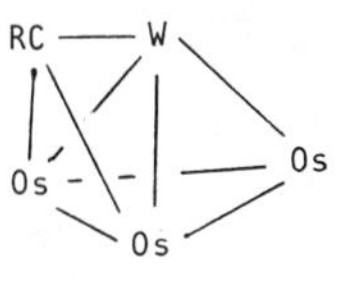

$Os_3WCp(CO)_{11}(CTol)$

82

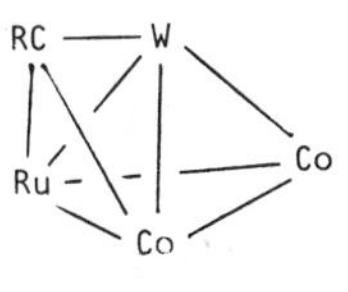

$RuCo_2WCp(CO)_{10}(CTol)$

83

5. *Mononuclear Fragment Reagents*

Since mechanistic information for cluster expansions is scarce it cannot be excluded that all such reactions described so far proceed via the addition of coordinatively unsaturated mononuclear complex fragments, even though they can be formulated differently. It is likely that most uncontrolled reactions changing the cluster nuclearity do proceed this way. However, there are a small number of simple cluster expansions which can be understood best by assuming intermediate fragments. This was already taken into account in the previous paragraph. It holds for the following reactions between clusters and simple complexes which do not bear an obvious center of reactivity.

The fragments $Fe(CO)_n$ ($n = 2$ or 3) and NiCp seem to be most easily attached to or removed from clusters since they appear in so many uncontrolled cluster syntheses. In a designed fashion $Fe(CO)_2$ units can be added to **84** and removed from the product **15a** in good yields (*61*). The interrelation of the two cluster types is obvious from their framework shapes. Similarly, cluster growth by $Fe(CO)_3$ fragment addition occurs in the sequence **85** → **86** → **87** (*162*). In both cases polynuclear iron carbonyls are the source of the $Fe(CO)_n$ fragments.

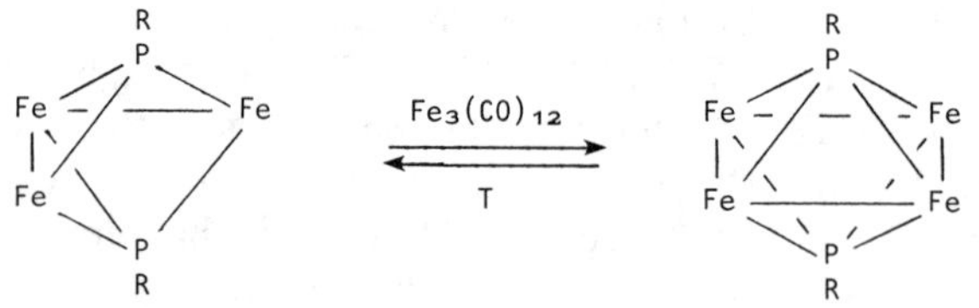

Fe₃(CO)₉(PR)₂

84

Fe₄(CO)₁₁(PR)₂

15a

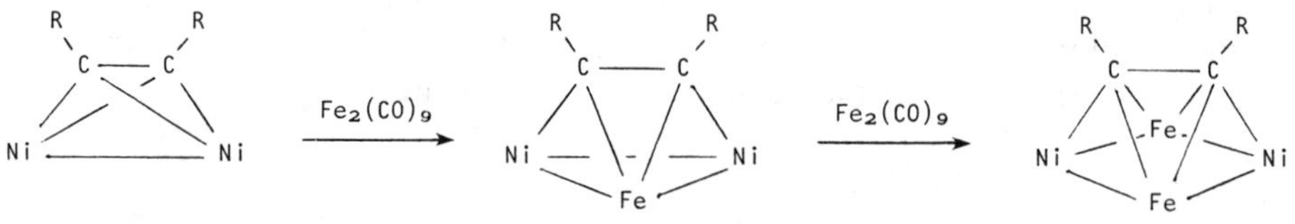

Ni₂Cp₂(R₂C₂)

85

Ni₂FeCp₂(CO)₃(R₂C₂)

86

Ni₂Fe₂Cp₂(CO)₆(R₂C₂)

87

The dinuclear complex (CpNiCO)₂ is a source of NiCp fragments. Their addition to clusters occurs with $Os_3(CO)_{12}$ to form **88** (*163*) or with the hydrocarbon bridged ruthenium clusters **89** to form **90** (*164*). The clusters

89 also add $Ru(CO)_3$ and FeCp fragments (*165*). A systematic consideration of the growth patterns of the higher osmium carbonyls leads to the conclusion that Os_1 and Os_2 fragments are involved in a rational way (*54, 62*). Such reactions are on the borderline between systematic and nonsystematic cluster chemistry. Their elucidation will require improvements in the understanding of metal–metal bond reactivity.

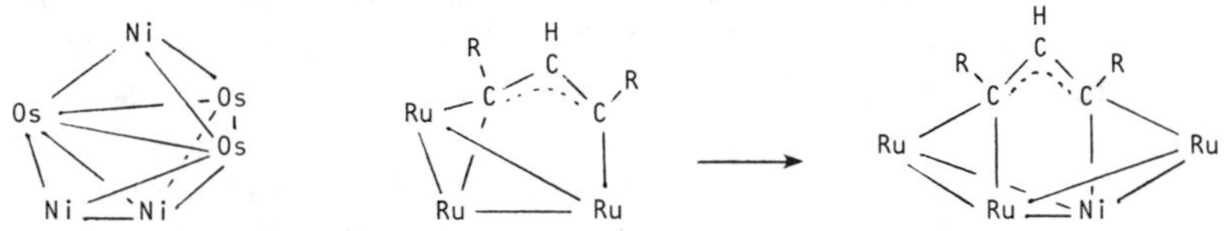

$$Os_3Ni_3Cp_3(CO)_9 \qquad HRu_3(CO)_9(C_6H_9) \qquad Ru_3Ni(CO)_8Cp(C_6H_9)$$

88   **89**   **90**

B. *Cluster Size Reduction*

This reaction, i.e., partial removal of metal units from clusters, should be favored by conditions opposing cluster expansion. Thus, an excess of donor ligands, conditions increasing the negative charge, formation of inert product clusters, or consumption of cluster fragments should direct interconversion toward smaller clusters. Some generalizations in this area can already be reached for ruthenium and osmium clusters (*54, 62*). Necessarily the trends in metal–metal bond strengths (*15*) and metal–metal bond scrambling equilibria (*166*) are of predictive power as to whether a certain cluster/fragment system will result in cluster expansion or degradation. One specific cause for cluster size reduction is the *in situ* chemical destruction of fragments which may result from cluster/fragment equilibria.

1. *Partial Degradation by Donor Ligands*

The total fragmentation of light transition metal clusters by donor ligands such as CO or PR_3 is commonplace; it is the reason for the use of such clusters as fragment sources. Partial metal framework opening by donor ligands was discussed in Section III,C. Removal of single fragments is also possible this way, using the appropriate starting clusters or reaction conditions.

A simple realization of this concept is the removal of weakly bound (i.e., light transition metal) capping units from a triangle of strongly bound (i.e., heavy transition metal) cluster atoms by means of CO. This way $H_2Os_3Co(CO)_{10}Cp$ is converted to $CpCo(CO)_2$ and $Os_3(CO)_{12}$ (*158*), or $H_2Os_3Rh(CO)_{10}(acac)$ is converted to $Rh(CO)_2(acac)$ and $H_2Os_3(CO)_{11}$ (*56*).

From each of the tetrahedral clusters $H_2FeRu_3(CO)_{13}$, $H_2FeRu_2Os(CO)_{13}$, $H_2FeRuOs_2(CO)_{13}$, and $HCoRu_3(CO)_{13}$ the light transition element is removed by CO leaving behind the corresponding $M_3(CO)_{12}$ compound (*167*).

Total degradation can also be avoided when the clusters involved belong to the Os, Ru, or Re cluster families with strong metal–metal bonds. The interconversion of the binary osmium carbonyls under CO has been investigated in detail (*75, 168, 169*). $Os_6(CO)_{18}$ in solution goes to $Os_5(CO)_{19}$, which in turn transforms to $Os_3(CO)_{12}$. Fragmentation products are $Os(CO)_5$ and $Os_2(CO)_9$, which can also be used in the opposite sense for osmium cluster expansion. Ruthenium clusters are more labile to total fragmentation, but in a controlled fashion $Ru_6C(CO)_{17}$ could be reduced to $Ru_5C(CO)_{15}$ (*111, 170*), and $HRu_4(CO)_{13}^-$ to $HRu_3(CO)_{11}^-$ (*171*) under CO. In rhenium cluster chemistry, $HRe(CO)_5$ seems to be a good leaving group. Its external attachment to the Os_3 cluster (see Section V,A,1) can be reversed with an excess of MeCN (*127*), and $H_4Re_4(CO)_{12}$ is fragmented under CO to $HRe(CO)_5$ and $H_3Re_3(CO)_{12}$ (*172*). So far, CO has been used to "extract" metal atoms from clusters. The applicability of other donor ligands for this purpose remains to be explored.

2. Oxidative Fragment Removal

Oxidation as a means of chemical fragment destruction is the method of choice to reduce carbido iron clusters in size. Normally ferric ion is used as an oxidant, and the clusters involved contain Fe_6C, Fe_5C, and Fe_4C skeletons (*134, 135, 173*). The starting clusters $Fe_6C(CO)_{16}^{2-}$, $Fe_5C(CO)_{14}^{2-}$, and $Fe_4C(CO)_{12}^{2-}$ are thus interrelated by Fe fragment additions and expulsions. Mixed iron clusters undergo the same degradations. For instance the hexanuclear systems **73** with $ML_n = Cr(CO)_3^{2-}$, $Mo(CO)_3^{2-}$, $W(CO)_3^{2-}$, and $Rh(CO)_2^-$ are fragmented to the appropriate pentanuclear systems **72** (*135*). And the Fe_5Au_2 cluster **91** is converted by oxidation to the Fe_4Au_2 cluster **78** (*146*).

The external rhenium unit is removed from **92** through iodine oxidation, and its place is taken by an iodide ligand in **93** (*174*). The reactivity of iodine towards clusters therefore can lie in one-electron oxidation (Section II), electrophilic attack (Section III,B), or partial degradation.

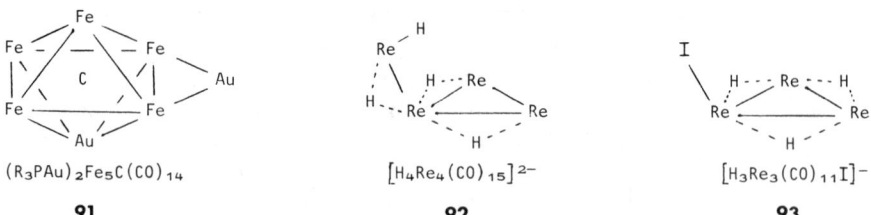

$(R_3PAu)_2Fe_5C(CO)_{14}$

91

$[H_4Re_4(CO)_{15}]^{2-}$

92

$[H_3Re_3(CO)_{11}I]^-$

93

3. Miscellaneous Fragmentation Reactions

Two further methods of fragment elimination seem to be generally applicable: controlled pyrolysis and hard base induced disproportionation. The thermal $Fe(CO)_2$ fragment removal from **15a** which completes the reversible interconversion of **84** and **15a** (*61*) has already been mentioned. Similarly the leaving group $HRe(CO)_5$ can be eliminated thermally from **92** yielding the stable unsaturated cluster $H_3Re_3(CO)_{10}^{2-}$ **13** (*175*).

Hydroxide induced disproportionation has been applied to osmium clusters whose stability again prevents total decomposition. In most cases one osmium atom is removed leaving behind an anionic cluster (*176, 177*). Thus, $Os_8(CO)_{22}$ converts into $Os_7(CO)_{20}^{2-}$, $Os_7(CO)_{21}$ goes to $Os_6(CO)_{18}^{2-}$, and $Os_6(CO)_{18}$ changes into $Os_5(CO)_{15}^{2-}$. All these species can be protonated to the corresponding hydrido clusters.

Controlled cluster degradation is more difficult than controlled expansion, due to the problem of preventing total degradation. For this reason few new clusters have been obtained as yet by this method. However, the mixed carbido iron clusters show possible future applications in providing a convenient entry via a cluster of an inexpensive metal which is partially degradated at the end of a synthesis sequence.

C. Metal Exchange

The combination of a cluster expansion reaction with removal of a mononuclear fragment results in a net metal exchange reaction retaining the overall cluster nuclearity. This simple process inevitably has a very complex mechanism if performed as a one-step synthesis. Nevertheless such reactions have been found for trinuclear clusters and systematized for three different kinds of reagents (*16, 178*).

The first observation enabling a designed metal exchange was made in trying to generate an external Co–M bond in the clusters **94** by CO elimination. Pyrolysis, however, resulted in elimination of the connecting Co–As unit and incorporation of the external M unit into the new trimetal clusters **95** (*179*). Subsequently, this reaction type was applied to $SFeCo_2(CO)_9$ and its $AsMe_2–M(CO)_3Cp$ attachment products **67** (*180*). Uncontrolled pyrolysis of these led to the metal exchange products **96**. However, changing the conditions allowed a stepwise reaction sequence which begins with a cluster expansion to form **25**, as described in Section V,A. Subsequently, the aggregation product **25** is partially opened and converted to **26**, as described in Section III,C. The open cluster **26** is then fragmented by CO with formation of **96** and elimination of $[(CO)_3Co–$

AsMe$_2$]$_x$ (*70*). The important steps in this sequence are the primary attachment of the incoming metal unit as in **94** or **67**, and the driving force supplied by the removal of the insoluble Co–As oligomer. Parts of the mechanism of this metal exchange process have thereby been elucidated and reconfirmed for similar sequences starting from RCCo$_3$, SRuCo$_2$, and RPFeCo$_2$ clusters (*71*). In each case it is an expansion/degradation sequence. Moreover, starting from the corresponding MCo$_2$ clusters, the chiral clusters **96** (*181*), **22** (*64*), and **97** (*182*) were obtained which could be separated into their pure enantiomers which have extreme optical properties.

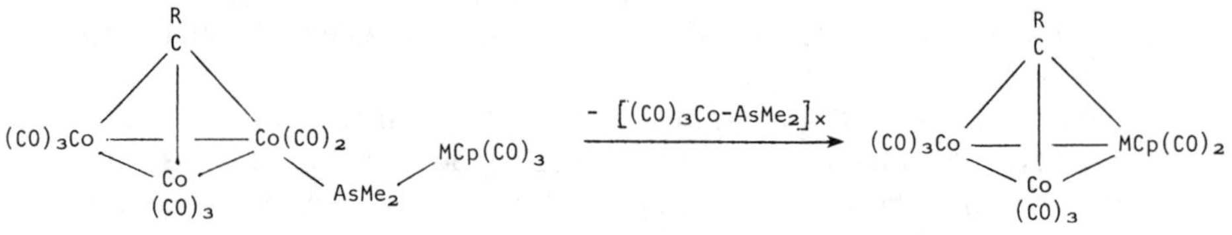

94 (M = Cr, Mo, W)   **95** (M = Cr, Mo, W)

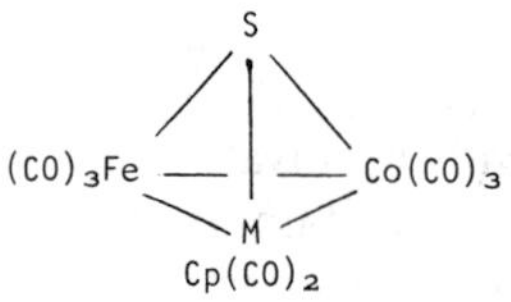

96 (M = Cr, Mo, W) 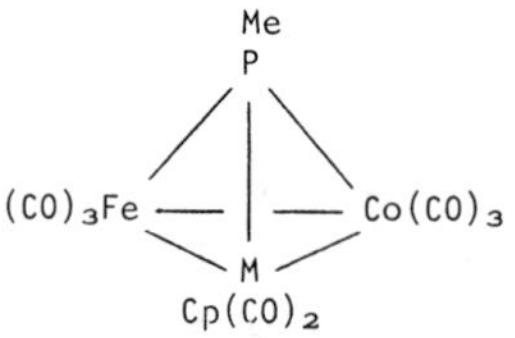   **22** (M = Mo, W) 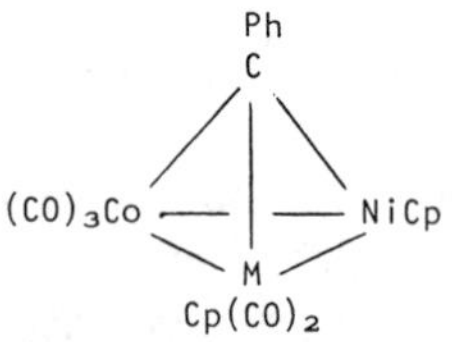   **97** (M = Mo, W)

The second type of metal exchange rests on the fact that mixtures of dinuclear metal–metal bonded complexes equilibrate easily with formation of mixed metal complexes (*166*). This equilibration can include clusters and is accompanied by CO transfer and replacement of Co(CO)$_3$ units of the clusters by other ML$_n$ units (*107, 182–184*). Useful dinuclear reagents for this purpose are [MoCp(CO)$_3$]$_2$, [WCp(CO)$_3$]$_2$, [FeCp(CO)$_2$]$_2$, and [NiCpCO]$_2$. They have been reacted with RCCo$_3$(CO)$_9$, RGeCo$_3$(CO)$_9$, SFeCo$_2$(CO)$_9$, SRuCo$_2$(CO)$_9$, (R$_2$C$_2$)RuCo$_2$(CO)$_9$, or RPFeCo$_2$(CO)$_9$, and mixed metal clusters thereby obtained. Typical exchange products are **98** and **99** which again have chiral frameworks. The mechanisms of these reactions could not be elucidated in as much detail as for the previous method, their yields are lower, sometimes product mixtures are formed, and the ability of the reagents to transfer ligands seems to play an important role (*182*).

$$
\begin{array}{ccc}
R\diagdown \ \ _C \underline{\quad} C\diagup^{R} & \underset{P}{\overset{t-Bu}{|}} & \overset{R}{\underset{C}{}} \\
(CO)_3Ru\underline{\quad}\diagup Co(CO)_3 & (CO)_3Fe\underline{\quad}\underline{\quad}Co(CO)_3 & (CO)_3Co\underline{\quad}\underline{\quad}Co(CO)_3 \\
Ni & Ni & Fe \\
Cp & Cp & H(CO)_3 \\
\mathbf{98} & \mathbf{99} & \mathbf{100}
\end{array}
$$

Just as simple dinuclear complexes can equilibrate (*166*), anionic mononuclear complexes can react with dinuclear complexes transferring their charge to the mononuclear fragments of those complexes ($2M^{\ominus} + M_2' \rightarrow M_2 + 2M'^{\ominus}$) (*185*). This metal exchange redox process should also be applicable to clusters and, as before, should be accompanied by ligand exchange in order to incorporate the previously anionic mononuclear units into the cluster. A few reactions have been observed which can be considered that way. Thus, one of the products from $Os_3(CO)_{12}$ and $Mn(CO)_5^-$ is $MnOs_2(CO)_{12}^-$, and from $Os_3(CO)_{12}$ and $Re(CO)_5^-$ after acidification $HReOs_2(CO)_{12}$ was obtained (*148*). A similar procedure was used systematically to prepare the clusters **100** from $RCCo_3(CO)_9$ and $Fe(CO)_4^{2-}$, expelling $Co(CO)_4^-$ (*186*), and the above-mentioned sequence **48** → **80** → **81** involves such a metal exchange step (*147*). Recently, it was found that electron transfer catalysis allows a nearly quantitative introduction of $CpM(CO)_2$ (M = Mo, W) units into $RCCo_3(CO)_9$ (*187*), $RGeCo_3(CO)_9$, and $RPFeCo_2(CO)_9$ (*188*), by reacting these clusters with $[CpM(CO)_3]_2$ or $CpM(CO)_3^-$ in the presence of catalytic amounts of benzophenone ketyl. Since such reactions depend on the redox potentials of the clusters and the mononuclear reagents it should be possible to develop a synthetic strategy by arranging cluster fragments and reagents according to their displaceability.

Cluster synthesis has not reached the stage of sophistication and predictability as the synthesis of hydrocarbon frameworks. Systematic cluster expansion, degradation, and metal exchange, however, are reactions which are much more likely to work for clusters than for hydrocarbons owing to the inherent lability of metal complexes. This is one of the clearest demonstrations of the uniqueness and the future potential of cluster chemistry.

VI

CLUSTER FRAMEWORK REARRANGEMENTS

The 18-electron rule generally governs the bonding in clusters with up to five metal atoms. The cluster core shape, i.e., the number of metal–metal bonds, can be deduced from the number of electrons provided by the metal atoms and the ligands, with the only reservation that the atoms of Rh, Ir, Pd, Pt, and Au sometimes require only 16 or 14 electrons. Clusters with five, six, and sometimes seven metal atoms can often be treated in terms of skeletal electron counting (*189, 190*), describing the frameworks by the nido, closo, and capping formalisms. Some clusters with six, several clusters with seven, and all clusters with more metal atoms (*122, 191*) do not show obvious relations between their structure and electron count. In these cases packing possibilities and ligand stereochemistries are optimized in a yet unpredictable way. The breakdown of the 18-electron rule beyond nuclearity five means that from here on a metal–metal connectivity no longer corresponds to a two-electron interaction, and localized bonding no longer exists. This, in turn, means that alternative metal frameworks with different connectivities can be of comparable stability and are able to exist for compounds with identical electron counts. For example at least four different types of closed metal polyhedra exist for hexanuclear clusters: **101** (octahedron), **102** (trigonal prism), **103** (capped trigonal bipyramid), and **104** (capped tetragonal pyramid), which can be adopted by clusters with the same electron counts. For each of these frameworks clusters with at least two different electron counts are known.

101   **102**   **103**   **104**

On this basis cluster framework rearrangements can be expected and understood for clusters with six or more metal atoms. So far they have been observed in chemical reactions of the osmium carbonyls (*54, 62*). Anionic $Os_6(CO)_{18}^{2-}$ and its protonation product $HOs_6(CO)_{18}^-$ are octahedral, as are $HRu_6(CO)_{18}^-$ and $H_2Ru_6(CO)_{18}$. The second protonation, however, is accompanied by an unpredictable framework change: $H_2Os_6(CO)_{18}$ has the core shape **104** (*192*). A similar rearrangement from a bicapped octahedron to a linked tetrahedral arrangement occurs upon conversion of $Os_8(CO)_{22}^{2-}$ to $HOs_8(CO)_{22}^-$ (*193*). Moreover, one and the same compound, $H_2Os_7(CO)_{20}$, exists in two isomers, one of which has been structurally characterized as containing an edge bridged bicapped tetrahedron (*128*).

Extensive framework rearrangements, i.e., cluster core fluxionality, has been found in rhodium and platinum clusters where it could be detected by metal atom NMR. The P-centered cluster $Rh_9P(CO)_{21}^{2-}$ has a capped square antiprism of rhodium atoms in the solid state, according to the 4:4:1 pattern in its ^{103}Rh NMR spectrum at $-90°C$. At 25°C it shows only one Rh NMR signal indicating total rhodium atom scrambling, which can also be concluded from the decet ^{31}P NMR signal at room temperature (194, 195). The same observations of low temperature rigidity and high temperature fluxionality were made for $Rh_{10}S(CO)_{22}^{2-}$, $Rh_{10}P(CO)_{22}^{3-}$, and $Rh_{10}As(CO)_{22}^{3-}$, all of which have a bicapped square antiprism of rhodium atoms in the solid state (196). A different kind of scrambling was found by ^{195}Pt NMR for the series of $Pt_{3n}(CO)_{6n}^{2-}$ clusters **105**–**108** with n = 2, 3, 4, or 5. The triangles of platinum atoms rotate independently of one another around the molecular threefold axis even at $-100°C$, and in mixtures of the Pt_9, Pt_{12}, and/or Pt_{15} compounds intermolecular exchange of $Pt_3(CO)_6$ units takes place (197). Whereas metal–metal connections are broken and re-formed in these cases, more subtle changes happen during the movement of the ligand sphere relative to the cluster core, the well-known ligand fluxionality. When, as has been discussed, the Fe_3 triangle moves inside the ligand polyhedron of $Fe_3(CO)_{12}$ (9), the unequally long Fe–Fe bonds of this triangle have to be interchanged simultaneously. And from ligand NMR spectroscopy a similar breathing of the metal core was deduced during the fluxional processes of $H_2FeRu_3(CO)_{13}$, $H_2FeRu_2Os(CO)_{13}$, $H_2FeRuOs_2(CO)_{13}$, and phosphine derivatives thereof (198, 199). Variable lengths of metal–metal bonds are an expression of their relative weakness (15). Cluster breathing, fluxionality, and framework changes are a unique extension of this property.

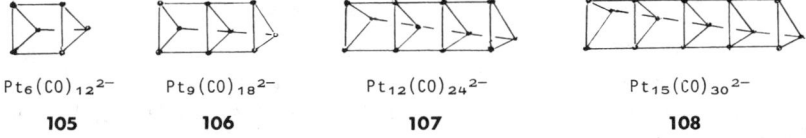

$Pt_6(CO)_{12}^{2-}$	$Pt_9(CO)_{18}^{2-}$	$Pt_{12}(CO)_{24}^{2-}$	$Pt_{15}(CO)_{30}^{2-}$
105	**106**	**107**	**108**

The framework rearrangements considered so far do not involve a change in the electron count. This normally means that the number of donor ligands does not change. Many ligands, however, still have unshared electron pairs, from which it follows that by changing their orientation they can donate further electrons. If this does not involve substitution of other ligands it changes the cluster electron count and thereby the core geometry. This has already been mentioned for the protonation induced Fe_4 rearrangements **9** → **10** and **11** → **12**, where CO ligands reorient to donate oxygen electron pairs. It is an underlying principle of the **43**/**44** interconversions where a bridging SR group changes between three- and five-electron donation. It is likely to be involved in the cluster openings by reagents which use their primary function to be externally attached to the cluster, e.g., before the formation of **36** or **59**. Also there seems to be one example of a reversible interconversion of this type without using a reagent. The cluster $Os_3(CO)_9S(NSiMe_3)$ with an open structure like **58** with two Os–Os bonds is in equilibrium at high temperatures with a species in which according to ^{13}C

NMR all three osmium atoms are chemically equivalent (*200*). This may involve changing the bridging sulfur atom from a four-electron donor to a two-electron donor in order to gain a third Os–Os bond and an equilateral metal triangle. These ligand induced cluster rearrangements are a further demonstration of metal–metal bond weakness as the reason for cluster mobility. For application-oriented purposes they may be the most important ones.

VII

LITTLE OR UNEXPLORED AREAS

In the preceding sections of this article, a systematization of basic cluster reactions has been attempted based on the present status of knowledge. It can be seen that the number of reactions reported is still relatively small for certain reaction types. The organization of the material described here, which mostly follows reactivity patterns, may not be the best one when more facts are available. An alternative way of presentation is outlined below, based on a cluster-centered viewpoint. It brings out which areas might show future developments or deserve research activities.

1. Reactions affecting single metal–metal bonds: From the knowledge of metal–metal bond reactivity (*15*) several reaction types which are common for dinuclear complexes can be expected to occur for single metal–metal bonds in clusters. Some of these, such as nucleophilic opening (cf. Section III,C) or oxidative additions with metal–metal bond breaking (cf. Section IV,C), have been observed. Others, such as thermal or photolytic bond breaking with subsequent reactions, reductive bond breaking, or insertion reactions, have yet to be developed.

2. Reactions affecting the cluster as a whole: Here the electron reservoir properties of clusters, the attack of reagents at several metal atoms, and changes in the ligand number without localized metal–metal bonding changes are to be considered, some of which, like the capping reactions (cf. Section IV,B), have been discussed above. Though different types of reactions of clusters with nucleophiles have been found, of the many electrophiles only the proton and the R_3PAu unit could be attached to the cluster core, despite the fact that many mono- and dinuclear complexes have been observed to be good nucleophiles toward other reagents. Similarly, the redox chemistry of clusters has not yet been developed past its physicochemical and exploratory stage. There are hardly any known nondestructive chemical redox reactions of clusters other than electron transfers. The modification of cluster–ligand bonding or ligand reactivity due to cluster redox reactions seems to be unknown. Unsaturated or "excess-electron" clusters are analogous to oxidized or reduced clusters. Few such clusters are known, but their number is growing slowly. From the examples given in Section III,B, it is clear that unsaturated clusters can be made by ligand elimination, and they should have a rich chemistry, as should excess-electron clusters.

3. Designed use of cluster fragments: It has been shown that dinuclear organometallic complexes are in equilibrium with their mononuclear fragments (*166*). That many clusters also detach or attach fragments is obvious from systematic cluster growth or shrinkage reactions (cf. Section V), as well as from the many degradations or uncontrolled nuclearity changes. The development of cluster chemistry has now reached a stage where the designed use of cluster fragments is becoming possible. Dinuclear fragments are involved in the growth patterns of the osmium carbonyls (*75, 169*), or in the conversion of $Co_6C(CO)_{15}^{2-}$ to $Co_8C(CO)_{18}^{2-}$ by $Co_4(CO)_{12}$ (*201*). Unsaturated mononuclear fragments like PtL_2 or $Cp(CO)_2W\equiv CR$ have been used very successfully in designed cluster syntheses (*202, 203*). Moreover, the generation of more than mononuclear fragments from clusters is becoming common, e.g., in the spontaneous fission of $Co_4(CO)_8(PPh_3)_4$ into two halves (*204*). The existence of intermediate dinuclear fragments is likely in the interconversion of $MeCCo_3(CO)_9$ to $(Me_2C_2)Co_4(CO)_{10}$ under the influence of $Mn(CO)_5$ radicals, the interconversion of $MePFeCo_2(CO)_9$ to $(MeP)_2Fe_2Co_2(CO)_{11}$ under the influence of CpNi fragments, and the interconversion of $RPCo_3(CO)_9$ to $(RP)_2Co_4(CO)_{10}$ upon chemical oxidation (*205*). In none of these cases, however, has the source of the dinuclear fragments been usable for more than one reaction.

Reactions between clusters also belong to this area. The easy interconversions in the $Pt_{3n}(CO)_{6n}^{2-}$ series are a fascinating example (*197, 206*). The synthesis of $Rh_9(CO)_{19}^{3-}$ from $Rh_4(CO)_{11}^{2-}$ and $Rh_5(CO)_{15}^{-}$ (*207*), of $Rh_{12}(CO)_{30}^{2-}$ from $Rh_6(CO)_{15}^{2-}$ and $Rh_6(CO)_{16}$ (*208*), and the splitting of $Rh_{12}(CO)_{30}^{2-}$ into two $Rh_5(CO)_{15}^{-}$ units under CO pressure (*209*) are other examples. Cluster growth and shrinkage in small steps are rather easy to understand; in big steps like these they should be much more efficient.

4. Hetero site reactivity: The simplest difference between a cluster and a mononuclear complex is that the cluster can do two or more things where a mononuclear complex can do one. Simple as it is, this difference has hardly ever been verified other than in multiple ligand substitutions. One verification is hetero site reactivity, i.e., different modes of reaction at different sites on one and the same cluster. Two examples of this appear to exist. Different phosphine ligands substitute CO on different metal atoms in $H_2FeRu_3(CO)_{13}$ depending on their size and basicity (*210*), and $Ru_2Co_2(CO)_{13}$ reacts with H_2 at the ruthenium atoms (cf. Section IV,A) and with acetylenes at the cobalt atoms (cf. Section IV,C) (*82*). More such reactions should be possible, especially for mixed metal clusters, and the often-stressed ability of clusters to activate more than one substrate for application purposes should drive the search for such reactivity patterns.

5. *Clusters with inorganic ligands:* As mentioned in the Section I, this review has attempted to present an inorganic chemist's view of cluster chemistry. Nevertheless, all clusters discussed have been organometallic species with the "application-oriented" organic ligand sphere being an integral part of the factors determining reactivity. The investigation of clusters having purely inorganic ligands should add considerable variety to the subject. So far for inorganic clusters, many of which are solid state compounds, only structure and bonding systematics have been developed. This synthesis is much more an art than is the preparation of organometallic clusters, and reactivity investigations are scarce, but some attractive possibilities are emerging (*211, 212*). It is to be expected that not only will organometallic cluster chemistry develop rapidly, but inorganic cluster chemistry will follow in its path.

ACKNOWLEDGMENTS

The work of the author's group described here was done by a team of hard-working and dedicated co-workers whose names are given in the references. It was made possible by the University of Freiburg, the Fonds der Chemischen Industrie, the Deutsche Forschungsgemeinschaft, and the Wissenschaftliche Gesellschaft Freiburg.

REFERENCES

1. K. F. Purcell and J. C. Kotz, "Inorganic Chemistry," Chapter 18. Saunders, Philadelphia, Pennsylvania, 1977.
2. F. A. Cotton and G. Wilkinson, "Advanced Inorganic Chemistry," 4th ed. Chapter 26. Wiley, New York, 1980.
3. G. Schmid, R. Pfeil, R. Boese, F. Bandermann, S. Meyer, G. H. M. Calis, and J. W. A. van der Felden, *Chem. Ber.* **114**, 3634 (1981).
4. D. Seyferth, *Adv. Organomet. Chem.* **14**, 98 (1976).
5. H. Vahrenkamp, *Struct. Bonding (Berlin)* **32**, 1 (1977).
6. A. J. Deeming, *in* "Transition Metal Clusters" (B. F. G. Johnson, ed.), p. 391. Wiley, New York, 1980.
7. J. S. Bradley, G. B. Ansell, and E. W. Hill, *J. Am. Chem. Soc.* **101**, 7417 (1979).
8. W. A. Herrmann, *Adv. Organomet. Chem.* **20**, 159 (1982).
9. B. F. G. Johnson and R. E. Benfield, *in* "Transition Metal Clusters" (B. F. G. Johnson, ed.), p. 471. Wiley, New York, 1980.
10. M. Moskovits, *Acc. Chem. Res.* **12**, 229 (1979).
11. E. L. Muetterties, T. N. Rhodin, E. Band, C. F. Brucker, and W. R. Pretzer, *Chem. Rev.* **79**, 91 (1979).

12. G. Huttner, *Nachr. Chem., Tech. Lab.* **27**, 261 (1979).
13. E. L. Muetterties, *J. Organomet. Chem.* **200**, 177 (1980).
14. J. A. Connor, *in* "Transition Metal Clusters" (B. F. G. Johnson, ed.), p. 345, Wiley, New York, 1980.
15. H. Vahrenkamp, *Angew. Chem.* **90**, 403 (1978); *Angew. Chem., Int. Ed. Engl.* **17**, 379 (1978).
16. H. Vahrenkamp, *Philos. Trans. R. Soc. London* **308**, 17 (1982).
17. B. F. G. Johnson, J. Lewis, and D. Pippard, *J. Organomet. Chem.* **160**, 263 (1978).
18. G. R. John, B. F. G. Johnson, and J. Lewis, *J. Organomet. Chem.* **181**, 143 (1979).
19. W. L. Gladfelter and G. L. Geoffroy, *Adv. Organomet. Chem.* **18**, 207 (1980).
20. C. D. Garner, *in* "Transition Metal Clusters" (B. F. G. Johnson, ed.), p. 265, Wiley, New York, 1980.
21. R. H. Holm, *Philos. Trans. R. Soc. London* **308**, 159 (1982).
22. P. A. Dawson, B. M. Peake, B. H. Robinson, and J. Simpson, *Inorg. Chem.* **19**, 465 (1980).
23. T. J. Meyer, *Prog. Inorg. Chem.* **19**, 1 (1975).
24. T. Madach and H. Vahrenkamp, *Chem. Ber.* **114**, 505 (1981).
25. T. Madach and H. Vahrenkamp, unpublished.
26. U. Honrath, B. M. Peake, B. H. Robinson, J. Simpson, and H. Vahrenkamp, *Organometallics* (in press).
27. A. M. Bond, P. A. Dawson, B. M. Peake, P. H. Rieger, B. H. Robinson, and J. Simpson, *Inorg. Chem.* **18**, 1413 (1979).
28. A. M. Bond, B. M. Peake, B. H. Robinson, J. Simpson, and D. J. Watson, *Inorg. Chem.* **16**, 410 (1977).
29. C. E. Strouse and L. F. Dahl, *Discuss. Faraday Soc.* **47**, 93 (1969).
30. H. Beurich, T. Madach, F. Richter, and H. Vahrenkamp, *Angew. Chem.* **91**, 751 (1979); *Angew. Chem., Int. Ed. Engl.* **18**, 690 (1979).
31. B. M. Peake, P. H. Rieger, B. H. Robinson, and J. Simpson, *Inorg. Chem.* **20**, 2540 (1981).
32. R. Fisel and R. Hoffmann, quoted in ref. *31*.
33. M. S. Paquette and L. F. Dahl, *J. Am. Chem. Soc.* **102**, 6621 (1980).
34. H. Vahrenkamp, V. A. Uchtman, and L. F. Dahl, *J. Am. Chem. Soc.* **90**, 3272 (1968).
35. P. D. Frisch and L. F. Dahl, *J. Am. Chem. Soc.* **94**, 5082 (1972).
36. L. R. Byers, V. A. Uchtman, and L. F. Dahl, *J. Am. Chem. Soc.* **103**, 1942 (1981).
37. J. J. Maj, A. D. Rae, and L. F. Dahl, *J. Am. Chem. Soc.* **104**, 3054 (1982).
38. T. Toan, W. P. Fehlhammer, and L. F. Dahl, *J. Am. Chem. Soc.* **94**, 3389 (1972).
39. G. L. Simon and L. F. Dahl, *J. Am. Chem. Soc.* **95**, 2164 (1973).
40. T. Toan, B. K. Teo, J. A. Ferguson, T. J. Meyer, and L. F. Dahl, *J. Am. Chem. Soc.* **99**, 408 (1977).
41. T. H. Lemmen, J. A. Kocal, F. Yip-Kwai Lo, M. W. Chen, and L. F. Dahl, *J. Am. Chem. Soc.* **103**, 1932 (1981).
42. C. R. Eady, B. F. G. Johnson, and J. Lewis, *J. Chem. Soc., Chem. Commun.* p. 302 (1976).
43. A. P. Humphries and H. D. Kaesz, *Prog. Inorg. Chem.* **25**, 145 (1979).
44. A. J. Deeming, B. F. G. Johnson, and J. Lewis, *J. Chem. Soc. A* p. 2967 (1970); M. Green, R. M. Mills, G. N. Pain, F. G. A. Stone, and P. Woodward, *J. Chem. Soc., Dalton Trans* p. 1321 (1982).
45. R. G. Teller and R. Bau, *Struct. Bonding (Berlin)* **44**, 1 (1981).
46. B. Delley, M. C. Manning, D. E. Ellis, J. Berkowitz, and W. C. Trogler, *Inorg. Chem.* **21**, 2247 (1982).
47. D. F. Shriver, *J. Organomet. Chem.* **94**, 259 (1975).

48. F. Richter and H. Vahrenkamp, *Angew. Chem.* **90**, 474 (1978); *Angew. Chem., Int. Ed. Engl.* **17**, 444 (1978).
49. B. F. G. Johnson, J. Lewis, D. Pippard, and P. R. Raithby, *J. Chem. Soc., Chem. Commun.* p. 551 (1978).
50. H. A. Hodali, D. F. Shriver, and C. A. Ammlung, *J. Am. Chem. Soc.* **100**, 5239 (1978).
51. E. M. Holt, K. Whitmire, and D. F. Shriver, *J. Chem. Soc., Chem. Commun.* pp. 778, 780 (1980); P. A. Dawson, B. F. G. Johnson, J. Lewis, and P. R. Raithby, *ibid.* p. 781.
52. E. M. Holt, K. H. Whitmire, and D. F. Shriver, *J. Am. Chem. Soc.* **104**, 5621 (1982).
53. M. Manassero, M. Sansoni, and G. Longoni, *J. Chem. Soc., Chem. Commun.* p. 919 (1976).
54. B. F. G. Johnson and J. Lewis, *Adv. Inorg. Chem. Radiochem.* **24**, 225 (1981).
55. L. J. Farrugia, J. A. K. Howard, P. Mitrprachachon, F. G. A. Stone, and P. Woodward, *J. Chem. Soc., Dalton Trans.* p. 162 (1981).
56. L. J. Farrugia, J. A. K. Howard, P. Mitrprachachon, F. G. A. Stone, and P. Woodward, *J. Chem. Soc., Dalton Trans.* p. 171 (1981).
57. G. Ciani, G. d'Alfonso, M. Freni, P. Romiti, and A. Sironi, *J. Organomet. Chem.* **186**, 353 (1980).
58. A. J. Deeming and S. Hasso, *J. Organomet. Chem.* **114**, 313 (1976).
59. J. R. Shapley, J. B. Keister, M. R. Churchill, and B. G. DeBoer, *J. Am. Chem. Soc.* **97**, 4145 (1975).
60. H. Vahrenkamp and D. Wolters, *Organometallics*, **1**, 874 (1982).
61. H. Vahrenkamp, E. J. Wucherer, and D. Wolters, *Chem. Ber.* **116**, 1219 (1983).
62. J. Lewis and B. F. G. Johnson, *Pure Appl. Chem.* **54**, 97 (1982).
63. G. Huttner, J. Schneider, H. D. Müller, G. Mohr, J. von Seyerl, and L. Wohlfahrt, *Angew. Chem.* **91**, 82 (1979); *Angew. Chem., Int. Ed. Engl.* **18**, 76 (1979).
64. M. Müller and H. Vahrenkamp, *Chem. Ber.* **116**, 8 (1983).
65. J. Schneider, L. Zsolnai, and G. Huttner, *Chem. Ber.* **115**, 989 (1982).
66. M. Müller and H. Vahrenkamp, *Chem. Ber.* **116**, 2311 (1983).
67. R. Ball, M. J. Bennett, E. H. Brooks, W. A. G. Graham, J. K. Hoyano, and S. M. Illingworth, *J. Chem. Soc., Chem. Commun.* p. 592 (1970).
68. D. A. Lesch and T. B. Rauchfuss, *Organometallics* **1**, 499 (1982).
69. A. J. Deeming, B. F. G. Johnson, and J. Lewis, *J. Chem. Soc. A* p. 897 (1970).
70. F. Richter and H. Vahrenkamp, *Organometallics* **1**, 756 (1982).
71. F. Richter, H. Beurich, M. Müller, N. Gärtner, and H. Vahrenkamp, unpublished.
72. J. S. Field, R. J. Haines, and D. N. Smit, *J. Organomet. Chem.* **224**, C49, (1982).
73. J. S. Field, R. J. Haines, D. N. Smit, K. Natarajan, O. Scheidsteger, and G. Huttner, *J. Organomet. Chem.* **240**, C23 (1982).
74. G. R. John, B. F. G. Johnson, J. Lewis, W. J. Nelson, and M. McPartlin, *J. Organomet. Chem.* **171**, C14 (1979).
75. D. H. Farrar, B. F. G. Johnson, J. Lewis, P. R. Raithby, and M. J. Rosales, *J. Chem. Soc., Dalton Trans.* p. 2051 (1982).
76. R. J. Goudsmit, B. F. G. Johnson, J. Lewis, P. R. Raithby, and K. H. Whitmire, *J. Chem. Soc., Chem. Commun.* p. 640 (1982).
77. D. Braga, B. F. G. Johnson, J. Lewis, J. M. Mace, M. McPartlin, J. Puga, W. J. H. Nelson, P. R. Raithby, and K. H. Whitmire, *J. Chem. Soc., Chem. Commun.* p. 1081 (1982).
78. A. Winter, L. Zsolnai, and G. Huttner, *J. Organomet. Chem.* **232**, 47 (1982).
79. D. G. Farrar, P. G. Jackson, B. F. G. Johnson, J. Lewis, W. J. H. Nelson, M. D. Vargas, and M. McPartlin, *J. Chem. Soc., Chem. Commun.* p. 1009 (1981).
80. R. Poilblanc, *Inorg. Chim. Acta* **62**, 75 (1982).
81. R. B. Calvert and J. R. Shapley, *J. Am. Chem. Soc.* **100**, 7726 (1978).
82. E. Roland and H. Vahrenkamp, *Organometallics* **2**, 183 (1983).

83. H. C. Foley and G. L. Geoffroy, *J. Am. Chem. Soc.* **103**, 7176 (1981).
84. G. N. van Buren, A. C. Willis, F. W. B. Einstein, L. K. Peterson, R. K. Pomeroy, and D. Sutton, *Inorg. Chem.* **20**, 4361 (1981).
85. B. F. G. Johnson, J. Lewis, and D. A. Pippard, *J. Chem. Soc., Dalton Trans.* p. 407 (1981).
86. M. Tachikawa and J. R. Shapley, *J. Organomet. Chem.* **124**, C19 (1977).
87. C. R. Eady, B. F. G. Johnson, and J. Lewis, *J. Chem. Soc., Dalton Trans.* p. 838 (1977).
88. A. J. Deeming, R. Ettorre, B. F. G. Johnson, and J. Lewis, *J. Chem. Soc. A* pp. 1797, 2701 (1971).
89. C. C. Yin and A. J. Deeming, *J. Chem. Soc., Dalton Trans.* p. 1013 (1974).
90. G. Süss-Fink, *Z. Naturforsch., B: Anorg. Chem., Org. Chem.* **35B**, 454 (1980).
91. F. Iwasaki, M. J. Mays, P. R. Raithby, P. L. Taylor, and P. J. Wheatley, *J. Organomet. Chem.* **213**, 185 (1981).
92. K. Natarajan, L. Zsolnai, and G. Huttner, *J. Organomet. Chem.* **220**, 365 (1981).
93. G. Ciani, G. d'Alfonso, M. Freni, P. Romiti, and A. Sironi, *J. Organomet. Chem.* **226**, C31 (1982).
94. N. M. Boag, C. E. Kampe, Y. C. Lin, and H. D. Kaesz, *Inorg. Chem.* **21**, 1706 (1982).
95. A. J. Deeming and M. Underhill, *J. Chem. Soc., Dalton Trans.* p. 2727 (1973).
96. B. F. G. Johnson, J. Lewis, and D. A. Pippard, *J. Organomet. Chem.* **213**, 249 (1981).
97. R. Bau, B. Don, R. Greatrex, R. J. Haines, R. A. Love, and R. D. Wilson, *Inorg. Chem.* **14**, 3021 (1975).
98. T. A. Creswell, J. A. K. Howard, F. G. Kennedy, S. A. R. Knox, and H. Wadepohl, *J. Chem. Soc., Dalton Trans.* p. 2220 (1981).
99. G. Huttner, J. Schneider, G. Mohr, and J. von Seyerl, *J. Organomet. Chem.* **191**, 161 (1980).
100. K. Natarajan, O. Scheidsteger, and G. Huttner, *J. Organomet. Chem.* **221**, 301 (1981).
101. G. Ciani, G. d'Alfonso, M. Freni, P. Romiti, and A. Sironi, *J. Organomet. Chem.* **219**, C23 (1981).
102. R. Bonfichi, G. Ciani, G. d'Alfonso, P. Romiti, and A. Sironi, *J. Organomet. Chem.* **231**, C35 (1982).
103. H. Vahrenkamp, *Angew. Chem.* **87**, 363 (1975); *Angew. Chem., Int. Ed. Engl.* **14**, 322 (1975).
104. E. W. Abel, T. Blackmore, and R. J. Whitley, *Inorg. Nucl. Chem. Lett.* **10**, 941 (1974).
105. O. Gambino, G. Cetini, E. Sappa, and M. Valle, *J. Organomet. Chem.* **20**, 195 (1969); J. F. Blount, L. F. Dahl, C. Hoogzand, and W. Hübel, *J. Am. Chem. Soc.* **88**, 292 (1966).
106. A. D. Clauss, J. R. Shapley, and S. R. Wilson, *J. Am. Chem. Soc.* **103**, 7387 (1981).
107. E. Roland and H. Vahrenkamp, *Organometallics* **2**, 8 (1983).
108. R. D. Adams and Z. Dawoodi, *J. Am. Chem. Soc.* **103**, 6510 (1981).
109. B. F. G. Johnson, J. Lewis, and P. A. Kilty, *J. Chem. Soc. A* p. 2859 (1968); N. Cook, L. Smart, and P. Woodward, *J. Chem. Soc., Dalton Trans.* p. 1744 (1977).
110. R. D. Adams, N. M. Golembeski, and J. P. Selegue, *Organometallics* **1**, 240 (1982).
111. I. A. Oxton, D. B. Powell, D. H. Farrar, B. F. G. Johnson, J. Lewis, and J. N. Nicholls, *Inorg. Chem.* **20**, 4302 (1981).
112. J. R. Moss and W. A. G. Graham, *J. Organomet. Chem.* **18**, P24 (1969).
113. R. D. Adams and D. A. Katahira, *Organometallics* **1**, 53 (1982).
114. P. V. Broadhurst, B. F. G. Johnson, and J. Lewis, *J. Chem. Soc., Dalton Trans.* p. 1881. (1982).
115. A. J. Deeming, B. F. G. Johnson, and J. Lewis, *J. Organomet. Chem.* **17**, P40 (1969).
116. A. Winter, L. Zsolnai, and G. Huttner, *J. Organomet. Chem.* **234**, 337 (1982).
117. U. Krüerke and W. Hübel, *Chem. Ber.* **94**, 2829 (1961); L. F. Dahl and D. L. Smith, *J. Am. Chem. Soc.* **84**, 2450 (1962).

118. J. R. Fox, W. L. Gladfelter, G. L. Geoffroy, T. Tavanaiepour, S. Abdel-Mequid, and V. W. Day, *Inorg. Chem.* **20**, 3230 (1981).

119. D. H. Farrar, G. R. John, B. F. G. Johnson, J. Lewis, P. R. Raithby, and M. J. Rosales, *J. Chem. Soc., Chem. Commun.* p. 886 (1981).

120. S. Aime, private communication.

121. E. Sappa, A. Tiripicchio, and M. Tiripicchio Camellini, *J. Chem. Soc., Dalton Trans.* p. 419 (1978).

122. P. Chini, *J. Organomet. Chem.* **200**, 37 (1980).

123. J. S. Bradley, *J. Am. Chem. Soc.* **101**, 7419 (1979).

124. C. W. Bradford, W. van Bronswijk, R. J. H. Clark, and R. S. Nyholm, *J. Chem. Soc. A* p. 2889 (1970).

125. R. D. Adams, I. T. Horvath, and B. E. Segmüller, *Organometallics* **1**, 1537 (1982).

126. J. R. Shapley, G. A. Pearson, M. Tachikawa, G. E. Schmidt, M. R. Churchill, and F. J. Hollander, *J. Am. Chem. Soc.* **99**, 8064 (1977); M. R. Churchill and F. J. Hollander, *Inorg. Chem.* **16**, 2493 (1977).

127. M. R. Churchill, F. J. Hollander, R. A. Lashewycz, G. A. Pearson, and J. R. Shapley, *J. Am. Chem. Soc.* **103**, 2430 (1981); M. R. Churchill and F. J. Hollander, *Inorg. Chem.* **20**, 4124 (1981).

128. E. J. Ditzel, H. D. Holden, B. F. G. Johnson, J. Lewis, A. Saunders, and M. J. Taylor, *J. Chem. Soc., Chem. Commun.* p. 1373 (1982).

129. F. Richter and H. Vahrenkamp, *Angew. Chem.* **91**, 566 (1979); *Angew. Chem., Int. Ed. Engl.* **18**, 531 (1979).

130. W. Hieber and E. H. Schubert, *Z. Anorg. Allg. Chem.* **338**, 32 (1965).

131. M. J. Mays, P. R. Raithby, P. L. Taylor, and K. Henrick, *J. Organomet. Chem.* **224**, C45 (1982).

132. A. Fumagalli, G. Longoni, P. Chini, A. Albinati, and S. Brückner, *J. Organomet. Chem.* **202**, 329 (1980).

133. T. L. Hall and J. R. Ruff, *Inorg. Chem.* **20**, 4444 (1981).

134. M. Tachikawa, A. C. Sievert, E. L. Muetterties, M. R. Thompson, C. S. Day, and V. Day, *J. Am. Chem. Soc.* **102**, 1726 (1980).

135. M. Tachikawa, R. L. Geerts, and E. L. Muetterties, *J. Organomet. Chem.* **213**, 11 (1981).

136. J. W. Lauher and K. Wald, *J. Am. Chem. Soc.* **103**, 7648 (1981).

137. E. Roland, K. Fischer, and H. Vahrenkamp, *Angew. Chem.* **95**, 324 (1983); *Angew. Chem. Int. Ed. Engl.* **22**, 326 (1983).

138. P. Braunstein, J. Rose, Y. Dusausoy, and J. P. Mangeot, *C. R. Hebd. Seances Acad. Sci. Serie II* **294**, 967 (1982).

139. M. Green, K. A. Mead, R. M. Mills, I. D. Salter, F. G. A. Stone, and P. Woodward, *J. Chem. Soc., Chem. Commun.* p. 51 (1982).

140. L. W. Bateman, M. Green, J. A. K. Howard, K. A. Mead, R. M. Mills, I. D. Salter, F. G. A. Stone, and P. Woodward, *J. Chem. Soc., Chem. Commun.* p. 773 (1982).

141. M. I. Bruce and B. K. Nicholson, *J. Chem. Soc., Chem. Commun.* p. 1141 (1982).

142. B. F. G. Johnson, D. A. Kaner, J. Lewis, and P. R. Raithby, *J. Organomet. Chem.* **215**, C33 (1981).

143. B. F. G. Johnson, D. A. Kaner, J. Lewis, P. R. Raithby, and M. J. Taylor, *J. Chem. Soc., Chem. Commun.* p. 314 (1982).

144. B. F. G. Johnson, D. A. Kaner, J. Lewis, P. R. Raithby, and M. J. Taylor, *Polyhedron* **1**, 105 (1982).

145. B. F. G. Johnson, D. A. Kaner, J. Lewis, P. R. Raithby, and M. J. Rosales, *J. Organomet. Chem.* **231**, C59 (1982).

146. B. F. G. Johnson, D. A. Kaner, J. Lewis, and M. J. Rosales, *J. Organomet. Chem.* **238**, C73 (1982).

147. K. Fischer and H. Vahrenkamp, unpublished.
148. J. Knight and M. J. Mays, *J. Chem. Soc., Dalton Trans.* p. 1022 (1972).
149. G. L. Geoffroy and W. L. Gladfelter, *J. Am. Chem. Soc.* **99**, 7565 (1977).
150. F. Takusagawa, A. Fumagalli, T. F. Koetzle, G. R. Steinmetz, R. P. Rosen, W. L. Gladfelter, G. L. Geoffroy, M. A. Bruck, and R. Bau, *Inorg, Chem.* **20**, 3823 (1981).
151. P. C. Steinhardt, W. L. Galdfelter, A. D. Harley, J. R. Fox, and G. L. Geoffroy, *Inorg. Chem.* **19**, 332 (1980).
152. S. Martinengo and P. Chini, *Gazz. Chim. Ital.* **102**, 344 (1972).
153. L. J. Farrugia, J. A. K. Howard, P. Mitrprachachon, J. L. Spencer, F. G. A. Stone, and P. Woodward, *J. Chem. Soc., Chem. Commun.* p. 260 (1978).
154. E. W. Burckhardt and G. L. Geoffroy, *J. Organomet. Chem.* **198**, 179 (1980).
155. J. S. Plotkin, D. L. Alway, C. R. Weisenberger, and S. G. Shore, *J. Am. Chem. Soc.* **102**, 6156 (1980).
156. M. R. Churchill, C. Bueno, W. L. Hsu, J. S. Plotkin, and S. G. Shore, *Inorg. Chem.* **21**, 1958 (1982).
157. S. G. Shore, W. L. Hsu, C. R. Weisenberger, M. L. Castle, M. R. Churchill, and S. Bueno, *Organometallics*, **1**, 1405 (1982).
158. J. Lewis, R. B. A. Pardy, and P. R. Raithby, *J. Chem. Soc., Dalton Trans.* p. 1509 (1982).
159. M. R. Churchill, F. J. Hollander, J. R. Shapley, and D. S. Foose, *J. Chem. Soc., Chem. Commun.* p. 534 (1978).
160. M. Green, J. C. Jeffery, S. J. Porter, H. Razay, and F. G. A. Stone, *J. Chem. Soc., Dalton Trans.* 2475 (1982).
161. E. Roland and H. Vahrenkamp, *Angew. Chem.* **93**, 714 (1981); *Angew. Chem., Int. Ed. Engl.* **20**, 679 (1981).
162. M. I. Bruce, J. R. Rodgers, M. R. Snow, and F. S. Wong, *J. Organomet. Chem.* **240**, 299 (1982).
163. E. Sappa, M. Lanfranchi, A. Tiripicchio, and M. Tiripicchio Camellini, *J. Chem. Soc., Chem. Commun.* p. 995 (1981).
164. D. Osella, E. Sappa, A. Tiripicchio, and M. Tiripicchio, *Inorg. Chim. Acta* **34**, L289 (1979).
165. S. Aime and D. Osella, *Inorg. Chim. Acta* **57**, 207 (1982).
166. T. Madach and H. Vahrenkamp, *Chem. Ber.* **113**, 2675 (1980).
167. J. R. Fox, W. L. Gladfelter, and G. L. Geoffroy, *Inorg. Chem.* **19**, 2574 (1980).
168. D. H. Farrar, B. F. G. Johnson, J. Lewis, J. N. Nicholls, and P. R. Raithby, and M. J. Rosales, *J. Chem. Soc., Chem. Commun.* p. 273 (1981).
169. J. N. Nicholls, D. H. Farrar, P. F. Jackson, B. F. G. Johnson, and J. Lewis, *J. Chem. Soc., Dalton Trans.* p. 1395 (1982).
170. D. H. Farrar, P. F. Jackson, B. F. G. Johnson, J. Lewis, J. N. Nicholls, and M. McPartlin, *J. Chem. Soc., Chem. Commun.* p. 415 (1981).
171. J. C. Bricker, C. C. Nagel, and S. G. Shore, *J. Am. Chem. Soc.* **104**, 1444 (1982).
172. R. B. Saillant, G. Barcelo, and H. D. Kaesz, *J. Am. Chem. Soc.* **92**, 5739 (1970).
173. M. Tachikawa and E. L. Muetterties, *J. Am. Chem. Soc.* **102**, 4541 (1980).
174. G. Ciani, G. d'Alfonso, M. Freni, P. Romiti, and A. Sironi, *J. Organomet. Chem.* **220**, C11 (1981).
175. A. Bertolucci, M. Freni, P. Romiti, G. Ciani, A. Sironi, and V. G. Albano, *J. Organomet. Chem.* **113**, C61 (1976).
176. C. R. Eady, J. J. Guy, B. F. G. Johnson, J. Lewis, M. C. Malatesta, and G. M. Sheldrick, *J. Chem. Soc., Chem. Commun.* p. 807 (1976).
177. R. Glynn, B. F. G. Johnson, and J. Lewis, *J. Organomet. Chem.* **169**, C9 (1979).
178. H. Vahrenkamp, *in* "Transition Metal Chemistry" (A. Müller and E. Diemann, eds.), p. 35. Verlag Chemie, Weinheim, 1981.

179. H. Beurich and H. Vahrenkamp, *Angew. Chem.* **90**, 915 (1978); *Angew. Chem., Int. Ed. Engl.* **17**, 863 (1978); *Chem. Ber.* **115**, 2385 (1982).

180. F. Richter and H. Vahrenkamp, *Angew. Chem.* **90**, 916 (1978); *Angew. Chem., Int. Ed. Engl.* **17**, 864 (1978); *Chem. Ber.* **115**, 3224 (1982).

181. F. Richter and H. Vahrenkamp, *Chem. Ber.* **115**, 3243 (1982).

182. H. Beurich, R. Blumhofer, and H. Vahrenkamp, *Chem. Ber.* **115**, 2409 (1982).

183. H. Vahrenkamp, D. Steiert, and P. Gusbeth, *J. Organomet. Chem.* **209**, C17 (1981).

184. M. Müller and H. Vahrenkamp, *Chem. Ber.* **116**, 8 (1983).

185. R. E. Dessy and L. A. Bares, *Acc. Chem. Res.* **5**, 415 (1972).

186. R. A. Epstein, H. W. Withers, and G. L. Geoffroy, *Inorg. Chem.* **18**, 942 (1979).

187. B. H. Robinson, private communication.

188. P. Gusbeth, U. Honrath, and H. Vahrenkamp, unpublished.

189. K. Wade, *Adv. Inorg. Chem. Radiochem.* **18**, 1 (1976).

190. K. Wade, *in* "Transition Metal Clusters" (B. F. G. Johnson, ed.), p. 193. Wiley, New York, 1980.

191. P. R. Raithby, *in* "Transition Metal Clusters" (B. F. G. Johnson, ed.), p. 5. Wiley, New York, 1980.

192. M. McPartlin, C. R. Eady, B. F. G. Johnson, and J. Lewis, *J. Chem. Soc., Chem. Commun.* p. 883 (1976).

193. D. Braga, K. Henrick, B. F. G. Johnson, J. Lewis, M. McPartlin, W. J. Nelson, and M. D. Vargas, *J. Chem. Soc., Chem. Commun.* p. 419 (1982).

194. O. A. Gansow, D. S. Gill, F. J. Bennis, J. R. Hutchinson, J. L. Vidal, and R. C. Schoening, *J. Am. Chem. Soc.* **102**, 2449 (1980).

195. J. L. Vidal, W. E. Walker, R. L. Pruett, and R. C. Schoening, *Inorg. Chem.* **18**, 129 (1979).

196. L. Garlaschelli, A. Fumagalli, S. Martinengo, B. T. Heaton, D. O. Smith, and L. Strona, *J. Chem. Soc., Dalton Trans.* p. 2265 (1982).

197. C. Brown, B. T. Heaton, A. D. C. Towl, G. Longoni, A. Fumagalli, and P. Chini, *J. Organomet. Chem.* **181**, 233 (1979).

198. W. L. Gladfelter and G. L. Geoffroy, *Inorg. Chem.* **19**, 2579 (1980).

199. W. L. Gladfelter, J. R. Fox, J. A. Smegal, T. G. Wood, and G. L. Geoffroy, *Inorg. Chem.* **20**, 3223 (1981).

200. G. Süss-Fink, U. Thewalt, and H. P. Klein, *J. Organomet. Chem.* **224**, 59 (1982).

201. V. G. Albano, P. Chini, G. Ciani, M. Sansoni, D. Strumolo, B. Heaton, and S. Martinengo, *J. Am. Chem. Soc.* **98**, 5027 (1976).

202. F. G. A. Stone, *Acc. Chem. Res.* **14**, 318 (1981).

203. F. G. A. Stone, in press.

204. R. Hug and A. J. Poë, *J. Organomet. Chem.* **226**, 277 (1982).

205. H. Beurich, U. Honrath, M. Müller, and H. Vahrenkamp, unpublished.

206. J. C. Calabrese, L. F. Dahl, P. Chini, G. Longoni, and S. Martinengo, *J. Am. Chem. Soc.* **96**, 2614 (1974).

207. S. Martinengo, A. Fumagalli, R. Bonfichi, G. Ciani, and A. Sironi, *J. Chem. Soc., Chem. Commun.* p. 825 (1982).

208. V. G. Albano, A. Ceriotti, P. Chini, G. Ciani, S. Martinengo, and M. Anker, *J. Chem. Soc., Chem. Commun.* p. 859 (1975).

209. B. T. Heaton, L. Strona, J. Jonas, T. Eguchi, and G. A. Hoffman, *J. Chem. Soc., Dalton Trans.* p. 1159 (1982).

210. J. R. Fox, W. L. Gladfelter, T. G. Wood, J. A. Smegal, T. K. Foreman, G. L. Geoffroy, T. Tavanaiepour, V. W. Day, and C. S. Day, *Inorg. Chem.* **20**, 3214 (1981).

211. K. Jödden and H. Schäfer, *Z. Anorg. Allg. Chem.* **430**, 5 (1977).

212. R. E. McCarley, *Philos. Trans. R. Soc. London* **308**, 141 (1982).

ADVANCES IN ORGANOMETALLIC CHEMISTRY, VOL. 22

Metal Isocyanide Complexes

ERIC SINGLETON and
HESTER E. OOSTHUIZEN*

National Chemical Research Laboratory
Council for Scientific and Industrial Research
Pretoria, Republic of South Africa

I

INTRODUCTION

It is over eight years since the last comprehensive review appeared on metal isocyanide chemistry. In the interim, reviews have appeared on specific aspects of isocyanide chemistry. Lippard reviewed seven and eight coordination in molybdenum isocyanide compounds (*1*), Yamamoto reviewed metal(0)–isocyanide complexes (*2*), and in related reviews on carbene complexes by Cotton and Lukehart (*3*), Lappert *et al.* (*4*), and Casey (*5*), mention was made of carbenes synthesized from metal isocyanides.

The surveys by Treichel (1973) (*6*) and Bonati and Minghetti (1974) (*7*)

---

* Previously Hester E. Swanepoel.

represent the last reasonably comprehensive coverage of metal – isocyanide complexes. Since that time over seven hundred articles have appeared on some aspect of metal isocyanide chemistry. Any reviewer trying to review such a mass of material is automatically faced with the dilemma of correlation and critical evaluation versus coverage. From our experience, however, both requirements have equal importance, and as a consequence we have attempted to strike a balance in this article between evaluating the work and the present need to have some comprehensive reference source on isocyanides covering the years 1973 – 1981. To keep this review to a suitable length, however, references to studies where isocyanides played only a minor chemical role have been omitted.

Some features have emerged from the large number of studies that have been undertaken over the last eight years and consequently can be highlighted.

The striking advance since 1973 has been the synthesis of homoleptic isocyanides for vanadium and almost all of the post-group-VIA metals up to group IB (see Table I), and this has been helped in some respects by synthetic rationale. For example, the discovery of the simple reductive or nonreductive cleavage of multiple metal – metal bonds in suitable dinuclear complexes of molybdenum and tungsten has provided a high yield route to $W(CNPh)_6$ and $[M(CNR)_7]^{2+}$ (M = Mo, W) and related complexes (8 – 14). During these studies the series $[M(CNR)_7]^{2+}$ was completed with the isola-

TABLE I

FULLY CHARACTERIZED HOMOLEPTIC METAL – ISOCYANIDE COMPLEXES

Group V	Group VI	Group VII	Group VIII			Group IB
$V(CNR)_6^+$	$Cr(CNR)_6$	$Mn(CNR)_6^+$	$Fe(CNR)_5$	$Co_2(CNR)_8$	$Ni(CNR)_4$	$Cu(CNR)$
	$Cr(CNR)_6^+$		$Fe_2(CNR)_9$	$Co(CNR)_5^+$	$Ni_4(CNR)_7$	
	$Cr(CNR)_6^{2+}$			$Co(CNR)_5^{2+}$	$Ni(CNR)_4^{2+}$	
	$Cr(CNR)_7^{2+}$			$Co_2(CNR)_{10}^{2+}$		
	$Mo(CNR)_6$		$Ru(CNR)_5$	$Rh(CNR)_4^+$	$Pd_3(CNR)_6$	$Ag(CNR)$
	$Mo(CNR)_7^{2+}$		$Ru_2(CNR)_9$	$Rh_2(CNR)_8^{2+}$	$[Pd_2(CNR)_6]^{2+}$	
			$Ru_2(CNR)_{10}^{2+}$	$Rh_2(CNRNC)_4^{2+a}$	$[Pd_3(CNR)_8]^{2+}$	
					$[Pd(CNR)_4]^{2+}$	
	$W(CNR)_6$	$Re(CNR)_6^+$	$Os(CNR)_5$	$Ir(CNR)_4^+$	$Pt_3(CNR)_6$	$Au(CNR)$
	$W(CNR)_7^{2+}$		$Os_2(CNR)_{10}^{2+}$		$Pt_7(CNR)_{12}$	
					$Pt(CNR)_4^{2+}$	
					$[PdPt(CNR)_6]^{2+}$	
					$[Pt_2(CNR)_6]^{2+}$	

^{a} CNRNC, chelating diisocyano-ligand.

tion of $[Cr(CNR)_7]^{2+}$ (*15*), which represents the only example of a seven-co-ordinate isocyanide complex of a first-row transition metal.

The sodium-amalgam reduction of metal–halo-isocyanide complexes was used by Stone and co-workers to synthesize $M(CNR)_5$ (M = Fe, Ru) (*16*), and $M_2(CNR)_9$ (M = Fe, Ru) (*17*) complexes and later extended by Swanepoel (*18*), Green (*19*), and Yamamoto (*20*) to the synthesis of $Os(CNXylyl)_5$, $Co_2(CNBu^t)_8$, and $Pt_7(CNXylyl)_{12}$, respectively.

Ligand displacement from labile precursors has led to the homoleptic clusters $Ni_4(CNR)_7$, $Ni_4(CNR)_6$ and $Ni_8(CNBu^t)_{12}$ (*21, 22*), $Pt_3(CNR)_6$ (*23*), and $Pt_7(CNXylyl)_{12}$ (*20*), which represent rare examples of polynuclear homoleptic metal(0)–isocyanide clusters.

Naturally, the ideal source of starting materials for homoleptic metal isocyanide compounds is via metal carbonyl complexes, but previously only with the two carbonyls $Ni(CO)_4$ (*24*) and $Co_2(CO)_8$ (*25*) has direct substitution of all carbonyl groups been effected. Recently, however, re-markable discoveries by Coville and co-workers (*26–31*) on the transition-metal-catalyzed substitution of carbonyl groups in monomeric and cluster compounds have shown that $Fe(CNR)_5$, $Mo(CNR)_6$, and $Ir_4(CO)_5(CNR)_7$ (*32*) can be prepared in high yield by stepwise substitution from the parent carbonyl.

Some interesting chemistry has appeared relating to the ability of the isocyanide ligand to stabilize unusual oxidation states. A series of palladium metal–metal bonded complexes has been synthesized by redox reactions involving two metal complexes in different formal oxidation states (*33–35*). Similar ruthenium(I) and osmium(I) dimers have been prepared by an unusual homolytic fission of a ruthenium–carbon bond (*36*) or by single-electron oxidation of $Os(CNXylyl)_5$ (*18*).

Metal–metal interactions in rhodium(I) isocyanide oligomers have been extensively studied by Gray *et al.* (*37, 38*), and these compounds are showing rich photochemical properties. As a consequence, they are proving ideal for the study of the photochemical behavior of metal complexes possessing low-lying metal–ligand charge-transfer (MLCT) excited states.[1]

---

[1] Abbreviations: arene, η^6-benzene or substituted benzene derivative; bipy, 2,2′-bipyridyl; Bu^i, Bu^n, Bu^t, *iso-*, *n-*, or *tert*-butyl; COD, 1,5-cyclo-octadiene; Cp, η^5-C_5H_5; DAD, dimethyl-acetylene dicarboxylate; dam, 1,2-bis(diphenylarsino)methane; DBA, dibenzylideneacetone; DMF, *N,N*-dimethylformamide; dpe, 1,2-bis(diphenylphosphino)ethane; dpen, *cis*-1,2-bis(di-phenylphosphino)ethylene; dpm, 1,2-bis(diphenylphosphino)methane; ESR, electron spin res-onance; F_6-acac, hexafluoroacetylacetone; FN, fumaronitrile; MA, maleic anhydride; Me, methyl; MLCT, metal ligand charge transfer; phen, 1,10-phenanthroline; Pr^i, Pr^n, *iso-* or *n*-propyl; py, pyridine; RT, room temperature; TCNE, tetracyanoethylene; tetraphos, $(Ph_2PCH_2CH_2)_3P$; THF, tetrahydrofuran; Xylyl, 2,6-$Me_2C_6H_3$.

Transition metal complexes which have low-lying MLCT are playing a role in demonstrating the use of molecular excited states in light-energy conversion processes.

Facile isocyanide insertion reactions into metal–carbon, –nitrogen, –sulfur, –oxygen, –hydride, and –halide bonds have been found to readily occur. The insertion into metal–hydrides to give stable formimidines is particularly noteworthy since corresponding formyls (—CHO) are exceptionally difficult to synthesize and tend to be very unstable. There is a great deal of interest in carbon monoxide reductions, and the instability of the intermediate reduction products has made a study of the reduction process extremely difficult. Recently, however, the interaction of isocyanides with zirconium hydrides has allowed the isolation of the individual reduction steps of the isocyanide which has provided a model study for carbon monoxide reduction (*39*).

What are the portents for the future? It will be interesting to see if metal–isocyanides undergo photochemical dissociation without decomposition to the metal–cyanide. This could further help to establish the relationship between specific photochemical reactivity and electronic structure and coordination environment about the metal atom.

Further studies are anticipated on metal–metal bonded oligomers possessing low-lying MLCT excited states in efforts to maximize the efficiency of light-energy conversion processes. Also in this context, linear oligomers containing metals in different oxidation states or different metals could exhibit anisotropic electrical behavior as one-dimensional conductors, so it would be of interest to devise synthetic routes to compounds of this type. One method that has been recently applied in an attempt to increase electrical conductance in the crystal has been by coupling the organic anion 7,7,8,8-tetracyano-*p*-quinodimethane (TCNQ) with the cations $[Pd(CNMe)_4]^{2+}$ (*40*) and $[Rh(CNR)_4]^+$ (*41*), or by synthesizing the salts $[Pt(CNR)_4][Pt(CN)_4]$ (*42*). These attempts met with limited success.

There are no routes yet to homoleptic metal isocyanide anions. If one considers the interesting products obtained from methyl iodide additions to molybdenum (*43*) and manganese (*44*) carbonyl isonitrile anions, negatively charged isocyanide complexes should have some interesting chemistry. Also, now that a simple route to $[CpFe(CNR)_2]_2$ complexes has been devised (*45*), the synthesis of the anion $[CpFe(CNR)_2]^-$ could provide a route to a range of products including heterometal–metal bonded systems.

Isocyanide ligands appear to be radical probes and hence activate or initiate radical reactivity. For example, the addition of TCNE to $[Co(CNMe)_5]^+$ is radical in nature (*46*) and homolytic fission of the metal–carbon bond in $[(\eta^1,\eta^2\text{-}C_8H_{13})Ru(CNR)_4]^+$ complexes has been observed (*36*), and it is possible that radical stabilization can occur through metal–

isocyanide π^* interactions which will give highly reactive isocyanide ligands. Thus, the generation of isocyanide radical species through homolytic metal–metal bond fission or by oxidation or reduction processes on diamagnetic isocyanide complexes in the presence of suitable substrates could lead to interesting reactivity patterns.

More emphasis should be placed in future endeavors on routes to metal–isocyanide clusters, since those of nickel (21) and platinum (47–49) are proving to have a rich chemistry.

Finally, the recent synthesis of a series of two-dimensional isocyanide coordination polymers containing rhodium(I) could lead to the development of a new area of metal-containing polymers (50–52).

II

STRUCTURE AND BONDING

The generally accepted valence bond and molecular orbital (MO) approach to the bonding of metal isocyanides has been well described in Treichel's review (6), and has been used to rationalize (i) variations in IR stretching frequencies between bonded and nonbonded isocyanides, and (ii) the better π-acceptor qualities of aryl versus alkyl isocyanide groups (53, 54). In valence bond theory the canonical forms involved in metal isocyanide bonds are

$$\overset{-}{M}-C\equiv\overset{+}{N}-R \longleftrightarrow M=C=N\overset{\diagup R}{\underset{\cdot\cdot}{}} \longleftrightarrow \overset{+}{M}-C=\overset{-}{N}-R$$

$$\quad\quad A \quad\quad\quad\quad\quad\quad B \quad\quad\quad\quad\quad\quad C$$

However, up to Treichel's review no complexes containing bent isocyanide ligands had been observed, and it was suggested by Cotton *et al.* that an MO approach was more appropriate to isocyanide bonding since the MO theory of metal–carbon π–bond formation does not require the C–N–R angle to be other than 180° (24). The MO description of CO and CNR is given in Fig. 1, and to account for nonlinearity, π bonding would be required to occur preferentially to one of the two degenerate π^* orbitals. Equivalent retrodative π^* bonding to both π^* orbitals would be cylindrically symmetric and would not have any effect on isocyanide geometry.

One of the most significant advances in isocyanide bonding in the last eight years has been the isolation of the complexes $M(CNR)_2(L_2)_2$ {M = Mo, W; L_2 = dpe, $(PMe_2Ph)_2$} (55, 56) and $M(CNR)_5$(M = Fe, Ru; R = alkyl) (16) containing bent monohapto isocyanide ligands. In valence

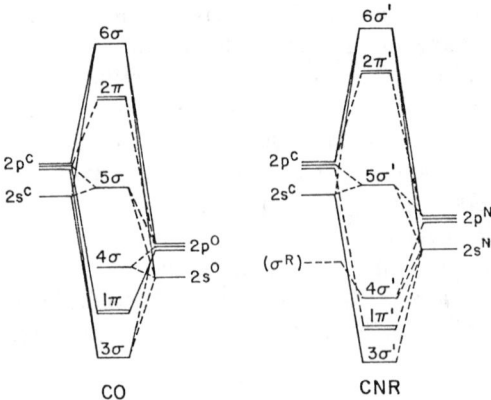

Fig. 1. The MO description of CO and CNR (6).

bond terms this means that a substantial contribution to the electronic structure is made by the bent canonical form B, but an adequate quantitative description has not been made at this stage.

Further evidence for the canonical form B comes from the reactivity patterns of $M(CNR)_2(L_2)_2$ with protons and alkylating agents (57, 58), which parallel the reactivities of bridging isocyanide ligands in Cp_2Fe_2 (μ-CNR)$_2$(CO)$_2$ in preferentially attacking the nitrogen atom. Surprisingly, no similar reaction patterns were observed with the bent isocyanide ligands in the $M(CNR)_5$ complexes.

Representative new types of bonding modes that have been observed are as follows: with the isocyanide symmetrically bridging three cobalt atoms through the isocyanide carbon (1) (59); with the isocyanide bridging two

(1)

metal atoms through lone-pair donation from both the isocyanide carbon and nitrogen atoms as in $Ni_4(CNR)_7$ (2) (60); in the manganese complex $Mn_2(\mu\text{-}CNC_6H_4Me\text{-}p)(dpm)_2(CO)_4$ (3) (61); and containing isocyanide li-

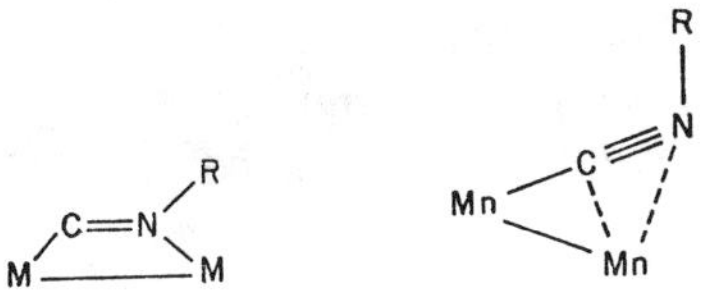

(2)   (3)

gands bonded to three osmium (4) (62) and five ruthenium (5) atoms (63).

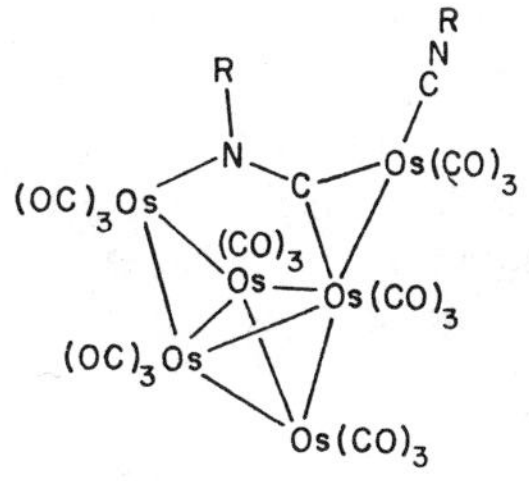

(R = C₆H₄Me-p)

(4) 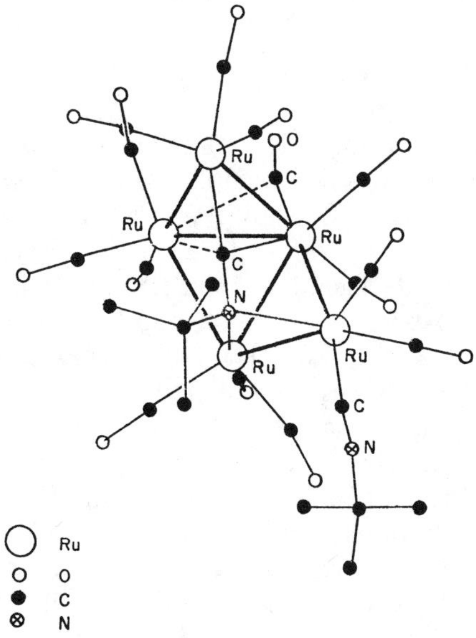   (5)

A comparison with carbonyl bonding modes can be made for symmetrical μ^3-bridging, and for asymmetrical μ^2-bridging (6) in $Mn_2(\mu\text{-}CO)(dpm)_2(CO)_4$ (64).

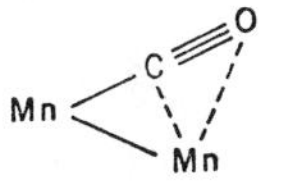

(6)

Dimeric and cluster compounds containing isocyanides exhibit dynamic behavior in solution. As there is special interest in dynamic processes owing to the possible role ligand migration may play in cluster catalysis, two systems are worthy of mention here. In the one, the substituted products $Cp_2Fe_2(CO)_{4-x}(CNR)_x(x = 1, 2)$ have played a pivotal role in formulating the Cotton–Adams rules for ligand migration in the $Cp_2Fe_2(CO)_4$ system (65), and in the other case the compound $Ir_4(CO)_{11}CNBu^t$ has provided an insight into the general question of exchange mechanisms in $M_4(CO)_{12}$ clusters (66).

III

SYNTHESIS OF ISOCYANIDE LIGANDS

A. From Metal Cyanides

Though alkylation of metal cyanides is one of the oldest routes to metal–isocyanide complexes, at the present time the usefulness of this method is confined to (i) partially characterizing new metal–cyanide complexes, (ii) providing access to complexes containing unstable or unusual isocyanide ligands which cannot be prepared by direct interactions of complex with isocyanide, and (iii) providing a route to chiral metal–isocyanide complexes. The following examples exemplify this.

During investigations into routes to new cyanide complexes of manganese the formation of the new cyanide salt $K_3[(C_5H_4Me)Mn(CN)_3]$ was inferred by the isolation of $K_2[(C_5H_4Me)Mn(CN)_2(CNEt)]$ and $K[(C_5H_4Me)Mn(CN)(CNEt)_2]$ from $[Et_3O]BF_4$ additions to an inseparable mixture of reaction products (67).

Complexes containing the unstable hydrogen isocyanide ligand have been isolated from phosphoric acid additions to ruthenium and manganese cyanide precursors and as a by-product from arene diazonium salt reactions on $Cr(CO)_5CN^-$. The products formed were $(C_5H_4Me)Mn(CO)_2CNH$ (67), $[CpRu(PPh_3)_2CNH]PF_6$ (68), and $Cr(CO)_5CNH$ (69). [Compare earlier preparations of $CpMn(CO)_2CNH$ (70) and $[Cp_2W(OEt)CNH]PF_6$ (71).]

Alkylation of the cyanide ligands in $Ag_4M(CN)_8$ with RX (X = halides) has provided the only route to the complexes $M(CN)_4(CNR)_4$ (M = Mo, W; R = alkyl, allyl, CH_2Ph, CPh_3) (72, 73), as direct reaction of isocyanides on the cyanide anions has given only reduced products (74).

Complexes containing unusual isocyanide ligands have evolved from attempts to alkylate the anions $[M(CN)_6]^{4-}$ (M = Fe, Ru, Os) with $[Et_3O]BF_4$ in acetone solution. The compounds isolated $[M\{CNCMe_2 CH_2COMe\}_6](BF_4)_2$ resulted from an initial acid-catalyzed aldol condensation of the acetone solvent followed by a nucleophilic attack of the carbonium ion of the dimer on the coordinated cyanide ligands. Analogous reactions of $K_4[Fe(CN)_6]$ with ethyl methyl ketone and cyclohexanone gave $[Fe\{CNC(R^1R^2)CHR^3COR^4\}_6](BF_4)_2$ [R^1 = Me, R^3 = H, R^2 = R^4 = Et; R^1R^2 = $(CH_2)_5$, R^3R^4 = $(CH_2)_4$] and with acetophenone produced $[Fe\{CNC(Me)(Ph)CH_2COPh\}_5CN]BF_4$ (75).

New routes to functionalized isocyanide ligands from metal cyanide precursors have appeared. Treatment of the anion $[CpMn(CO)_2CN]^-$ with R_2ECl (R = Et, OEt, Ph; E = P, As) gave the isocyanophosphine and arsine complexes $CpMn(CO)_2(CNER_2)$. The potential donor properties of these ligands were demonstrated by reactions of $CpMn(CO)_2THF$ with $CpMn(CO)_2CNPPh_2$ which produced the dimer $\{CpMn(CO)_2\}_2CNPPh_2$ (76). Similar interactions of (olefin)$Fe(CO)_2CN^-$ (olefin = e.g., C_8H_{14}, C_4H_6, C_8H_8) with R_3ECl (R = Me, Et; E = Si, Ge, Sn, Pb) gave (olefin)$Fe(CO)_2(CNER_3)$ (77).

The interaction of an oxidizable anionic metal cyanide with a diazonium salt has led to the synthesis of a series of novel α-functionalized isocyanide complexes. Arene diazonium chloride $[4\text{-}XC_6H_4N_2]Cl$ (X = H, Cl, Br) additions to $[M(CO)_5CN]^-$ (M = Cr, Mo, W) in THF, CH_2Cl_2, $CHCl_3$, C_6H_5Cl, and CH_3Cl gave the complexes $M(CO)_5\{CN\overline{CH(CH_2)_3O}\}$ (7), $Cr(CO)_5CNCHCl_2$, $Cr(CO)_5CNCCl_3$, $Cr(CO)_5CNCOC_6H_4Cl$, and $Cr(CO)_5$ $CNCOCH_2Cl$, respectively. A mechanism for their formation is given in Scheme 1. This radical alkylation of cyano complexes opens up routes to isocyanides bearing strongly electronegative, and chemically exploitable,

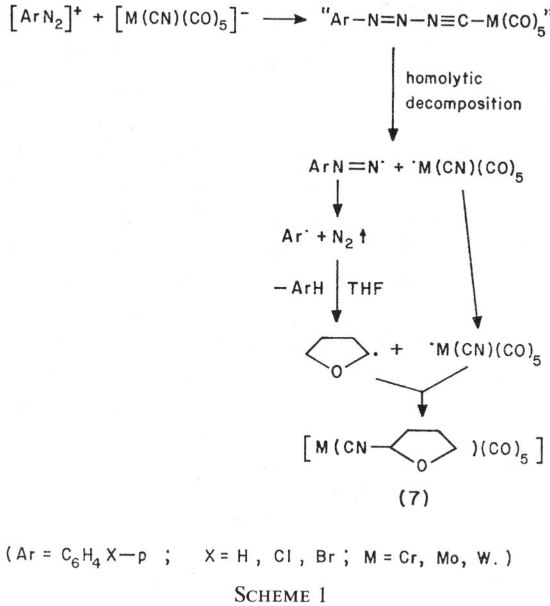

$$[ArN_2]^+ + [M(CN)(CO)_5]^- \longrightarrow \text{"} Ar-N=N-N\equiv C-M(CO)_5\text{"}$$

$$\downarrow \begin{array}{l} \text{homolytic} \\ \text{decomposition} \end{array}$$

$$ArN\equiv N\cdot + \cdot M(CN)(CO)_5$$

$$\downarrow$$

$$Ar\cdot + N_2\uparrow$$

$$-ArH \,\Big|\, THF$$

$\cdot \; + \; \cdot M(CN)(CO)_5$

$$[M(CN- \text{⬠}O)(CO)_5]$$

(7)

(Ar = C_6H_4X—p ;   X = H , Cl , Br ;   M = Cr, Mo, W.)

SCHEME 1

substituents which are inaccessible by the usual methods (69). An interesting reaction of these isocyanides was observed with the formation of $(OC)_5CrCNCCo_3(CO)_9$ (1) from $Cr(CO)_5CNCCl_3$ and $Co_2(CO)_8$, in which the CNC-skeleton functions as a μ_4-bridge between the chromium atom and the three cobalt atoms. The related product $(OC)_9Co_3CNCCl_2$ was formed from Cl_3CNCCl_2 and $Co_2(CO)_8$ and this reacted further with $Pt(C_2H_4)$- $(PPh_3)_2$ to give $(OC)_9Co_3CNCPtCl_2(PPh_3)$ (59).

Other complexes containing N-α-functionalized isocyanide ligands, e.g., (arene)$Cr(CO)_2CNCOPh$, were obtained from K[(arene)$Cr(CO)_2CN$] and benzoyl chloride. Two pairs of diastereoisomers $(C_6H_5CO_2Me)$- $Cr(CO)(CNCOPh)\{PPh_2(neomenthyl)\}$ and $(m\text{-}MeOC_6H_4CO_2Me)Cr(CO)$- $(CNCOPh)\{P(OPh)_3\}$ were prepared as synthetic applications of this reaction (78, 79). $(C_6H_5CO_2Me)Cr(CO)_2(CNCOPh)$ has been structurally refined (80). $Cr(CO)_5(CNCOR)$ (R = Bu^t, Ph) was shown to reconvert to $[Cr(CO)_5CN]^-$ and RCO_2Et in ethanol in the presence of a base (81).

The use of the chiral alkyl iodides sec-butyl iodide and α-phenylethyl iodide on $CpFe(CO)(CNR)CN$ (R = Me, Et) has produced configurationally nonlabile diastereoisomers (82).

Other examples of isocyanide complexes formed from cyanide alkylations are $[CpFe(CO)(CNEt)(PPh_3)]^+$ (83), trans-$[Pt(CNMe)_2(PMe_2Ph)_2]$- $(SFO_3)_2$ (84), and $PtI_2(CNMe)_2$ (85).

B. Generation at the Metal–Atom Center

A number of routes have been employed for the synthesis of metal isocyanide complexes by generating the isocyanide ligand on the metal atom.

Complexes containing the N-isocyanoiminotriphenylphosphine ligand, e.g., $M(CO)_5CNNPPh_3$ (M = Cr, Mo, W), are useful intermediates to iso-diazomethane complexes, e.g., $M(CO)_5CNNH_2$ (M = Cr, Mo, W) by hydrolytic cleavage, and to $M(CO)_5CNNCRR^1$ (M = Cr, Mo; R = R^1 = Me, R = Ph, R^1 = H) by reacting with acetone and phenaldehyde, respectively (86).

The formation of isocyanides from a series of intermediates of the type 8–12 is reported. The ylides 8 [M = $Fe(CO)_4$, R = $C_6H_4Me\text{-}o,m,p$; M = $PdCl_2(PhN=PPh_3)$, R = Ph] were proposed as intermediates in the

$$M \overset{\overset{\overset{\ominus}{O}}{|}}{=\!\!\!=}C — N \overset{\overset{\oplus}{PPh_3}}{\underset{|}{}} — R$$

(8)

$$M — \overset{\overset{O}{||}}{C} — NHR$$

(9)

$$M \rightarrow C \overset{NR}{\underset{NR'}{\diagup\!\!\!\diagdown}} \ominus$$

(10)

$$\underset{(11)}{M-C\underset{NHR'}{\overset{NHR}{\big\langle}}} \qquad \underset{(12)}{M\underset{\overset{||}{O}}{\overset{\overset{NR}{\overset{||}{C}}}{\big\langle}}\underset{C}{\overset{C}{\big\rangle}NR'}}$$

interactions of $Fe(CO)_5$ with $arylN=PPh_3$ to give $Fe(CO)_4(CNAryl)$ (87), and from reactions of $PhN=PPh_3$ with $PdCl_2$ and CO which gave $(PhN=PPh_3)PdCl_2CNPh$, respectively (88). Carbamoyl species **9**, [M = $CpFe(CO)_2$, R = Bu^t], which formed from addition of Bu^tNCO to $CpFe(CO)_2H$, spontaneously decomposed to $Cp_2Fe_2(CO)_3(CNBu^t)$ (89), whereas **9** [M = $CpFe(CO)_2$, R = Me] only converted to $[CpFe(CO)_2CNMe]Cl$ with $COCl_2$ and NEt_3 (90). The reaction of a carbonyl metal anion with carbodiimides, $RN=C=NR^1$, produced **10** [M = $Cr(CO)_5$, R = R^1 =Ph; M = $CpFe(CO)_2$, R = Ph, R^1 = alkyl, allyl; R = R^1 = Ph, C_6H_{11}], which on protonation produced **11** [M = $Cr(CO)_5$, R = R^1 = Ph; M = $CpFe(CO)_2$, R = Ph, R^1 = alkyl, allyl; R = R^1 = Ph, C_6H_{11}]. Both **10** and **11** converted to the isocyanide compounds $Cr(CO)_5CNPh$, $[CpFe(CO)(CNPr^i)(CNPh)]PF_6$, and $[CpFe(CO)(CNBu^t)-(CNPh)]PF_6$ with $COCl_2/NEt_3$ (91, 92). The intermediate **12** {M = $CpFe(CO)$} was implicated in the conversion of **10** [M = $CpFe(CO)_2$; R^1 = Ph; R = Bu^t, Pr^i] to the bis-isocyanide iron complexes, and to $Cp_2Fe_2(CO)_{4-x}(CNPh)_x$ (x = 2, 3) with excess $PhN=C=NPh$ (92).

The products formed from the carbene (**11**) were dependent upon the nature of the alkyl group (93). Good leaving groups were found to be a factor in converting the carbenes $[CpFe(CO)_2\{C(YR)YR\}]^+$ (YR = SePh, SPh, OPh, SMe) to the isocyanide complex $[CpFe(CO)_2(CNCH_2Ph)]^+$ with benzylamine (94, 95), whereas $[CpFe(CO)_2(CNR)]^+$ (R = Me, CH_2Ph) formed in the melt from $[CpFe(CO)_2\{C(NHR)(OMe)\}]SO_3CF_3$ by intramolecular extrusion of methanol (95). Treatment of $PtCl_2\{C(OEt)NHPh\}PEt_3$ with $AlCl_3/NEt_3$ gave $PtCl_2(CNPh)PEt_3$ (93).

Treatment of the iron carbene $Fe(tetraphenylporphyrin)(CCl_2)$ with primary amines provided a novel route to the isocyanide derivatives $Fe(tetraphenylporphyrin)(NH_2R)(CNR)$ (R = Pr^i, Bu^n, Bu^t) (96).

The compounds $[CpFe(CO)_2CNR]PF_6$, $Cp_2Fe_2(CO)_3CNR$ (R = Me, Ph) and $Cr(CO)_5CNR$ (R = Me, C_6H_{11}, Ph) were prepared from desulfurization reactions of the corresponding isothiocyanate with the carbonyl anion (97). Other desulfurization reactions of isothiocyanates with the complexes $CpCo(PMe_3)_2$ (98), $Ru(CO)_2(PPh_3)_3$, and $RhCl(PPh_3)_3$ (99) gave $CpCo(PMe_3)CNR$ and the dithiocarbonimidato derivatives $Ru(CO)$-

$(CNR)(S_2CNR)(PPh_3)_2$ (R = Me, Et, Ph) and $RhCl(CNR)(S_2CNR)$-$(PPh_3)_2$ (R = alkyl, aryl), respectively.

The reaction of a series of thiocarbonyl compounds with primary amines has given isocyanide complexes via the intermediate carbene (13) (100).

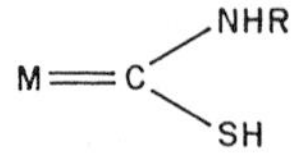

$$M{=\!=}C\!\!\begin{array}{c} {}^{\nearrow NHR} \\ {}_{\searrow SH} \end{array}$$

(13)

Kinetic studies on these reactions have shown a second-order dependence on amine concentration consistent with the addition of the amine to the thiocarbonyl carbon atom in a rate-determining step (101). Complexes isolated from this and related studies were $M(CO)_5CNR$ (M = Cr, Mo, W; R = alkyl) (101–103), $[RuCl_2(CO)_2(CNR)L]$ (L = CO, PPh_3; R = C_6H_{11}, aryl) (104), and $[CpFe(CO)(CNR)L]PF_6$ (L = phosphine; R = Me, C_6H_{11}) (105).

The dimeric products $[RuI(SH)(CO)(CNR)(PPh_3)]_2$ were formed by a novel rearrangement on reacting the dithiomethyl ester complex $RuI(CS_2Me)(CO)(PPh_3)_2$ with primary amines in refluxing benzene (Scheme 2) (106).

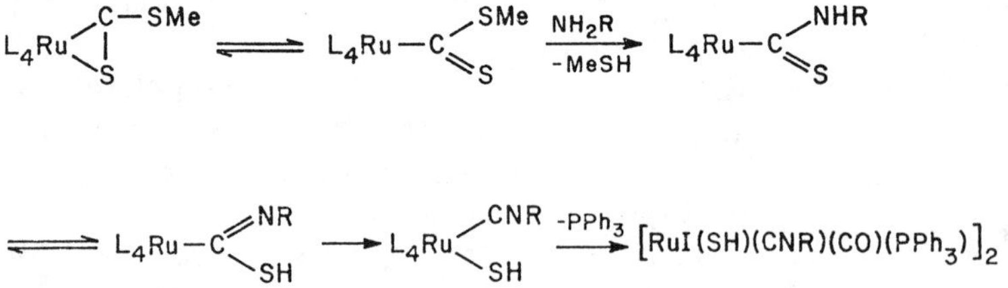

$$L_4Ru\!\!\begin{array}{c}{}^{\nearrow C-SMe}\\{}^{|}\\{}_{\searrow S}\end{array} \rightleftharpoons L_4Ru-C\!\!\begin{array}{c}{}^{\nearrow SMe}\\{}_{\searrow\!\!\!=S}\end{array} \xrightarrow[-MeSH]{NH_2R} L_4Ru-C\!\!\begin{array}{c}{}^{\nearrow NHR}\\{}_{\searrow\!\!\!=S}\end{array}$$

$$\rightleftharpoons L_4Ru-C\!\!\begin{array}{c}{}^{\nearrow NR}\\{}_{\searrow SH}\end{array} \longrightarrow L_4Ru\!\!\begin{array}{c}{}^{\nearrow CNR}\\{}_{\searrow SH}\end{array} \xrightarrow{-PPh_3} [RuI(SH)(CNR)(CO)(PPh_3)]_2$$

SCHEME 2

Oxidative addition of Cl_2CNPh with the low-valent metal complexes $Fe(CO)_5$, $[Fe(CO)_4]^{2-}$, $Pt(PPh_3)_4$, and $RhCl(PPh_3)_3$ produced $Fe(CO)_4CNPh$, $PtCl_2(CNPh)PPh_3$, and $RhCl_3(CNPh)(PPh_3)_2$ (107, 108).

IV

SYNTHESIS OF ISOCYANIDE COMPLEXES

A. Homoleptic Systems

No homoleptic isocyanide complexes of niobium, tantalum, titanium, zirconium, and halfnium have yet been synthesized. The vanadium cation

[V(CNBu^t)$_6$]$^{2+}$ has been prepared either by treating *mer*-VCl$_3$(CNBu^t)$_3$ with isocyanide in ethanol (*109*) or by oxidizing [V(CO)$_6$]$^-$ in the presence of CNBu^t (*110*).

A number of new synthetic routes to isocyanide complexes of chromium, molybdenum, and tungsten have been investigated. For example, new routes to the long-established species Cr(CNR)$_6$ have utilized the reductive elimination of isopropyl groups from Cr(Pr^i)$_4$ (*111*) or the displacement of naphthalene from bis(η^6-naphthalene)-chromium(0) by CNR(R = Bu^n, C$_6$H$_{11}$) (*112, 113*). The 17- and 16-electron salts [Cr(CNR)$_6$](PF$_6$)$_n$ (n = 1, 2; R = aryl) were prepared by successive AgPF$_6$ oxidations of Cr(CNR)$_6$ (*114*), or, for [Cr(CNR)$_6$](PF$_6$)$_2$ (R = Bu^t, C$_6$H$_{11}$), directly from blue chromium(II) solutions by additions of CNR and KPF$_6$ (*15*). Addition of neat CNR to [Cr(CNR)$_6$](PF$_6$)$_2$ (R = Bu^t, C$_6$H$_{11}$, Ph) produced the 18-electron salt [Cr(CNR)$_7$](PF$_6$)$_2$ for R = Bu^t and C$_6$H$_{11}$, and the reduced complex [Cr(CNPh)$_6$]PF$_6$ for R = Ph (*15*). The isolation of [Cr(CNR)$_7$](PF$_6$)$_2$ has provided a rare example of a seven-coordinate homoleptic compound of the first-row transition elements (the first one containing isocyanide ligands) and has completed the first homologous series of homoleptic seven-coordinate complexes of any transition metal group.

A number of attempts have been made over the years to develop reproducible synthetic routes to six- and seven-coordinate isocyanide complexes of molybdenum and tungsten. Two of the older methods, namely, the reaction of the hexacarbonyls with halogens in the presence of an isocyanide (*115, 116*) or reactions of the salt Ag$_4$Mo(CN)$_8$ with isocyanides (*74*), have given six- and seven-coordinate products. Recently, however, the discovery of the reductive or nonreductive cleavage of multiple metal–metal bonds in dinuclear group VIA compounds by isocyanides has provided a facile route to the synthesis of a variety of homoleptic and related isocyanide complexes of these metals in reasonable yields.

This facile cleavage of metal–metal bonds has been explained in terms of the ability of isocyanides to compete effectively through π interactions for the electron density in the metal HOMOs, which are the metal–metal bonding orbitals. A variety of precursors have been used, including chloro complexes of molybdenum (*8–12*) and tungsten (*8*), carboxylato complexes of molybdenum (*9–11*), and tungsten complexes containing the anions of alkylated hydroxypyridine (*13*) and pyrimidine (*14*). Complexes synthesized were [M(CNR)$_7$]$^{2+}$ {R = alkyl, M = Mo (*8, 9*), W (*8, 13*); R = aryl, M = Mo, W (*14*)} and W(CNPh)$_6$ (*14*), as well as the intermediates [MoX(CNBu^t)$_6$]X (X = Cl, O$_2$CCF$_3$) (*10,11*), [Mo(CNR)$_5$(L$_2$)$_x$]$^{n+}$ [L$_2$ = O$_2$CMe, n = 0, x = 2 (*10, 11*); L$_2$ = dpm, dpe, (PR$_3^1$)$_2$, n = 2, x = 0 (*12*)] and [Mo(CNR)$_6$(PR$_3^1$)]$^{2+}$ (*12*). The chloro-complex MoCl$_4$ also provided a route to [Mo(CNR)$_7$]$^{2+}$ (R = alkyl) on treatment with CNR in ether. This reaction appears to be solvent specific as [MoCl(O)(CNR)$_4$]$^+$ (see Section

IV,D,2) is produced from methanol (117). The synthesis of the missing congeners $Mo(CNBu^t)_6$ (14), $[M(CNPh)_7]^{2+}$ (M = Mo, W) (14), and $[Cr(CNR)_7]^{2+}$ (R = Bu^t, C_6H_{11}) (15) has placed some doubt on the significance of electronic effects on the formation of these compounds. Their isolation should now allow an evaluation of the factors controlling the stabilities of the various isocyanide complexes.

Only the manganese and rhenium complexes $[M(CNR)_6]^+$ (M = Mn, Re) are known in group VII, and these have been prepared earlier. A new route to $[Re(CNBu^t)_6]^+$ via cleavage of the metal–metal bond in $Re_2Cl_2(O_2CMe)_4$ is reported (11) [cf. $Re_2Cl_8^{2-}$ and CNMe gave $ReCl_5(CNMe)^-$ (118)].

An interesting aspect of the recent isocyanide chemistry of iron and ruthenium has been the formation of the highly reactive electron-rich metal(0) complexes $M(CNR)_5$ (M = Fe, R = alkyl, xylyl; M = Ru, R = Bu^t) from reductions of $trans$-$MCl_2(CNR)_4$ with Na/Hg in the presence of CNR (16). A similar reduction of the osmium polymer $[CODOsCl_2]_x$ ($x > 2$) with Na/Hg in the presence of $CNBu^t$ gave only $CODOs(CNBu^t)_3$ (16), but mixtures of $trans$-$OsCl_2(CNXylyl)_4$ and xylyl isocyanide are smoothly reduced to $Os(CNXylyl)_5$ by Na/Hg (18). UV irradiation of $Fe(CNR)_5$ (R = Et, Pr^i) or Na/Hg reductions of $trans$-$RuCl_2(CNPr^i)_4$ in the presence of $CNPr^i$ gave the dinuclear isocyanide analogues of the carbonyl compounds, e.g., $M_2(CNR)_9$ (M = Fe, Ru). Corresponding irradiations of solutions of $Fe(CNBu^t)_5$ gave only $trans$-$Fe(CN)_2(CNBu^t)_4$ or (η^4-C_8H_8)$Fe(CNBu^t)_3$ in the presence of cyclooctatetraene (17).

The dimeric ruthenium(I) and osmium(I) salts $[M_2(CNXylyl)_{10}](PF_6)_2$ have been isolated respectively from the melt of $[(\eta^1,\eta^2$-$C_8H_{13})$-$Ru(CNXylyl)_4]PF_6$ and from single-electron oxidations on $Os(CNXylyl)_5$ with tropylium hexafluorophosphate (36). A brief mention of the formation of the salt $[Ru_2(L_2)_4](PF_6)_2$ (L_2 = 2,5-dimethyl-2,5-diisocyanohexane) has appeared (37).

A number of publications have appeared on the synthesis of cationic cobalt(I) complexes by known routes. The cations are of the well-established type $[Co(CNR)_5]^+$ (R = aryl) (119–122). Reactions of aromatic isocyanides with $Co_2(CO)_8$ in refluxing toluene have given the fully substituted cobalt(0) dimer $Co_2(CNR)_8$ (R = xylyl, $C_6H_2Me_3$-2,4,6, C_6H_2Br-4-Me_2-2,6) (25, 123).

There are few other examples of complete substitution of carbonyl groups from a homoleptic metal–carbonyl complex by isocyanide ligands [cf. $Ni(CO)_4$ (24), $Fe(CO)_5$ (26), and $Mo(CO)_6$ (124)]. The corresponding butyl isocyanide derivative $Co_2(CNBu^t)_8$ was formed by reduction of $[Co(CNBu^t)_5]PF_6$ with potassium amalgam (19).

Full papers have appeared on the synthesis, under mild conditions, of the rhodium and iridium cations $[M(CNR)_4]^+$ (M = Rh, Ir; R = alkyl, aryl) (125–129) from a variety of sources, e.g., $[RhClL_2]_2$ [L_2 = COD (126),

$(CO)_2$ (128), $(C_2H_4)_2$ (129)] and $[IrCl(CO)_3]_n$ (127). Stepwise addition of $CNBu^t$ to $[RhCl(CO)_2]_2$ has allowed the isolation of the intermediates $RhCl(CO)_2CNBu^t$, $RhCl(CO)(CNBu^t)_2$, and $RhCl(CNBu^t)_3$ in $[Rh(CNBu^t)_4]^+$ formation (130).

The $[M(CNR)_4]^+$ cations associate in solution giving intense violet, blue, or green oligomers. The dimer $[Rh_2(CNPh)_8](BPh_4)_2$ has been fully characterized (131, 132) and other dimeric rhodium(I) and iridium(I) compounds have been synthesized using a range of chelating diisocyanides. They include $[M_2(L_2)_4]^{2+}$ [M = Rh, L_2 = $CN(CH_2)_nNC$; n = 3,4,5,6 (133–135); M = Rh, Ir; L_2 = 2,5-dimethyl-2,5-diisocyanohexane (37), meso-1,3-diisocyanocyclohexane (136)] (see Sections IV,D,2 and V,C).

To date, only nickel, palladium, and platinum have formed homoleptic metal(0) isocyanide clusters. This may be due to two factors, either the greater tendency to nucleation in the group or simply the ready access to the highly labile metal(0) precursors $M(COD)_2$ [M = Ni (137), Pt (138)].

Muetterties and co-workers (21, 22, 60, 139) found that addition of deficiencies of isocyanides to $Ni(COD)_2$ gave the clusters $Ni_4(CNR)_7$ (R = Bu^t, C_6H_{11}) and $Ni_4(CNBu^t)_6$. With isopropyl and benzyl isocyanides, dark brown and red solids of compositions $\{Ni_4(CNPr^i)_6\}_x$ and $\{Ni_4(CNCH_2Ph)_4\}_x$ (x unknown) were obtained. Corresponding reactions in the presence of L [L = $P(OMe)_3$, CNMe] gave $Ni_7(CNBu^t)_6L$.

The complexes $M_3(CNR)_6$ [M = Pd, R = Bu^t (22, 60), M = Pt, R = alkyl (23, 140)], containing a triangle of palladium and platinum atoms, were synthesized by treating $[CpPdC_3H_5]_2$ or $Pt(COD)_2$ with isocyanides. The multinuclear isocyanide cluster $Pt_7(CNXylyl)_{12}$ was formed from reactions of $Pt(COD)_2$ and xylyl isocyanide in a 1:2 ratio or from reduction of $PtCl_2(CNXylyl)_2$ with sodium amalgam (20).

The first well-defined dicationic isocyanide nickel(II) complex $[Ni(CNBu^t)_4](ClO_4)_2$ has been prepared from nickel perchlorate, $CNBu^t$, and ethanol (141). This compound completes the series $[M(CNR)_4]^{2+}$ (M = Ni, Pd, Pt); the latter two having been known for some time. The scarcity of isocyanide complexes of nickel(II) may be a consequence of the high catalytic activity of nickel complexes for isocyanide polymerization. The formulation of the salts $[Ni_3(CNPh)_{11}](ClO_4)_6$ and $[Ni_4(CNPh)_{14}](ClO_4)_8$, isolated from reactions of CNPh with $Ni(ClO_4)_2$, was proposed primarily from their magnetic behavior. A linear arrangement of nickel atoms was suggested (142).

Chemical routes to palladium(I) and platinum(I) isocyanide complexes have been devised. Addition of CNMe to an aqueous solution of $[MCl_4]^{2-}$ (M = Pd, Pt) resulted in the rapid formation of $[M(CNMe)_4]^{2+}$ cations, which were isolated as the hexafluorophosphate salts. On standing, however, the cations $[M(CNMe)_4]^{2+}$ slowly converted over 18 hr to $[M_2(CNMe)_6]^{2+}$

via a mechanism which involved the oxidation and hydrolysis of a methyl isocyanide ligand [Eqs. (1)–(3)] (34).

$$[M(CNMe)_4]^{2+} + H_2O \rightarrow [(MeNC)_3M\{C(O)NHMe\}]^+ + H^+ \quad (1)$$

$$[(MeNC)_3M\{C(O)NHMe\}]^+ + H_2O \rightarrow [(MeNC)_3M] + H^+ + CO_2 + MeNH_2 \quad (2)$$

$$[(MeNC)_3M] + [M(CNMe)_4]^{2+} \rightarrow [M_2(CNMe)_6]^{2+} + CNMe \quad (3)$$

Earlier work by Otsuka et al. (33) had shown that coupling of Pd(CNBu^t)$_2$ with PdX$_2$(CNBu^t)$_2$ (X = Cl, Br, I) readily produced Pd$_2$X$_2$(CNBu^t)$_4$. With this as an example, Balch and co-workers confirmed the above mechanism by obtaining the Pd(I) dimer from the reaction of a solution of Pd$_2$(DBA)$_3$·CHCl$_3$ in acetonitrile with [Pd(CNMe)$_4$]$^{2+}$ (34). Other workers have synthesized the compounds [PdX(CNR)$_2$]$_2$ and [Pd$_2$(CNR)$_6$]Y$_2$ (X = Cl, Br; Y = Cl, Br, PF$_6$; R= Me, Bu^t) by similar coupling reactions of PdCl$_2$(NCPh)$_2$ and bis- or tris(dibenzylideneacetone)palladium(0) in the presence of excess CNR (35, 143).

By varying the ratios of Pd(CNMe)$_x$ with respect to [Pd(CNMe)$_4$]$^{2+}$ or by adding Pd(CNMe)$_x$ to [Pd$_2$(CNMe)$_6$]$^{2+}$ in acetone solution, the linear trimer [Pd$_3$(CNMe)$_8$](PF$_6$)$_2$ is formed (144).

B. Carbonyl Substitution Reactions

The ideal route to homoleptic isocyanide metal(0) complexes is via substitution of carbonyl ligands in metal–carbonyl compounds. This is simply because the carbonyls constitute the largest readily available source of transition metal compounds in low oxidation states. If one also considers the importance of metal carbonyl chemistry in synthetic organometallic chemistry and as catalysts in organic synthesis and industrial processes, the reasons for synthesizing and studying corresponding compounds containing the more versatile isocyanide ligand become apparent. However, the subtle differences between the π-acceptor and σ-donor properties of isocyanides and carbon monoxide have, until very recently, precluded the carbonyl substitution pathway as a viable route to highly substituted and homoleptic isocyanide–metal complexes. Traditionally, substitution reactions of metal carbonyls have been carried out thermally, usually in high boiling solvents and with long reaction times, or in sealed Carius tubes. A number of alternative routes have recently been investigated, but not all these routes have as yet been applied to isocyanide substitutions. For example, photoinduced (145) and reagent-induced substitution, photoinduced (146, 147) and electrochemical-induced radical catalysis (148), the displacement of a weakly coordinated ligand or ligands (e.g., olefins), or the chemical modification of an already coordinated group (e.g., CO, CS, CN^-) have been investigated with limited success. Although photoinduction has been used to prepare $CpMn(CNR)_3$ (R = Me, Ph, C_6H_4Cl-p, C_6Cl_5) from $CpMn(CO)_3$ (149), it has a limited application (presumably because of isonitrile polymerization) in isocyanide substitution reactions, as has the displacement of weakly coordinated ligands and the chemical modification of a coordinated group. However, electrochemical-induced substitution pathways via highly labile radical cations may yet provide a facile route to isocyanide compounds. Examples of chemical reagents that have been used to enhance substitutions are $RhCl(PPh_3)_3$ (150) (via carbonyl group abstraction), amine oxides (carbonyl oxidation) (151), borohydrides (kinetic labilization through transient hydride or formyl formation) (152), and hydroxide ions (in phase transfer catalysis) (153).

One of the most remarkable recent advances in metal carbonyl substitution chemistry has been the discovery by Coville and co-workers of the homogeneous and heterogeneous catalytic labilization of the metal–carbon bond in metal–carbonyl complexes (26–31). Considering that restrictions to catalysis involving metal carbonyl species can, in some instances, be related to the strength of the metal–carbon bond, these discoveries could have far-reaching implications. To exemplify these catalytic substitution processes, comparisons in the systems $M(CO)_6$ (M = Cr, Mo, W), $CpMoI(CO)_3$, $CpFeI(CO)_2$, $Fe(CO)_5$, $Fe(CO)_4$(olefin), and $Ir_4(CO)_{12}$ will be made.

Thermal reactions of $M(CO)_6$ with neat isocyanides or with high boiling solvents such as decalin over extended periods (16 hr to days) has given mainly the monosubstituted product $M(CO)_5CNR$ [M = Cr, Mo, R = Bu^t (154); M = Cr, Mo, W, R = $M^1R_3^1$, M^1 = Si, Ge, R^1 = Me, Bu^t, Ph (155)]. Varying amounts of cis-$Mo(CO)_4(CNMe)_2$ (156) have been observed from thermal reactions but in only one instance has a direct thermal synthesis of a trisubstituted product been achieved. Thus, treatment of $W(CO)_6$ with neat $CNBu^t$ gave fac-$W(CO)_3(CNBu^t)_3$ after about four days (115).

Recently, phase transfer catalysis involving chemical modification of a carbonyl group has been used to achieve the substitution products $M(CO)_5CNR$ (M = Cr, Mo, W; R = Bu^t, C_6H_4Me-p) and cis-$M(CO)_4$ $(CNR)_2$ (M = Mo, W; R = Bu^t, C_6H_4Me-p) in 1 hr. The two-phase system used was benzene/50% aqueous sodium hydroxide, and the products formed as mixtures in 37–80% yield (152).

The most successful and widely used synthesis of higher substituted derivatives of $M(CO)_6$ (M = Cr, Mo, W) is via the thermal displacement of mono- or polyhapto ligands from the appropriate carbonyl complex. Thus, from $C_7H_8M(CO)_4$ (M = Cr, Mo, W; C_7H_8 = norbornadiene, cycloheptatriene), fac-M(CO)$_3$(NCMe)$_3$ (M = Cr, Mo), and (C$_6$H$_8$)$_2$M(CO)$_2$ (M = Mo, W), substituted products of general formula cis-M(CO)$_4$(CNR)$_2$, fac-M(CO)$_3$(CNR)$_3$ (M = Cr, Mo, W; R = alkyl, aryl) (*154, 157, 158*), and cis-M(CO)$_2$(CNBu^t)$_4$ (M = Mo, W) were obtained (*154*).

In contrast, stepwise substitution reactions on M(CO)$_6$ (M = Cr, Mo, W) have been achieved with a series of heterogeneous catalysts including co-balt(II) chloride (*27*), activated charcoal (*159*), and platinum metals dispersed on oxide or carbon supports (*31*), to give mono-, di-, tri-, and complete substitution (*124*) in yields > 90%. Representative reaction times are given in Table II (*159*). The efficiency of the method was further demonstrated by the stepwise synthesis of the mixed isocyanide complexes cis-Mo(CO)$_4$(CNMe)(CNBu^t) and fac-Mo(CO)$_3$(CNMe)(CNBu^t)$_2$ from Mo(CO)$_6$ in < 25 min in 85 and 95% yields, respectively (*159*).

Thermal substitution reactions of CpMoY(CO)$_3$ (M = Mo, W; Y = halide) are difficult to accomplish. Though the products CpMoY (CO)$_{3-x}$(CNR)$_x$[x = 1,2; R = Me, Y = Cl (*160, 161*); R = Bu^t, Y = Cl (*162*); x = 1, R = Bu^t, Y = Br, I (*163*)] have been isolated after long reaction times; they form as mixtures with the salt [CpMo(CNR)$_4$]Y, with the proportion of the salt increasing with increasing ratios of isocyanide to starting complex. Recently, however, Coville has shown that stepwise sub-stitution is effected with short reaction times (< 30 min) using [CpMo(CO)$_3$]$_2$ as a catalyst to give CpMoI(CO)$_{3-x}$(CNR)$_x$ (x = 1,2; M = W, R = Bu^t; M = Mo, R = Bu^t, CH$_2$Ph, xylyl; x = 3, M = Mo, R = xylyl). These compounds represent the only trisubstituted halide complexes and substituted tungsten halide compounds known with isocyanides. A free radical chain mechanism involving initial homolytic fission of the Mo–Mo bond in the dimer is proposed (*164*).

Fe(CO)$_5$ is fairly inert to substitution reactions, usually requiring Carius tube reactions or high boiling solvents, both with extended reaction times, to reach, in the case of isocyanides (*165*), disubstitution. As a testimony to the unreactivity of Fe(CO)$_5$ only three publications have appeared on isocya-nide reactions with complexes related to this molecule since 1974. Thus, photolysis of Fe(CO)$_3${P(OAryl)$_3$}$_2$ in the presence of CNMe gave Fe(CO)$_2${P(OAryl)$_3$}$_2$(CNMe) (*166*), substitution reactions on Fe(CO)$_4$MA with CNCH$_2$Ph produced Fe(CO)$_{4-x}$(CNCH$_2$Ph)$_x$MA (x = 1–3) (*167*), and the reaction of Fe$_3$(CO)$_{12}$ with CN(CH$_2$)$_n$NC (n = 2,6) produced exclusively the dimer [Fe(CO)$_4$]$_2$CN(CH$_2$)$_n$NC (*168*).

The discovery that additions of catalytic amounts of CoCl$_2$, activated carbon, or metals on metal oxide or carbon supports are extremely effective in labilizing carbonyl groups in Fe(CO)$_5$ is thus particularly noteworthy. Stepwise substitution products Fe(CO)$_{5-x}$(CNR)$_x$ (x = 1–5) have been obtained in high yield with short reaction times using these catalysts (see Table III) (*159*).

TABLE II

REACTION CONDITIONS FOR THE CATALYZED SYNTHESIS OF THE COMPLEXES $[M(CO)_{6-n}(CNCH_2Ph)_n]$
($M = Cr$, Mo, W; $n = 1-3$) WITH Pd/C IN REFLUXING TOLUENE

Complexes	Reaction time[a] (min)	Complexes	Reaction time[a] (min)	Complexes	Reaction time[a] (min)
$[M(CO)_5(CNCH_2Ph)]$[b]		cis-$[M(CO)_4(CNCH_2Ph)_2]$[c]		fac-$[M(CO)_3(CNCH_2Ph)_3]$[d]	
M = Cr	4	M = Cr	15	M = Cr	60
M = Mo	2	M = Mo	2	M = Mo	6
M = W	2	M = W	4	M = W	15

[a] As determined by IR spectroscopy.
[b] Yields between 87 and 93%.
[c] Yields between 92 and 98%.
[d] Yields between 93 and 98%.

TABLE III

REACTION TIMES FOR THE FORMATION OF THE COMPLEXES $[Fe(CO)_{5-n}(CNR)_n]$ ($n = 3-5$, R = Me, C_6H_{11}, Bu^t, CH_2Ph, Ph, $C_6H_3Me_2$-2,6, $C_6H_2Me_3$-2,4,6)

Complexes	Reaction time[a] (min)	Complexes	Reaction time[a] (min)	Complexes	Reaction time[a] (min)
$[Fe(CO)_2(CNR)_3]$[b]		$[Fe(CO)(CNR)_4]$[c]		$[Fe(CNR)_5]$[c]	
R = $C_6H_3Me_2$-2,6	25	R = $C_6H_3Me_2$-2,6	5	R = $C_6H_3Me_2$-2,6	10
R = $C_6H_2Me_3$-2,4,6	5	R = $C_6H_2Me_3$-2,4,6	5	R = $C_6H_2Me_3$-2,4,6	10
R = Ph	5	R = Ph	5	R = Ph	10
R = Bu^t	35	R = Bu^t	720		
R = Me	180				
R = C_6H_{11}	35				

[a] As estimated by IR spectroscopy.
[b] Yields between 50 and 78%.
[c] Yields between 50 and 60%.

Rhodium(I) or polymer supported rhodium(I) compounds catalyzed the formation of $Fe(CO)_{4-x}(CNR)_xL$ ($x = 1-3$; R = Bu^t, xylyl; L = MA, citraconic anhydride, acrylamide) (29, 30), and the dimer $[CpFe(CO)_2]_2$ catalyzed the stepwise substitution of carbonyl groups in $CpFeI(CO)_2$ to give $CpFeI(CO)_{2-x}(CNR)_x$ ($x = 1,2$; R = Bu^t, xylyl) in 60–80% yields. A nonchain free-radical mechanism was proposed for the latter reaction (28). The compounds $CpFeX(CO)_{2-x}(CNR)_x$ ($x = 1,2$; X = halide, $SiMe_3$) are known for a range of alkyl and aryl isocyanides (169–171).

This catalytic labilization of carbonyl groups has been extended to the replacement of carbonyls in metal carbonyl clusters. Metal cluster complexes are at present the subject of extensive studies, partly because of their possible relevance as models for chemisorbed metal surfaces and because of their catalytic activity. The majority of these clusters contain carbonyl ligands, and these have been prepared from the vast number of metal carbonyl precursors generally available by a variety of synthetic methods usually without recourse to designed or rational procedures. In metal isocyanide chemistry, however, suitable precursors are lacking, and as a consequence, there are few routes to homoleptic metal–isocyanide clusters, and few isocyanide clusters are known (see Section IV,A).

Uncatalyzed substitution of carbonyls by isocyanide ligands in carbonyl clusters has found limited scope, but has in some cases revealed interesting bonding modes and highlighted the versatility of the isonitrile ligand (see Section IV,D,2). Substitution reactions on the clusters $M_3(CO)_{12}$ by isocyanides in high boiling solvents has given up to trisubstitution for iron and ruthenium (172, 173) and tetrasubstitution for osmium (174). In the osmium case the precursors $Os_3(CO)_{12-x}(NCMe)_x$ [$x = 1$ (175), 2 (176)] were used to synthesize mono- and disubstituted isocyanide products. Corresponding reactions with the hydrides $H_4Ru_4(CO)_{12}$ and $(\mu\text{-}H)_2Os_3(CO)_{10}$ gave the complexes $H_4Ru_4(CO)_{12-x}(CNBu^t)_x$ ($x = 1,2,4$) (173), $(\mu\text{-}H)HOs_3(CO)_{10}(CNBu^t)$, and $(\mu\text{-}H)_2Os_3(CO)_9(CNBu^t)$ (176a).

Up to five carbonyl groups have been replaced by isocyanides in $Co_4(CO)_{12}$ to give $Co_4(CO)_{12-x}(CNR)_x$ ($x = 1-5$, R = Bu^t; $x = 1-4$, R = Me, C_6H_{11}) (177), whereas in the corresponding iridium carbonyl reasonable yields of mono- to tetrasubstituted products $Ir_4(CO)_{12-x}(CNR)_x$ ($x = 1-4$; R = Me, Bu^t) were only obtained with Me_3NO (66). With 5% palladium on carbon as a catalyst stepwise substitution of up to seven carbonyl groups in $Ir_4(CO)_{12}$ was effected in refluxing toluene with reaction times of 10–15 min (32).

Other interesting carbonyl substitutions have been reported. The lability of ammonia in $[V(CO)_5NH_3]^-$ has been utilized as a route to the isocyanide carbonyl anions $[V(CO)_5CNR]^-$ (R = Me, Bu^t, C_6H_{11}, Ph), one of the few anionic complexes containing isocyanide ligands known (178). Mixing

$Cr(CNPh)_6$ with $Cr(CO)_6$ has given the missing chromium congeners $Cr(CO)(CNPh)_5$ and cis-$Cr(CO)_2(CNPh)_4$ (*179*).

Substitution reactions of $CpMo(CO)_3R$ (R = Me, CH_2Ph) and $CpMo(CO)_2R$ (R = NO, $N_2C_6H_4Me$-p) with $CNBu^t$ have given $CpMo(CO)_2(CNBu^t)COR$ (R = Me, CH_2Ph) and $CpMo(CO)_{2-x}(CNBu^t)_xR$ ($x = 1$, 2; R = NO, $N_2C_6H_4Me$-p) (*163*). The reactions of $CNBu^t$ on $CpMo(CO)_3\{C(Cl)C(CN)_2\}$ were accompanied by an interesting rearrangement and insertion which gave $CpMo(CNBu^t)_2\{\eta^2\text{-}CH(CN)\!\!=\!\!C(CN)_2\}Cl$ (isolated in two isomeric forms) and **14**. Other products isolated during this

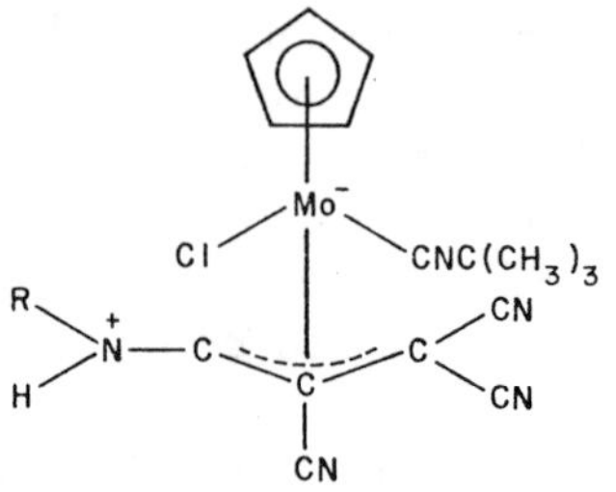

(14)

study were $CpMo(CO)_2(CNBu^t)\{C(X)\!\!=\!\!C(CN)_2\}$ (X = Cl, CN) and two isomers of $CpMo(CNBu^t)_2\{trans\text{-}C_2(CN)_2H_2\}Cl$ (*162*).

Substitution products of $M_2(CO)_{10}$ (M = Mn, Re) containing isocyanides have been obtained thermally and by nucleophilic attack on $Mn(CO)_{5-x}(CNR)_xX$ (X = halide) by $[Mn(CO)_5]^-$. Products characterized were $M_2(CO)_{10-x}(CNR)_x$ {M = Mn, $x = 1$–4, R = alkyl, aryl (*172, 180, 181*); M = Re, $x = 1$, R = Me (*181*)}. A Raman investigation of these compounds suggests weakening of the metal–metal bond on substitution (*181*). The manganese complexes represent some of the very few examples of equatorial subsitutions.

The three series $M(CO)_{5-x}(CNR)_xX$ [$x = 1$–5; M = Mn, X = halide, $SnCl_3$ (*182, 183*); M = Re, X = halide (*184*)], $[M(CO)_{6-x}(CNR)_x]^+$ {$x = 1$–6; M = Mn (*179, 182, 183*); M = Re (*184*)}, and $CpMn(CO)_{3-x}(CNR)_x$ ($x = 1$–3) (*180*) have been completed for alkyl and aryl isocyanides. The product $Re(CO)_2(CNMe)_3Br$ was isolated in three isomeric forms, fac, mer-trans, and trans, an important precedent in substitutions of $M(CO)_5Br$ (*184*).

The substituted products $Cp_2Fe_2(CO)_{4-x}(CNR)_x$ (R = Me, Bu^t; $x = 1$–3) have been prepared from the parent carbonyl as part of detailed NMR studies of their fluxional behavior in solution (see Section IV,D,1) (*185–190*). Compounds of general formula $(dienyl)_2M_2(CO)_{4-x}(CNR)_x$ [dienyl = Cp, C_5H_4Me, C_9H_7; M = Fe, Ru; $x = 1$, 2; R = alkyl, aryl]

($191-195$) and $\{Cp_2Fe_2(CO)_3\}_2CN(CH_2)_nNC$ ($n = 2, 3, 4, 6$) (168) have also been reported as products of direct substitution reactions. It is surprising that in these studies $Cp_2Fe_2(CO)(CNR)_3$ has been obtained only for alkyl isocyanides with aryl isocyanides, giving at the most disubstitution and then in poor yields {e.g., for CNPh yields of $<1\%$ are reported (192)}. Yields of between 27 and 73% of $Cp_2Fe_2(CO)_{4-x}(CNPh)_x$ ($x = 1-3$) have been obtained, however, from interactions of $[CpFe(CO)_2]^-$ with $[CpFe(CO)_{3-x}(CNPh)_x]^+$ ($x = 1-3$). The first example of the fully substituted dimer $Cp_2Fe_2(CNPh)_4$ was obtained by treating $[CpFe(CNPh)_3]^+$ with one equivalent of the isocyanide acceptor $Na_2Cr(CO)_5$ (196). The stability of this latter complex prompted a reinvestigation of the reaction between $[(C_5H_4R)Fe(CO)_2]_2$ and the aryl isonitriles, CNPh and xylyl isocyanide. Contrary to previous reports, the complexes $(C_5H_4R^1)_2Fe_2(CNR)_4$ ($R^1 = H$, Me, CO_2Me; R = Ph, xylyl) formed in 70–80% yield within 1 hr in refluxing toluene. The reaction rates followed the sequence $C_5H_4CO_2Me > Cp > C_5H_4Me$, suggesting a rate enhanced by increasing electrophilicity of the metal atom centre (45). The low yields in previous reactions was attributed to workup problems, specifically to complex decomposition during chromatographic separations.

Pyrolysis of the carbonyl isocyanide complexes $Ru_3(CO)_{11}CNBu^t$ (63) and $Os_3(CO)_{12-x}(CNBu^t)_x$ ($x = 1, 2$) (197) has produced the high nuclear clusters $Ru_5(CO)_{14}(CNBu^t)_2$, $Os_6(CO)_{18-x}$ $(CNBu^t)_x$ ($x = 1-5$) and a product tentatively formulated as $Os_6(CO)_{17}(CNBu^t)_2$. This latter product may be regarded as a substitution product of the unknown $Os_6(CO)_{19}$ or may be related to $Os_6(CO)_{18}(CNC_6H_4Me\text{-}p)_2$ (198) by addition of a metal–metal bond.

The cobalt(I) and cobalt(III) cyclopentadienyl derivatives $C_5R_5^1$ $Co(CO)_{2-x}(CNR)_x$ ($x = 1, 2$; R = Bu^t; $R^1 = H$, Me) (199), $CpCo(CNR)I_2$, $[CpCo(CNR)_2I]PF_6$, and $[CpCo(CNR)_3](PF_6)_2$ (R = Me, $C_6H_4OMe\text{-}p$) (200) were prepared from $C_5R_5^1Co(CO)_2$ and $CpCo(CO)I_2$, respectively.

C. General

The high basicity of the cation $[Cp_2V]^+$, generated from Cp_2VCl in aqueous solution, has been shown by the rapid formation of $[Cp_2VL_2]^+$ {$L_2 = (CO)_2$, $(CNC_6H_{11})_2$, dpe} with π-acceptor ligands ($201, 202$).

Treatment of the olefin–hydrides $Cp_2Ta(H)L$ (L = C_3H_6, 1-butene) with the aryl or alkyl isocyanides CNR (R = xylyl, C_6H_{11}, Me, Bu^t) effected insertion of the olefin into the Ta–H bond to give $Cp_2TaR^1(CNR)$ ($R^1 = C_3H_7$, C_4H_9). The observation that these alkyls do not convert to iminoacyl compounds, has been ascribed to the "carbenelike" character of the Ta–C bond as a result of the strong π-donor properties of Cp_2TaR. Further evidence for this bonding mode came from the reaction with $AlEt_3$ which produced the two isomers **15** and **16** in the ratio of $2:1$ (*203*).

$$(15) \qquad\qquad (16)$$

Addition of $CNBu^t$ to the halocomplex $MoCl_4(THF)_2$ in nonpolar solvents gave $[MoCl_3(CNBu^t)_3]Cl$ (*204*), whereas $MoCl_4$ and CNR in methanol formed $[MoCl(O)(CNR)_4]^+$ (R = alkyl). The bromine analog of the latter salt was prepared from $Mo(CO)_6$, bromine, and CNMe in methanol. These compounds represent rare examples of an oxomolybdenum(IV) species (*117*).

Corresponding treatment of WCl_6 with $CNBu^t$ in nonpolar solvents has given $trans$-$WCl_4(CNBu^t)_2$ (*204*).

The interaction of $CpMoI(CO)_2CNMe$ with $Na[CpMo(CO)_3]$ gave the monosubstituted dimer $Cp_2Mo_2(CO)_5CNMe$, which is the only example of this type known (*205*).

Replacement of ligands in $C_3H_5MoCl(CO)_2(NCMe)_2$ by isocyanides has given the substituted products $C_3H_5MoCl(CO)_2(CNR)_2$ (R = alkyl) and $C_3H_5MoCl(CO)(CNBu^t)_3$, and the reduced products $[MoCl(CNBu^t)_4]_2$ and cis-$Mo(CO)_2(CNR)_4$ (R = Me, Et). No rationale for the loss of allyl and allyl chloride in the latter two cases was proposed (*206*). These reactions are rare examples of the formation of low-oxidation state metal–isocyanide complexes via reductive elimination of allyl or allyl chloride from metal–allyl species. The potential applications of mono-, bis-, and tris-π-allylic systems as precursors to low-oxidation state compounds remain to be explored. Substitution and simultaneous reduction of $Mo(SBu^t)_4$ also occurred on reaction with $CNBu^t$ to give $Mo(SBu^t)_2(CNBu^t)_4$ (*207*) (see Section IV,D,2).

Isonitriles displace dinitrogen quantitatively from cis-$M(N_2)_2$ $(PMe_2Ph)_4$ (M = Mo, W) (*56*) and $trans$-$M(N_2)_2(L_2)_2$ {M = Mo, L = $PMePh_2$ (*56*); M = Mo, W, L_2 = dpe (*55*)} in benzene solution under UV irradiation. The products formed from the cis isomers were cis-$M(CNR)_2(PMe_2Ph)_4$, $trans$-$M(CNR)_2(PMe_2Ph)_4$, mer-$M(CNR)_3(PMe_2Ph)_3$, and cis-$M(CNR)_4$-$(PMe_2Ph)_2$, and from the trans isomers, mer-$Mo(CNR)_3(PMePh_2)_3$, $trans$-$Mo(CNR)_4(PMePh_2)_2$ (*56*), and $trans$-$M(CNR)_2(dpe)_2$ (*55*), with a range of alkyl and aryl isocyanides. The mechanism of the sequential displacement of N_2 from cis-Mo(N_2)$_2$(PMe$_2$Ph)$_4$ by CNMe involves dissociation via the intermediate cis-Mo(N_2)(CNMe)(PMe$_2$Ph)$_4$. For CNBu^t substitution, the only detectable intermediate appears to be Mo(CNBu^t)(PMe$_2$Ph)$_4$ and no intermediate was detected in reactions of $trans$-Mo(N_2)$_2$ (dpe)$_2$ with CNMe. An order of cis ligand labilization of N_2 was given as CNBu^t > CNMe > N_2 > NCR (208).

Qualitative molecular orbital calculations on MY$_2$L$_4$ (M = Mo, W; Y = N_2, CNMe; L = tertiary phosphine or arsine) have shown that isonitriles and dinitrogen behave as strong π acceptors (209). The π-acceptor properties, together with the electron donor properties of the ML$_4$ site, account for the carbene-like character of the CNC bond, resulting in particularly low ν(NC) values, reduced CNC bond angles, and fairly short Mo–C distances (see Section IV,D,2). The low temperature electronic absorption spectra of [Mn(CNPh)$_6$]$^{n+}$ (n = 1, 2) and M(CNPh)$_6$ (M = Cr, Mo, W) (210) and the far-IR and Raman spectra of [Mn(CNR)$_6$]$^+$ (R = Me, Ph, C$_6$H$_4$Cl-p) (211) have been recorded and assigned.

Extensions of the limited number of halo-isocyanide–rhenium complexes have been reported. Treatment of [Re$_2$Cl$_8$]$^{2-}$ with CNMe in methanol produced [ReCl$_5$(CNMe)]$^-$, one of the few examples of an anion containing an isocyanide ligand (118). The reaction of K$_2$ReI$_6$ with an excess of CNR in acetone or ethanol gave the violet paramagnetic Re$_3$I$_6$(CNC$_6$H$_{11}$)$_6$ (17) (L$'$ = CNC$_6$H$_{11}$) (212) or yellow diamagnetic ReI(CNC$_6$H$_4$Me-p)$_5$

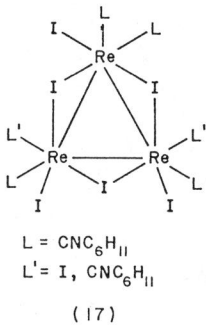

L = CNC$_6$H$_{11}$
L$'$ = I, CNC$_6$H$_{11}$

(17)

(213). With a deficiency of isocyanide and short reaction times in ethanol diamagnetic ReI$_3$(CNC$_6$H$_4$Me-p)$_4$ precipitated from solution, and ReI$_3$(CNC$_6$H$_4$Me-p)$_3$ was obtained from the mother liquors (213). Two products arise from reactions of Re$_3$I$_9$ with isocyanides. The trimer Re$_3$I$_6$(CNC$_6$H$_{11}$)$_3$ (18) results from dichloromethane solution whereas Re$_3$I$_9$(CNC$_6$H$_{11}$)$_3$ (17) (L$'$ = I) is the product formed from ethanol (212). The type of products obtained from the bromo analog K$_2$ReBr$_6$ depends on both the isocyanide and the reaction conditions; ReBr$_3$(CNC$_6$H$_4$Me-p)$_3$ is

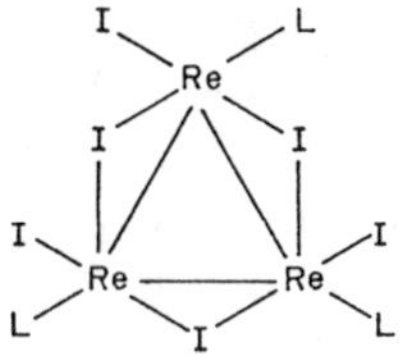

L = CNC$_6$H$_{11}$

(18)

the sole product from ethanol with this isocyanide (213), whereas [ReBr$_2$(O)(CNC$_6$H$_{11}$)$_2$]$_2$ (19) is formed under the same conditions with

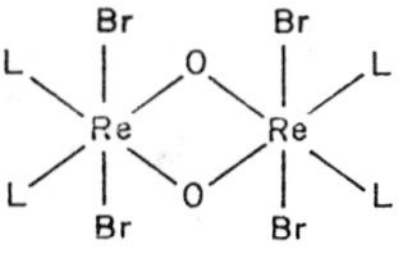

L = CNC$_6$H$_{11}$

(19)

CNC$_6$H$_{11}$. Neat cyclohexylisocyanide and K$_2$ReBr$_6$ produced trans-ReBr$_2$(CNC$_6$H$_{11}$)$_4$ (212), whereas ReBr$_3$(CNR)$_4$ resulted from CNR (R = Me, C$_6$H$_4$Me-p) and ReBr(CO)(CNR)$_4$ and bromine (214).

Isocyanides have been used to probe the bonding sites in hemoproteins and related compounds (215–219). Additions of CNCH$_2$Ph, PBu$_3^n$, and P(OEt)$_3$ to Fe(II)cap (cap = dianion of capped porphyrine) has produced five-coordinate low-spin iron(II) compounds by π-acid destabilization of the dz^2 orbital of the iron(II) porphyrin (215).

Few publications relating to new isocyanide complexes of iron, ruthenium and osmium in the (II) or (III) oxidation state have appeared. Reactions of MX$_3$L$_n$ (M = Ru, X = Cl, Br, L = AsPh$_3$, PPh$_3$, n = 2, 3; M = Os, X = Cl, L = AsPh$_3$, n = 3) with CNC$_6$H$_4$Me-p have given MX$_3$(CNC$_6$H$_4$Me-p)$_x$L$_{3-x}$ [M = Ru, X = Cl, Br, L = AsPh$_3$, x = 2 (220), L = PPh$_3$, x = 1 (221); M = Os, X = Cl, L = AsPh$_3$, x = 2 (222)], which in the osmium case was isolated in three isomeric forms. The isomerization of trans-RuCl$_2$(CNBu^t)$_2$(PPh$_3$)$_2$ occurs under UV irradiation via the photo elimination of PPh$_3$ (223), whereas for trans-RuX$_2$(CNEt)$_2$(EPh$_3$)$_2$ (X = Cl, Br; E = P, As, Sb) isomerization to the cis isomer is effected in high boiling solvents (224).

Treatment of the η^6-arene compounds [C$_6$H$_6$RuCl$_2$]$_2$ with cyclohexyliso-cyanide gave only C$_6$H$_6$RuCl$_2$(CNC$_6$H$_{11}$), whereas trans-RuCl$_2$(CNR)$_4$ (R = aryl) was formed with aryl isocyanides (225). These latter compounds were synthesized earlier by Malatesta *et al.* (*226*) as mixtures of the cis and trans isomers from $RuX_3 \cdot xH_2O$ and the appropriate isocyanide in boiling alcohol, and recently as intermediates in the reduction of mixtures of $RuCl_3 \cdot xH_2O$ and isocyanide with Na/Hg to ruthenium(0) isocyanide compounds (*16*). Considering the widespread interest in clusters and the possible routes to isocyanide clusters by reduction of species like $FeX_2(CNR)_2$ and $MX_2(CNR)_4$ (M = Fe, Ru) it is surprising that few simple routes to metal–halo isocyanide complexes in this group have appeared. In fact until recently there were no osmium(II) halo isocyanide complexes known. Recent investigations have found, however, high yield syntheses to the osmium chlorocompounds $[OsCl_2(CNR)_2]_x$ ($x > 2$), $CODOsCl_2(CNR)_2$ and cis- and trans-$OsCl_2(CNR)_4$ (R = Bu^t, xylyl) from the polymer $[CODOsCl_2]_x$ ($x > 2$) and from $CODOsCl_2(NCMe)_2$ (*32*).

Useful compounds have recently been synthesized from reactions of $CpRuCl(PPh_3)_2$ with CNR (R = alkyl, aryl) (*68*). The products $[CpRu(CNR)_x(PPh_3)_{3-x}]PF_6$ ($x = 1$, 2) and $CpRuCl(CNR)_x(PPh_3)_{2-x}$ ($x = 1$, 2) might be precursors to some interesting reaction products, especially $CpRuCl(CNR)_2$, which could provide a route to the as yet unknown $[CpRu(CNR)_2]_2$.

Metal–metal bond cleavage in $Ru_2Cl(O_2CMe)_4$ is effected by $CNBu^t$ to yield trans-$Ru(O_2CMe)_2(CNBu^t)_4$. Bond weakening was ascribed to the ability of $CNBu^t$ to delocalize electron density not only from the antibonding HOMOs but also from the metal–metal bond orbitals (*11*). Attempts have been made to effect reductive elimination of CH_3COOH from $RuH(O_2CMe)(PPh_3)_3$ by $CNBu^t$ without success. The complex isolated was $RuH(O_2CMe)(CNBu^t)_2(PPh_3)_2$, and considering in general the tendency of aryl isocyanides to preferentially stabilize lower oxidation states, when compared with alkyl isocyanides the choice of $CNBu^t$ may not have been propitious (*227*).

Displacement of labile groups from ruthenium(II) and osmium(II) cations has led to a series of very interesting reaction products. For example, during an investigation into the solution dynamics of the highly reactive salt $[CODRuH(NH_2NMe_2)_3]PF_6$, the dienyl complexes $[(\eta^1,\eta^2-C_8H_{13})Ru(CNR)_4]PF_6$ (R = Bu^t, xylyl) were isolated as intermediates in the conversion of the hydride to the η^3-allyl compounds (*228*). These isocyanide complexes undergo some unusual decomposition reactions in solution and in the solid state (*36*). Refluxing $[(\eta^1,\eta^2-C_8H_{13})Ru(CNBu^t)_4]PF_6$ in dry acetone for 12 hr gave cyclooctene and 4-methylpent-3-ene-2-one. In deuteroacetone, cyclooctene-d_1 was formed, together with a range of deuterated and partially deuterated aldol condensation products. The evidence pointed to a homolytic cleavage of the Ru–C bond. This conclusion was further enhanced by the isolation of the dimeric ruthenium(I) salt $[Ru_2$-

$(CNXylyl)_{10}](PF_6)_2$ from corresponding reactions of $[(\eta^1,\eta^2\text{-}C_8H_{13})Ru\text{-}(CNXylyl)_4]PF_6$. When $[(\eta^1,\eta^2\text{-}C_8H_{13})Ru(CNBu^t)_4]PF_6$ was refluxed in acetone containing water, methanol, ethanol, or n-propyl mercaptan the dimeric ruthenium(II) complexes (20) (X = O,S) were obtained.

L = CNBu^t

(20)

In contrast, melting the salt $[(\eta^1,\eta^2\text{-}C_8H_{13})Ru(CNXylyl)_4]PF_6$ produced four organic products (Fig. 2), together with the dimer $[Ru_2(CNXylyl)_{10}](PF_6)_2$. No mechanism has yet been proposed for these reactions but once again a radical process appears to be implicated (36). The ruthenium methyl complex $[CODRu(Me)Cl(NCMe)]_2$ has also provided a route to reactive ruthenium(I) complexes; the dimer $Ru_2Cl_2(CNXylyl)_6$ (21) (L =

$$L_3Ru \overset{\overset{\displaystyle Cl}{\diagdown}\ \ \overset{\displaystyle Cl}{\diagup}}{\underset{}{=\!=\!=\!=\!=\!=}} RuL_3$$

(21)

CNXylyl) formed in high yield from reactions with CNXylyl in acetone (18).

Other ruthenium and osmium compounds that have been investigated as precursors to isocyanide complexes are $[CODOsCl(N_2H_4)_3]^+$, $[CODM(N_2H_4)_4]^{2+}$ (M = Ru, Os) and $[CODRu(NH_2NHMe)_4]^{2+}$, and these formed mer-$[M(CNBu^t)_3(N_2H_4)_3]^{2+}$ and $trans$-$[M(CNR)_4L_2]^{2+}$ (M = Ru, Os; L = N_2H_4, NH_2NHMe; R = Bu^t, CH_2Ph, xylyl) from alcohols, and the hydrazones mer-$[Ru(CNBu^t)_3(NH_2NCMe_2)_3]^{2+}$, $trans$-$[Ru(CNBu^t)_4\text{-}(NH_2NCMe_2)_2]^{2+}$, and $[CODOs(CNR)_2(NH_2NCMe_2)_2]^{2+}$ (R = Bu^t, xylyl) from acetone, and CNR (36, 229).

Reactions of isocyanides with metal–metal multiple-bonded dimers of molybdenum, rhenium, ruthenium, and rhodium have effected cleavage of

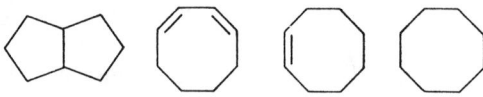

FIG. 2.   The four organic products produced from melting the salt $[(\eta^1,\eta^2\text{-}C_8H_{13})Ru(CNXylyl)_4]PF_6$.

the metal–metal bond in all cases except for rhodium. It is proposed that the binuclear axial adducts $Rh_2(O_2CR^1)_4(CNR)_2$ [R^1 = H, Me, Et; R = C_6H_{11} (230); R^1 = Me; R = Bu^t (11)], formed from $Rh_2(O_2CR^1)_4$, retain a metal–metal bond because the π-acceptor properties of the isocyanide strengthen this bond by effectively delocalizing electron density from HOMO's which are metal–metal antibonding orbitals (11).

Full papers have appeared on the formation and reactivity of the compounds $ML(CNR)_2$ (M = Ni, Pd, Pt; L = O_2, azobenzene, olefin, diazofluorene, acetylene) (231–237) (see also Sections IV,D,2 and V,D). Complexes of the type $Ni(olefin)(CNBu^t)_2$ have been prepared for a large range of olefins (234, 237). The isocyanide stretching frequencies have been measured and related to the electron-withdrawing properties of the olefin. Other unsaturated molecules such as imines, diazenes, ketones, nitroso compounds, and acetylenes have been similarly studied. The effect of substituent change has been found to be cumulative and an empirical relationship has been developed to predict $v(NC)$ (237).

Other related complexes that have been characterized are $ML(CNBu^t)_2$ [M = Ni, Pd, L = 9-diazofluorene; M = Ni, L = diphenyldiazomethane (22) (238)], $ML(CNBu^t)(PPh_3)$ (M = Ni, Pd; L = 9-diazofluorene) (238,

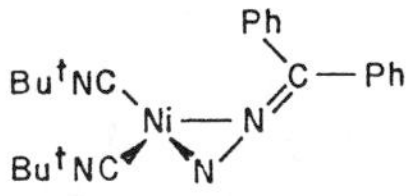

(22)

239) and $M\{C(CF_3)_2NN{=}C(CF_3)_2\}(CNR)_2$ (M = Ni, Pd, R = Bu^t; M = Pd, R = C_6H_{11}) (240). The platinum compounds $PtL(CNBu^t)_2$ {L = olefin (241), acetylene (48)} were prepared from PtL(COD) and CNBu^t, or from $Pt_3(CNBu^t)_6$ and the olefin. $Pt(Ph_2C_2)(CNBu^t)_2$ further reacts with $Pt(C_2H_4)(PPh_3)_2$ to give 23 (241a). The displacement of ethylene from

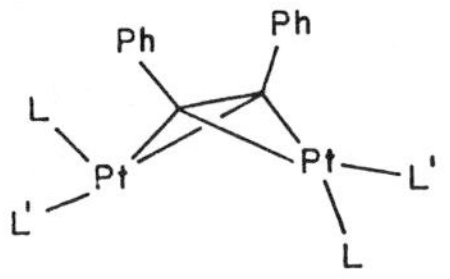

L = CNBu^t
L' = PPh$_3$

(23)

$Pt(C_2H_4)(PPh_3)_2$ (85) or the reductive elimination of 1-cyanoalkane from $PtH\{(CH_2)_nCN\}(PPh_3)_2$ (n = 1, 3) (242) with isocyanides gave $Pt(CNR)_2(PPh_3)_2$ (R = alkyl, aryl). In the latter reaction the intermediate $PtH(CH_2CN)(CNR)(PPh_3)$ was isolated.

Addition of the chelating diisocyanide p-1,8-diisocyanomethane (DMB) to nickel chloride or bromide in methanol has given the salt $[Ni_2X(DMB)_4](PF_6)_3$ (X = Cl, Br), on addition of a counteranion (*37*).

Novel dinuclear platinum(I) hydrides were obtained by treating $[Pt_2(dpe)_2H_3]X$ with CO or CNR (R = Me, Bu^t, C_6H_4Me-p). For R = Me, or C_6H_4Me-p, IR data of the violet compound indicate a bridging isocyanide group [ν(CN) 1645–1660] and, in the absence of ν(Pt–H), a bridging hydride group (**24**). For bulky R = Bu^t however, the complex isolated is yellow and has ν(CN) at 2160 and ν(Pt–H) at 1995 cm^{-1} indicating that both groups are terminal (**25**) (*243*).

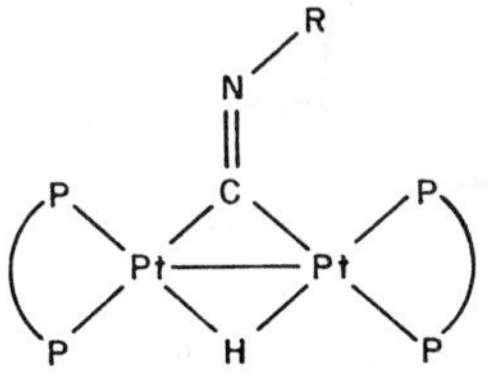

P——P = dpe   L = CNBu^t

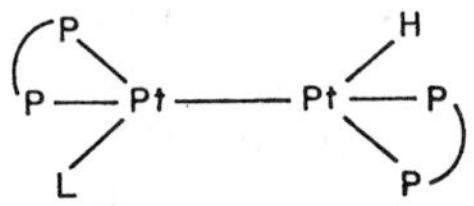

(24)   (25)

A range of nickel(II), palladium(II), and platinum(II) and (IV) complexes of general formulas $[MX(CNR)L_2]^+$ (M = Ni, Pd, Pt; X = halide, Me; L_2 = COD; L = phosphine, arsine, CNR) (*244–246*), $[MXY(CNR)_2]^{n+}$ (M = Pt; X = Y = Me, CF_3, halide, SCN, X = Me, Y = Cl, n = 0; X = Y = phosphine, n = 2; (*246–248*), $[PtR_2^1(CNR)_2L_2]^+$ (R^1 = Me, CF_3; L = phosphine) (*248, 249*) and $[PtMe_3(CNR)_xL_{3-x}]^+$ (x = 1, 2; L = phosphine) (*250, 251*) have been prepared with alkyl and aryl isocyanide ligands. The five-coordinate salt $[PtI(CNMe)_2(PR_3)_2]BF_4$ was formed by adding $[Bu_4^tN]I$ to $[Pt(CNMe)_2(PR_3)_2](BF_4)_2$ (*252*).

Trimetallic complexes of platinum(II) have been synthesized by interacting the neutral halides $PtX_2(CNR)_2$ (R = Bu^t, C_6H_{11}; X = Cl, I) with a series of carbonylmetallate anions to give the neutral compounds *trans*-M-Pt(CNR)$_2$-M {M = Co(CO)$_4$, Co(CO)$_3$PPh$_3$, Fe(CO)$_3$NO, Mn(CO)$_5$, CpM^1(CO)$_3$; M^1 = Cr, Mo, W} (*253*).

D. Physical Properties

1. NMR Studies

Some generalizations can be made from the ^{13}C NMR spectra recorded for metal–isocyanide complexes. In many cases difficulties are experienced in observing the resonance of the isonitrile carbon ($254-257$), but the use of relaxation agents, long accumulation times, and repetition times of 1 sec have partially overcome this problem (258). The isonitrile ligating carbon resonance, which should appear as a $1:1:1$ triplet from $^{13}C-^{14}N$ spin couplings, is usually observed as a broad singlet or a triplet with a more intense central peak stemming from quadrupolar relaxation of the coupling. $J(^{13}C-^{14}N)$ couplings have been observed in the range $12-30$ Hz (254), trans-$J(^{31}P-^{13}C)$ couplings in $[PtCl(CNR)(PEt_3)_2]^+$ were of the order of 15 Hz (258), and $J(Pt-^{13}C)$ were found to be between 1720 and 1742 Hz (258).

The ^{13}C isonitrile carbon chemical shift exhibits considerably greater sensitivity to changes in bonding and molecular structure (254, 256) than carbonyls, but is generally observed in the range $105-175$ ppm (254, 258) [cf. free isonitriles resonate between 154 and 165 ppm (254, 259)].

Dimeric and cluster compounds containing isocyanide ligands exhibit dynamic behavior in solution. The dimer $Cp_2Mo_2(CO)_5CNMe$ exists as a mixture of nonbridged isomers and permutamers which are rapidly interconverted by unimolecular processes at RT. The isocyanide passes rapidly from one metal atom to the other via carbonyl or carbonyl–isocyanide doubly bridged species (205).

Variable temperature NMR studies on $Mn_2(CO)_7(CNMe)_3$ have shown that the isonitrile ligands rapidly exchange between the two metal atoms. This rearrangement is energetically very similar to that in $Cp_2Fe_2(CO)_2(CNMe)_2$ and implies that geometric variations might not be important factors in determining the way that CO and CNR move to and from bridging positions (260).

Studies on the fluxional nature of $[CpFe(CO)_2]_2$ and isocyanide substituted derivatives continue to attract much attention. Considerable evidence now supports a mechanism proposed by Adams and Cotton (65) which involves initial pairwise opening of the bridge to yield an intermediate which may undergo rotation about the Fe–Fe bond. The dimer $Cp_2Fe_2(CO)_4$ exists as a cis–trans mixture in solution and NMR studies have shown that while cis–trans interconversions and bridge-terminal CO exchange in the cis isomer is slow on the NMR time scale, bridge-terminal exchange in the trans isomer is rapid. Warming results in the onset of cis–trans interconversion and bridge-terminal CO exchange in the cis isomer at equivalent rates. In terms of the Adams–Cotton mechanism, bridge-terminal exchange in the trans isomer may take place without Fe–Fe bond rotation on the unbridged intermediate, whereas bridge-terminal exchange in the cis isomer requires Fe–Fe bond rotation which also results in cis–trans interconversion. The significantly higher activation energy for the latter process is ascribed to the need for bond rotation.

Strong support for this mechanism has come from variable temperature

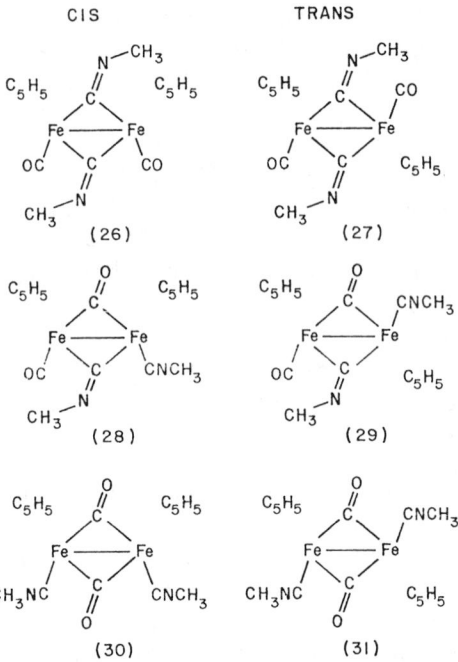

FIG. 3.   The six possible isomers of $Cp_2Fe_2(CO)_2(CNMe)_2$. Reprinted with permission from *J. Am. Chem. Soc.* **95,** 6592 (1973). Copyright 1973 American Chemical Society.

NMR studies on $Cp_2Fe_2(CO)_2(CNMe)_2$ (*186*). As a consequence of these rules the six possible isomers of this complex (excluding optical isomers) (Fig. 3) should interconvert in two definite sets (**26, 28,** and **30; 27, 29,** and **31**) in which it is not possible to convert members of the one set into members of the other set (assuming that there is no intermediate nonbridged structure with both isonitriles on the same metal atom). The NMR observations were in keeping with these predictions.

For compounds containing bridging isonitrile groups a further source of isomerism and stereochemical nonrigid behavior is the orientations of the alkyl or aryl groups, which can give rise to syn (**32**) and anti (**33**) isomers (*65*).

A detailed investigation of such isomers and their rates of interconversion has been carried out on $Cp_2Fe_2(CO)_{4-x}(CNMe)_x$ ($x = 2, 3$) (*185*). In both cases it was shown that the expected isomers do occur and can be observed at low temperature. The barriers to inversion at nitrogen were relatively low and the step itself was not rate determining.

In the complex $Cp_2Fe_2(CO)_3CNBu^t$ there is no detectable concentration of an isonitrile-bridged isomer, and the spectroscopic evidence is consistent

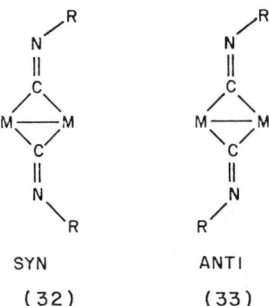

SYN   ANTI

(32)   (33)

with the molecule existing in solution primarily as a rapid interconverting mixture of cis and trans isomers, with the exchange of the isonitrile ligand between metal atoms being the slower process ($187-189$).

For $Cp_2Fe_2(CO)_2(CNBu^t)_2$ two types of isomerization take place in solution, through a selectivity in pairwise bridge opening. The low temperature isomerization is accounted for by the Adams–Cotton rules but in the second higher temperature isomerization, the isomers with the two closed sets, defined by Adams and Cotton, interconvert by asymmetric bridge cleavage giving two isonitriles on one metal atom (190).

Other complexes synthesized which also exist in solution as interconverting isomers were $(dienyl)_2M_2(CO)_{4-x}(CNR)_x$ (M = Fe, Ru; $x = 1, 2$; dienyl = Cp, C_5H_4Me; R = alkyl, aryl) ($191-195$) and the isomer distribution from these series was found to be related to the electron withdrawing properties of R (increase in electronegativity favors bridging) and the bulk of R (increase in size favors terminal bonding) (194).

Variable-temperature NMR studies on $M_2(CNR)_9$ (M = Fe, R = Et, Pr^i; M = Ru, R = Pr^i) have revealed dynamic intramolecular behavior involving bridge-terminal isocyanide exchange, this process occurring via synchronous pairwise exchange with inversion at nitrogen (17). With $Co_2(CNBu^t)_8$, no evidence in solution for nonbridged species in the intramolecular exchange process was observed (19).

A ^{13}C NMR study of the compounds $Os_3(CO)_{12-x}(CNR)_x$ ($x = 1, 2$, R = Me, C_6H_4OMe-p, Bu^n; $x = 1-4$, R = Bu^n, Bu^t) has found axial and equatorial isomers at low temperatures in solution, the ratio of which depends upon the size of the isocyanide group. Thus, for the monosubstituted derivative, the axial isomer is present in significant concentrations in solution at low temperatures except for $CNBu^t$, for which an equilibrium mixture of the axial and equatorial isomer is obtained at $-60°C$. As substitution increases the proportion of the equatorial isomer increases except for $CNBu^t$, for which in the tetrasubstituted derivative an all-axial isomer is found. Above $0°C$ these isomers rapidly interconvert (174).

The IR and ^{13}C NMR spectra of $Ir_4(CO)_{12-x}(CNR)_x$ ($x = 1-4$; R = Me), indicate that most of these molecules adopt structures related to $Ir_4(CO)_{12}$ in that they contain no bridging carbonyl or isocyanide groups (66) whereas with $Ir_4(CO)_{12-x}(CNBu^t)_x$ ($x = 3-7$) all have either bridging carbonyl or isonitrile ligands (32, 66). A variable-temperature ^{13}C NMR study on $Ir_4(CO)_{11}CNBu^t$ has established that a carbonyl scrambling process occurs through at least two distinct routes. The lower energy process yields two sets of carbonyl environments of relative intensities 10:1. These phenomena can be described by an idealized $T_d \rightleftharpoons C_{3v}$ process that avoids an isocyanide bridge intermediate or transition state and by which carbonyl "a" (Fig. 4) (L = $CNBu^t$) remains magnetically distinct. This is the formal reverse of the mechanism proposed (261) for $Rh_4(CO)_{12}$. The higher energy exchange process equilibrates all carbonyl ligands and could be described as an analogous traverse, but one in which a bridging isocyanide state is allowed (66).

The $Ni_4(CNR)_7$ molecules are stereochemically nonrigid in solution, undergoing a two step intramolecular rearrangement between $-10°$ and 80°C, and probably a fast dissociative process, $Ni_4(CNR)_7 \rightleftharpoons Ni_4(CNR)_6 + CNR$, between 80° and 120°C. The first intramolecular process averages ligand environments between the unique apical site and either the terminal base or the bridging sites, and in the second, higher temperature process, terminal-bridge exchange of ligands analogous to that operative in $Rh_4(CO)_{12}$ (261) is proposed. At 0°C the ^{1}H NMR spectrum of $Ni_4(CNBu^t)_7$ consists of three methyl resonances with intensities 3:3:1, consistent with the solid-state structure (see Section IV,D,2) (60).

The platinum cluster $Pt_3(CNBu^t)_6$ has been found to undergo dynamic

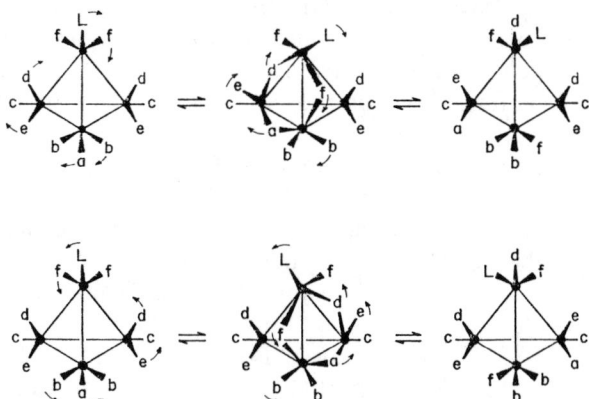

FIG. 4. Proposed ligand migration process for $Ir_4(CO)_{11}CNBu^t$. Reprinted with permission from *Chem. Rev.* **78,** 642 (1978). Copyright 1978 American Chemical Society.

behavior via an intermolecular process involving terminal and bridging isocyanide site exchange catalyzed by free ligand (23).

2. X-Ray Structures

The product of the reaction of VCl_3 and $CNBu^t$ in ethanol has been shown crystallographically to be mer-$VCl_3(CNBu^t)_3$ (109), and not the product derived from insertions into V–Cl bonds as previously proposed (see Section VI,A,4) (204). $[V(CNBu^t)_6][V(CO)_6]$ has also had its structure elucidated (110).

The X-ray structural determination of $Mo(CN)_4(CNMe)_4$ confirmed Orgel's prediction (262) that d^2 MX_4Y_4 molecules should adopt a dodecahedral geometry (72). Attempts to rationalize the factors governing molecular geometry in seven-coordinate systems of molybdenum and tungsten have led to a series of X-ray structural determinations. The structures are intermediate between two types, a 4:3 piano stool geometry or a monocapped trigonal prismatic structure (1). The structure of $WI_2(CO)_2(CNBu^t)_3$ approaches closely the piano stool geometry (115), whereas those of $[(Ph_3B)_2CN][Mo(SnCl_3)(CNBu^t)_6]$ (263) and $[Mo(CNMe)_7](BF_4)_2$ (9) lie between these two geometries and can be considered distorted examples of either. The structures of the salts $[MoX(CNBu^t)_6]X$ [X = Br (116), I (264)] and $[M(CNBu^t)_7]X_2$ [M = Mo, X = PF_6 (265); M = W, X_2 = W_6O_{19} (8)] approximate to a capped trigonal prism of C_{2v} symmetry. At this stage all that can be said about the various heptacoordinate geometries is that they depend upon both the anion and the nature of the isocyanide alkyl substituent.

Shake-up satellite structure in the X-ray photoelectron spectra of $[Mo(CNR)_7](PF_6)$ (R = Me, Bu^t, C_6H_{11}) has been observed. The similarity of the nitrogen $1s$ and carbon $1s$ Is/Ip ratios to those of $Mo(CO)_6$ oxygen $1s$ and carbon $1s$ Is/Ip ratios argues for a similarity in bonding, as a decrease in the metal–carbon bond length (i.e., stronger M–C bonding) will influence both the satellite position relative to the primary peak and the Is/Ip intensity ratio (266).

The X-ray analysis of $Cp_2Mo_2(CO)_5CNMe$ (205), $[MoCl(O)(CNMe)_4]I_3$ (267), and $CpMoI(CO)_2(CNPh)$ (268, 269) have found the compounds to have trans rotamer, trans, and trans diagonal structures, respectively. A detailed structural characterization of $Mo(SBu^t)_2(CNBu^t)_4$ has found the molecule to be substantially deformed from the ideal octahedral geometry so that the S–Mo–S and C–Mo–C angles in the equatorial plane are 115.3° and 73.7°. A molecular orbital analysis of the model $Mo(SH)_2(CNH)_4$ has traced the deformation to the d^4 electron count. The lowest lying unoccupied MO consists of an S p–S p bonding combination and an Mo d orbital. The resulting imbalance in S–S bonding leads to an opening up of the S–Mo–S angle (*207*).

The crystal structures of [Mn(CNR)$_6$]I$_3$ [R = Et (*270*), Ph (*271*)] and Mn(CO)$_3$(CNR)$_2$Br [R = Me (*272*), Ph (*273*)] have been resolved. The monosubstituted product Mn$_2$(CO)$_5$(CNR)(dpm)$_2$ (R = Me, CH$_2$Ph, C$_6$H$_4$Me-*p*) formed from Mn$_2$(CO)$_5$(dpm)$_2$ and CNR converts to Mn$_2$(CO)$_4$(CNR)(dpm)$_2$ on heating. The structural elucidation of the *p*-tolyl isocyanide derivative (**34**) has confirmed a four-electron doubly bridged

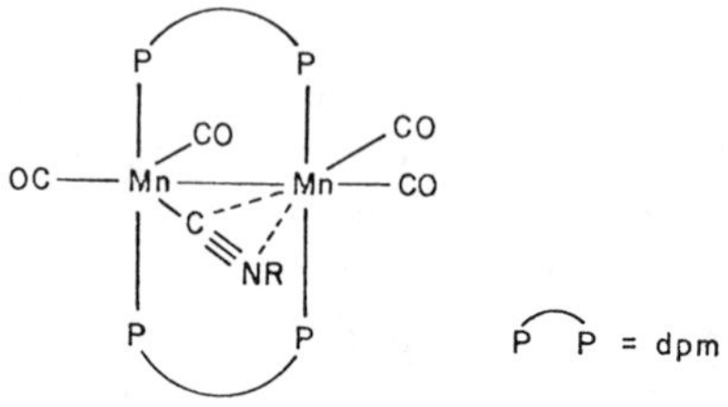

(34)

isocyanide ligand, analogous to the carbonyl bonding mode (**6**) in the corresponding carbonyl product (*61*).

The reaction between Re$_2$(CO)$_4$(PhCCPh)$_4$ and CNR (R = Bu^n, Bu^t, C$_6$H$_4$OMe-*p*, CH$_2$SO$_2$C$_6$H$_4$Me-*p*) was accompanied by coupling of the acetylenes into a chain, as shown by the X-ray structure of the product Re$_2$(CO)$_4$(PhCCPh)$_3$(CNCH$_2$SO$_2$C$_6$H$_4$Me-*p*)$_2$ (**35**) (*274*). The salt

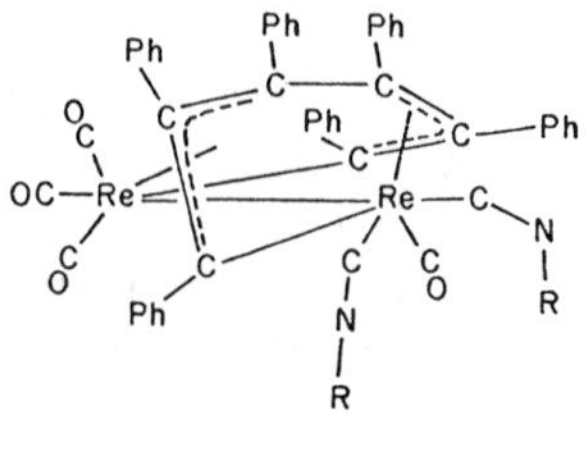

R = CH$_2$SO$_2$C$_6$H$_4$Me -*p*

(35)

[NBu$_4^n$][ReCl$_5$(CNMe)] has been structurally characterized (*118*). The elucidation of the structure of ReBr$_3$(CNC$_6$H$_4$Me-*p*)$_4$ has found the complex to have a capped octahedral geometry about the metal with three isocyanides and three bromide ligands in fac-octahedral arrangement and the fourth isocyanide situated on the face bounded by the three isocyanide groups (*214*).

The molecular structures of the diiron metal–metal bonded dimers $Fe_2(CNEt)_9$ (**36**) (*17*), *cis*-$Cp_2Fe_2(\mu$-CO$)_2$(CO)CNBu^i (*275*), *cis*-$Cp_2Fe_2(\mu$-

(36)

CO$)_2$(CO)CNBu^t (*187*), $Cp_2Fe_2(\mu$-CNR$)_2$(CO)$_2$ [R = Me(*cis-anti*) (*276*), Ph(*trans-anti*) (*277*)], *cis*-$Cp_2Fe_2(\mu$-CO)$(\mu$-CNPr^i)(CO)(CNPr^i) (*194*), and $Cp_2Fe_2(\mu$-CNR$)_2$(CNR$)_2$ [R = Ph (*278*), xylyl (*279*)] have been resolved. The dimensions within the cyclopentadienyl derivatives agree remarkably well.

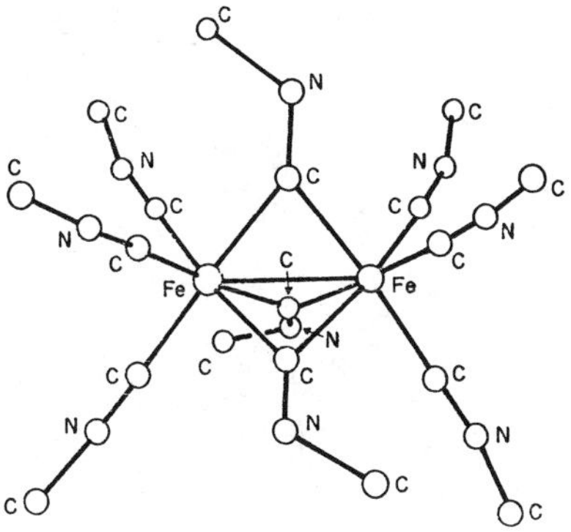

Three structural determinations have been reported on monomeric complexes containing bent isocyanide ligands. Ru(CNBu^t)$_4$PPh$_3$ has a trigonal–bipyramidal structure with PPh$_3$ and two bent isocyanide ligands (mean C–N–C 130(2)°) occupying the equatorial sites, and two essentially linear isocyanides occupying the axial positions (*16*). The iron complex Fe(CNBu^t)$_5$ has a similar structure with the CNC angle averaging 134° for two radial isocyanides and 154° for one axial isocyanide (*16*). A preliminary report on the structure of *trans*-Mo(CNMe)$_2$(dpe)$_2$ has found the CNC angle to be 156(1)° (*280*). In the iron and ruthenium complexes the bent isocyanide ligands account for the stretching vibrations observed in the region 1815–1870 cm^{-1} in their IR spectra.

The ruthenium(I) dimer [Ru$_2$(CNXylyl)$_{10}$](BPh$_4$)$_2$ is composed of two perfectly eclipsed Ru(CNXylyl)$_5$ units joined by a Ru–Ru bond of 3.00 Å (**37**). In spite of this long bond the molecule is remarkably stable, reacting over a few hours with CF$_3$COOH to give [Ru(O$_2$CCF$_3$)(CNXylyl)$_5$]PF$_6$ (*36*). The structure of the ruthenium compound [(η^1,η^2-C$_8$H$_{13}$)Ru(CNBu^t)$_4$]PF$_6$ (**38**) has also been resolved (*36*).

The product of the reaction of [CODOs(N$_2$H$_4$)$_4$](BPh$_4$)$_2$ with CNBu^t in acetone has been shown to be [CODOs(CNBu^t)$_2$(NH$_2$NCMe$_2$)$_2$]-(BPh$_4$)$_2$·(acetone)$_2$ (**39**) (*281*).

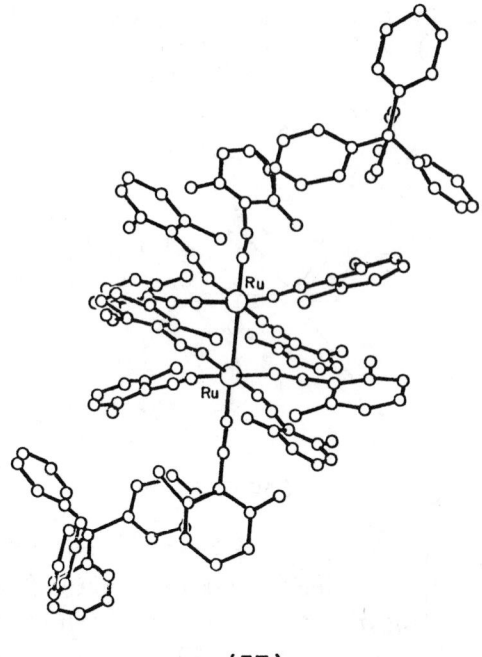

(37)

(38)

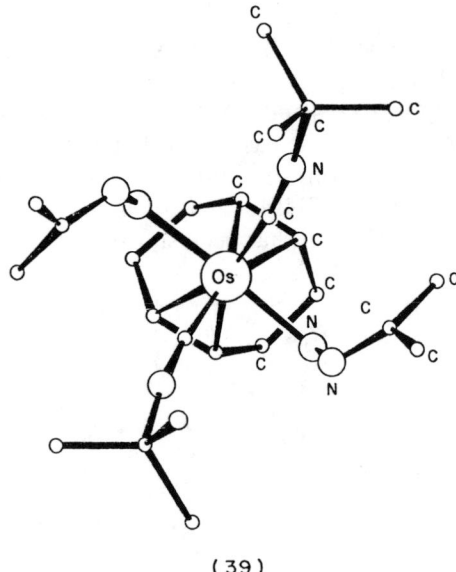

(39)

Structural determinations of $Os_6(CO)_{16}(CNBu^t)_2$ (*282*) and $Os_6(CO)_{18}(CNC_6H_4Me-p)_2$ (*62*) have shown the former to have the same Os_6 bicapped tetrahedron of the parent compound $Os_6(CO)_{18}$, whereas the latter has a rearranged metal skeleton with one isocyanide bridging three osmium atoms (**40**). With the ruthenium product $Ru_5(CO)_{14}(CNBu^t)_2$ (*63*), isolated

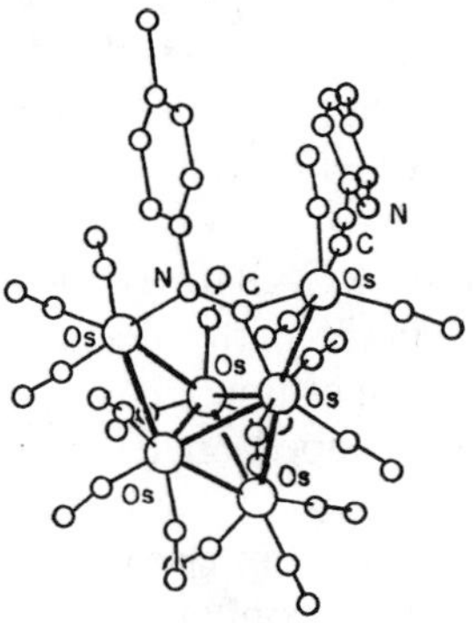

(40)

from $Ru_3(CO)_{11}CNBu^t$ by moderate heating, the molecule contains an open "swallow" cluster with one isocyanide acting as a six-electron donor and bonded to all five ruthenium atoms (**5**). The complexes serve as further examples that the stereochemistry of the cluster unit is sensitive to the number of electrons donated (*173*). The crystal structure determinations of the products (μ-H)(H)Os$_3$(CO)$_{10}$(CNBu^t) and (μ-H)$_2$Os$_3$(CO)$_9$(CNBu^t), formed from (μ-H)$_2$Os$_3$(CO)$_{10}$ and CNBu^t, have shown the former to have an axial isocyanide group, whereas in the latter the isocyanide is in an equatorial position (*177*). Both structures relate to the parent carbonyl, as does that of (μ-H)$_4$Os$_4$(CO)$_{11}$CNMe (*283*).

The structure of Co$_2$(CNBu^t)$_8$ has been shown to be **41** (*19*).

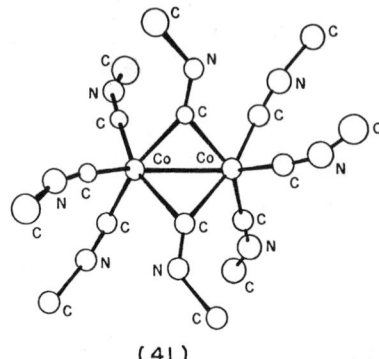

(41)

The stereochemistry of the pentakisisocyanidecobalt(I) and (II) complexes is apparently a function of crystallization procedures. To date, four isomeric structures have been identified for isocyanide complexes of cobalt(I) (*121*) and three for those of cobalt(II) (*284*). Crystal structure determinations of [Co(CNPh)$_5$]ClO$_4$·CHCl$_3$ (*285*) and [Co(CNPh)$_5$]-(ClO$_4$)$_2$·$\frac{1}{2}$ClCH$_2$CH$_2$Cl (*284*) have shown the coordination around the cobalt to be square pyramidal, whereas with [Co(CNC$_6$H$_4$Me-p)$_5$][Co(NMA)$_3$] (NMA = nitromalonaldehyde) a trigonal bipyramidal structure was found for the cation (*286*).

The ability of the cations [M(CNR)$_4$]$^+$ (M = Rh, Ir) to self-associate is a function of steric crowding in the molecule, and for bulky R groups monomeric species predominate. An estimate has been made of the steric size of isocyanides in terms of fan-shaped angles and as part of this study the structure of RhCl(CNC$_6$H$_2$Bu$_3^t$-2,4,6)$_3$ has been elucidated (*126*). The structural determinations of a series of dimeric rhodium(I) isocyanide salts have been completed. An eclipsed configuration was found for [Rh$_2$(CN(CH$_2$)$_3$NC)$_4$](BPh$_4$)$_2$·NCMe (**42**)[2] (*287*), whereas [Rh$_2$(CNPh)$_8$]-(BPh$_4$)$_2$·NCMe had a staggered configuration (**43**)[3] (*131, 132*) and

[2] Reprinted with permission from *Inorg. Chem.* **19**, 2463 (1980). Copyright 1980 American Chemical Society.

[3] Reprinted with permission from *Inorg. Chem.* **17**, 831 (1978). Copyright 1978 American Chemical Society.

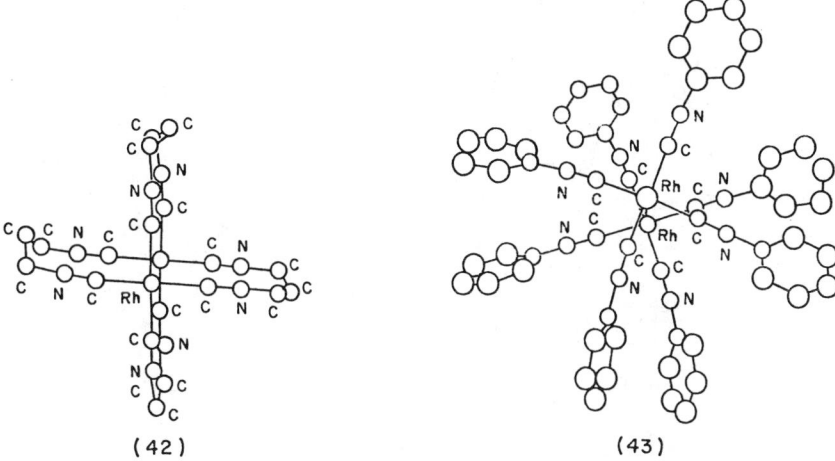

(42)   (43)

$[Rh_2(L_2)_4](PF_6)_2 \cdot (NCMe)_x$ [L_2 = 2,5-dimethyl-2,5-diisocyanohexane, $x = 2$ (287); L_2 = 1,8-diisocyanomethane, $x = 0$ (288)] have partially staggered configurations. Rhodium–rhodium distances varied from 3.19 to 3.26 Å. The molecular structure of $Ir_4(CO)_{11}(CNBu^t)$ has been resolved (289).

X-ray structural determinations of $NiL(CNBu^t)_2$ [L = O_2 (235), azobenzene (232), diazofluorene (236), diphenylacetylene (231)] have confirmed the square-planar geometry predicted by INDOR calculations on $NiL(CNH)_2$ [L = O_2 (290), HN=NH (291), HC≡CH (291)]. The cation in the salt $[Ni_4(OMe)_4(O_2CMe)_2(TMB)_4](BPh_4)_2$ (TMB = 2,5-dimethyl-2,5-diisocyanohexane) contains a cubane arrangement of Ni(II) and methoxide ions; the acetates bridge two nickel ions on opposite faces of the cube and the isocyanide ligands bridge the remaining four faces (**44**).[4] The change from paramagnetism to diamagnetism with variations in temperature for this salt has been related to the Ni–O–Ni angle in that the larger angles are coupled antiferromagnetically and that the ferromagnetic interaction is associated with the smaller angles (292).

The metal(0) isocyanide clusters $Ni_4(CNBu^t)_7$ (**45**)[5] (60), $Pt_3(CNBu^t)_6$ (**46**) (23) and $Pt_7(CNXylyl)_{12}$ (**47**) (20) have been characterized crystallographically. In the nickel complex the nickel atoms define the vertices of a highly compressed C_{3v} tetrahedron, with each nickel having a terminal isocyanide and three basal nickel atoms joined by three four-electron donor isocyanide ligands. The unusual feature of the structure of $Pt_7(CNXylyl)_{12}$ is that one

[4] Reprinted with permission from *Inorg. Chem.* **20**, 2394 (1981). Copyright 1981 American Chemical Society.
[5] Reprinted with permission from *J. Am. Chem. Soc.* **97**, 2572 (1975). Copyright 1975 American Chemical Society.

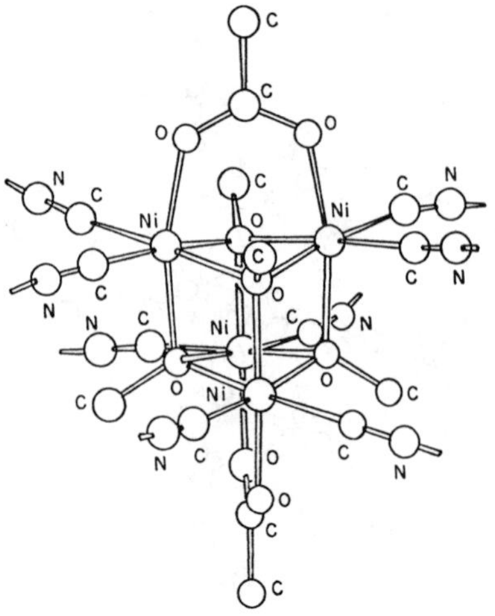

(44)

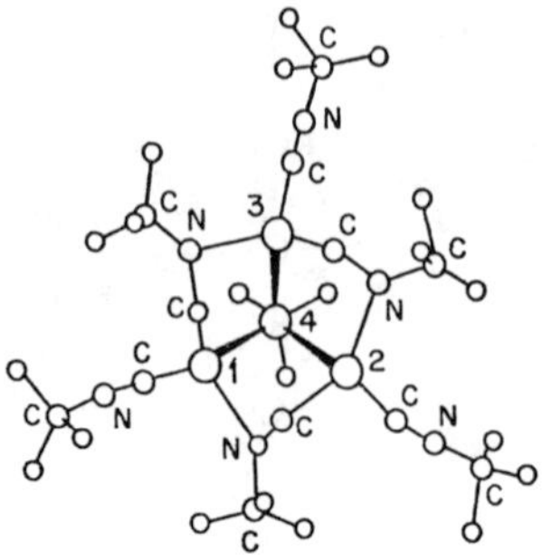

(45)

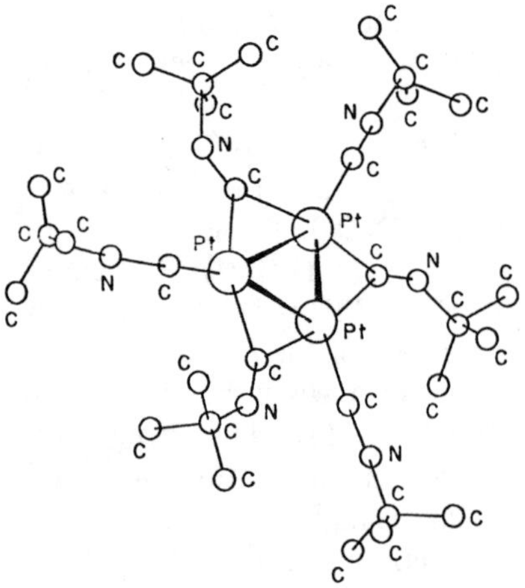

(46)

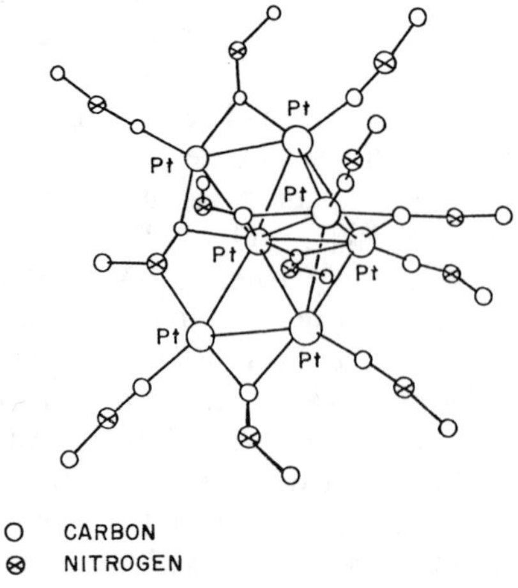

(47)

○   CARBON
⊗   NITROGEN

isocyanide ligand is coordinated to three platinum atoms in a manner similar to that found in $Os_6(CO)_{18}(CNC_6H_4Me-p)_2$ (62). The molecular geometry of $[Pd_2(CNMe)_6](PF_6)_2$ has been found to be **48**,[6] with a Pd–Pd bond length of 2.53 Å and a dihedral angle of 86.2° between the two coordination planes (293). A metal–metal stretching frequency of 156 cm^{-1} and Pt–Mo bond lengths of 2.89 Å are indicative of relatively strong

[6] Reprinted with permission from *J. Am. Chem. Soc.* **97**, 1961 (1975). Copyright 1975 American Chemical Society.

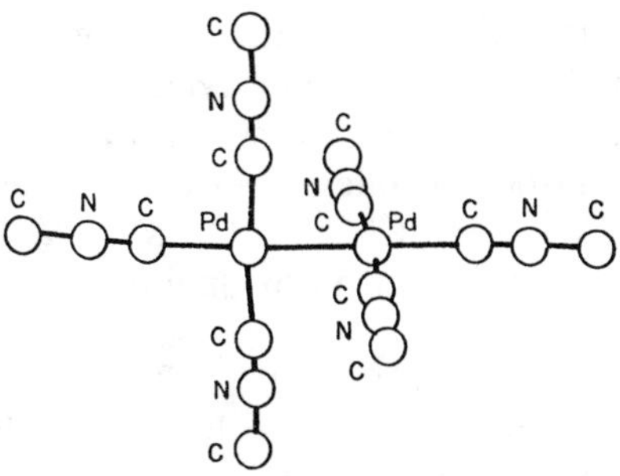

(48)

Pt–Mo bonding in the trinuclear complex *trans*-Pt{C(OEt)-NHC$_6$H$_{11}$}(CNC$_6$H$_{11}$)[Mo(CO)$_3$Cp]$_2$ (49) (*294*). The crystal and molecular

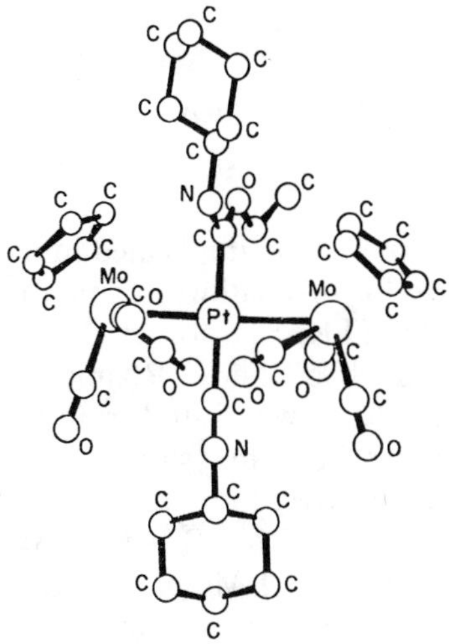

(49)

structures of *trans*-PdI$_2$(CNBu^t)$_2$ (*295*), *cis*-PtCl$_2$(CNPh)$_2$ (*296*), and *cis*-PtCl$_2$(CNEt)PEt$_2$Ph (*297*) have been resolved.

V

REACTIONS OF METAL ISOCYANIDE COMPLEXES

A. Substitution Reactions

A rare example of photodissociation of an isocyanide ligand has appeared with the formation of M(CNR)$_5$py (M = Cr, Mo, W; R = Ph, C$_6$H$_3$Pr^i-2,6) from $M(CNR)_6$ by irradiation in pyridine solution. The proposed mechanism is dissociative for chromium, but associative for molybdenum and tungsten via a bimolecular excited-state substitution pathway. Irradiation of $M(CNR)_6$ (R = 2,6-diisopropylphenyl) in $CHCl_3$ yields the one electron oxidation products $[M(CNR)_6]Cl$ (M = Cr, Mo, W), whereas $M(CNPh)_6$ (M = Mo, W) gives the two electron oxidation products $[M(CNPh)_6Cl]Cl$ (38, 298).

The reaction of the dimers $Co_2(CNR)_8$ with carbon monoxide effected fission of the metal–metal bond to give $[Co(CNR)_5][Co(CO)_4]$ (R = Bu^t, xylyl) (19, 25, 123). With diphenylacetylene, **50** is produced (19).

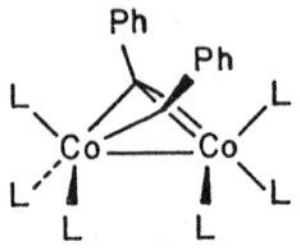

L = CNBu^t

(50)

Radical intermediates have been implicated in the addition of TCNE to $[Co(CNMe)_5]^+$. Evidence for this pathway came from the detection of the TCNE radical anion by ESR, and by the high yield preparation of $[Co(CNMe)_4TCNE]^+$ from $[Co(CNMe)_5]^{2+}$ and K[TCNE] (46). No evidence for radical intermediates was found from kinetic measurements of addition of TCNE to similar rhodium systems (299). Free rotation of TCNE coupled with Berry-type pseudorotations in the TCNE adducts $[M(CNR)_4TCNE]^+$ (M = Co, Rh; R = Me, Bu^t, C_6H_4OMe-p) and $[Rh(CNR)_2L_2(TCNE)]^+$ [R = C_6H_4Me-p, C_6H_4OMe-p; L = $P(OMe)Ph_2$, $P(OMe)_2Ph$, $P(OPh)_3$] were observed during NMR studies (300–302).

A number of monomeric complexes have been prepared which may be considered as substitution products of the $[Co(CNR)_5]^+$ and $[M(CNR)_4]^+$ cations. Their general formulations are $[Co(CNR)_{5-x}L_x]^+$ [x = 1(303–306), 2(303–313), 3(303, 308, 313), 4(303)], $[M(CNR)_{5-x}L_x]^+$ [x = 2, 3, M = Rh (300, 314), Ir (314–316); x = 4, M = Ir (317)] and $[Rh(CNR)_2L_2]^+$ (300, 314) for a range of alkyl and aryl isocyanides, and L_x = a series of mono-, bi-, tri-, and tetradentate tertiary phosphorus compounds, or monoolefins {for cobalt (46, 318), rhodium (300, 319) or iridium (316)}. They have been prepared either from substitution reactions on the homoleptic isocyanide cation or by adding mixtures of isocyanide and ligand to some suitable labile metal precursor.

Addition of dpm to $[Rh(CNR)_4]^+$ or addition of mixtures of dpm and CNR to $[CODRhCl]_2$ has given the dimeric complexes $[Rh_2(CNR)_4$-

(dpm)$_2$]$^{2+}$ (**51**) [R = Bu^n (*320, 321*), Me, Bu^t, C$_6$H$_{11}$ (*320*)]. Other routes to

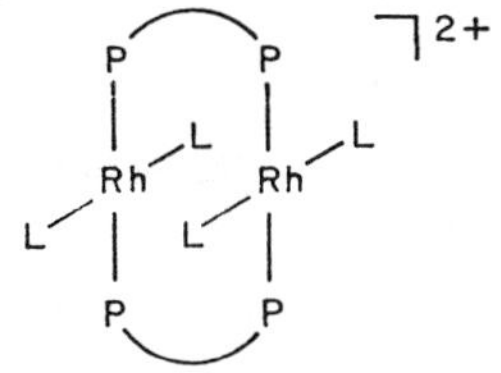

P‿P = dpm ,   L = CNR

(51)

similar compounds have involved substitution reactions on A-frame dimers. For example [Rh$_2$Cl$_2$(CO)$_2$(dam)$_2$], [Rh$_2$Cl(CO)$_2$(L$_2$)$_2$]$^+$, and [Ir$_2$Cl(CO)$_3$(L$_2$)$_2$]$^+$ gave [Rh$_2$(CNR)$_4$(dam)$_2$]$^{2+}$ (R = Me, Bu^t, Bu^n, C$_6$H$_{11}$), [Rh$_2$Cl(CO)$_{2-x}$(CNBu^t)$_x$(L$_2$)$_2$]$^+$ (*x* = 1, 2; L$_2$ = dpm, dam), [IrCl(CO)$_{3-x}$(CNBu^t)$_x$(L$_2$)$_2$]$^+$ (*x* = 1, 2; L$_2$ = dpm, dam), and [Ir$_2$(CO)-(CNBu^t)$_4$(L$_2$)$_2$]$^{2+}$ (L$_2$ = dpm, dam), respectively, on treatment with CNR in alcohols. The structures proposed for the dimers **52–54** reflect the differ-

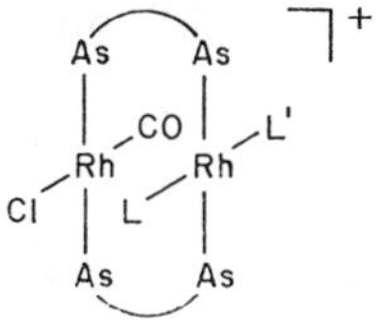

L' = CO, Bu^{t}NC

L = CNBu^t

(52)

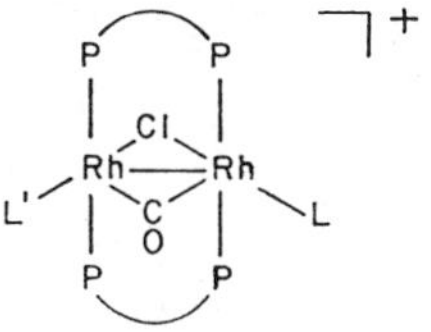

L' = CO, Bu^{t}NC

L = CNBu^t

(53)

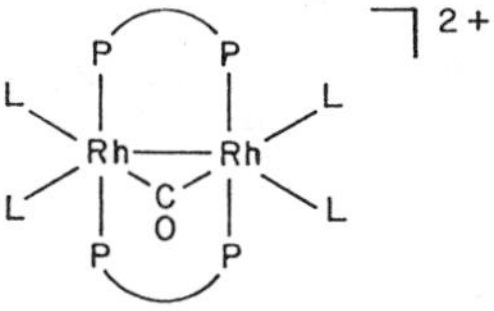

L = CNBu^t

(54)

ences in lability of the bridging carbonyl group between A-frame complexes of dpm and dam (*321, 322*).

The metal–metal bond in the dimers [MM^1(CNMe)$_6$]$^{2+}$ is quite robust and the complexes readily undergo substitution reactions with PPh$_3$ and halides or pseudohalides to give the dimers $[MM^1(CNMe)_{6-x}(PPh)_x]^{2+}$ (M = M^1 = Pd, x = 1, 2; M = M^1 = Pt, x = 1; M = Pd, M^1 = Pt, x = 1, 2) and $[Pd_2(CNMe)_4X_2]$ (X = Cl, I, SCN) (*323, 324*). In contrast, the metal–metal bond in the diphosphine-bridged dimers Pd_2X_2 (dpm)$_2$ (X = Cl, Br, I, SCN) is readily cleaved by a variety of simple addenda molecules. With isocyanides and isocyanates, both produced the same product **55** (R = Me, C_6H_{11},

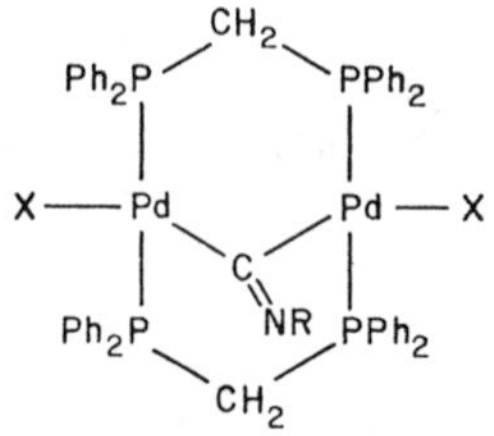

(55)

Ph, C_6H_4Me-p) (*325–327*). Further reactions on **55** with CNR or by addition of dpm to $[Pd_2(CNR)_6](PF_6)_2$ produced $[Pd_2(\mu\text{-}CNR)(CNR)_2$- $(dpm)_2](PF_6)_2$ (R = Me, C_6H_4Me-p) which for R = Me was characterized structurally (*327a*). The salt $[Pd_2\{\mu\text{-}C_2(CO_2Me)_2\}(CNMe)_2(dpm)_2](PF_6)_2$ (**56**) was also prepared during these studies (*325*). Addition of PPh_3 to the

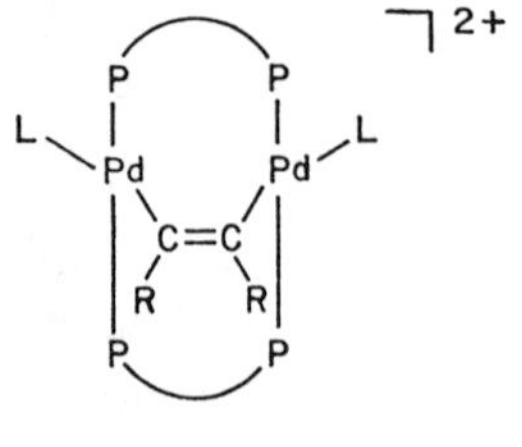

L = CNMe ,   P—P = dpm ,   R = CO_2Me

(56)

linear tripalladium cations $[Pd_3(CNMe)_8]^{2+}$ has given the disubstituted product $[Pd_3(CNMe)_6(PPh_3)_2]^{2+}$ (*144*).

Reactions of Ni(COD)$_2$ with isocyanides in the presence of aryl acetylenes gave the products $Ni_4(CNR)_4(ArC_2Ar)_3$ (R = Bu^t, C_6H_{11}, Ar = Ph; R = Bu^t, Ar = C_6H_4Me-p) together with Ni(ArC$_2$Ar)(CNR)$_2$. NMR studies on the addition of aryl acetylenes to Ni$_4$(CNR)$_7$ indicated only the formation of the trisacetylene adduct, with no evidence for intermediates being observed in this addition. However, addition of 3-hexyne to Ni$_4$(CNBu^t)$_7$

provided NMR evidence of three clusters $Ni_4(CNBu^t)_{7-x}(EtC_2Et)_x$ with $x = 1$, 2, or 3 in solution, although no complexes were isolated from this reaction. The alkyl acetylenes underwent facile exchange with bonded alkyl acetylenes in the cluster in contrast to $Ni_4(CNBu^t)_4(PhC_2Ph)_3$ which was nonlabile in solution and inert to intermolecular exchanges with diphenylacetylene (21).

B. *Chemical and Electrochemical Oxidation – Reduction Reactions*

Electrochemical investigations have been reported on a range of homoleptic and mixed carbonyl – isocyanide complexes, in attempts to rationalize substituent effects on the isocyanide with the electronic structure of the metal. The ease of electrochemical oxidations correlate with (i) the decrease in the number of carbonyls in the coordination sphere and (ii) the replacement of aryl with alkyl isocyanides (179, 182, 183, 312, 328 – 331). $E_{\frac{1}{2}}$ values have also been found to differ substantially for different isomers (182, 183). Cyclic voltammetry measurements on chromium isocyanides have given a direct measure of the thermodynamic stability of pairs of alkyl or aryl isocyanides of the group VI elements, showing the oxidation $Cr^{2+} \rightarrow Cr^{3+}$ is more favored for alkyl isocyanides whereas the reductions $Cr^{2+} \rightarrow Cr^+$ and $Cr^+ \rightarrow Cr^0$ are more favored for aryl isocyanides (15, 328). Other workers have confirmed the three sequential one electron oxidation processes $Cr^0 \rightleftharpoons Cr^+ \rightleftharpoons Cr^{2+} \rightleftharpoons Cr^{3+}$ (179), and a good correlation of these oxidation processes with the Hammett σ parameters assigned to substituent groups on the aryl ring has been found (332).

With the $[Co(CNR)_3(PR_3^1)_2]^+$ cations the primary reversible electron transfer is followed by a fast chemical reaction which brings about a change in configuration (and dimerization). It was not found possible to isolate intermediate products from these oxidations (333).

In other cases, however, isolation of oxidized species from electrochemical oxidations has been possible. For example, the salts $[Mn(CNMe)_6](PF_6)_2$, $[Mn(CO)(CNMe)_5](PF_6)_2$ (182), and $[Co(CNBu^t)_3(PPh_3)_2](PF_6)_2$ (312) were isolated from electrochemical studies on the corresponding manganese(I) and cobalt(I) systems. These oxidized species can be highly labile and hence constitute precursors to a range of other complexes. Thus, the salts $[Mo(bipy)_2(CO)_2(NCMe)](BF_4)_2$ and $[\{Mo(bipy)_2(CO)_2\}_2](BF_4)_2$, generated by redox reactions on *cis*-$Mo(bipy)_2(CO)_2$, provided a facile route to the isocyanide compounds *cis*-$Mo(CO)_2(CNC_6H_4Me$-$p)_4$, *cis,cis*-$Mo(bipy)(CO)_2(CNEt)_2$, $[Mo(bipy)_2$-

$(CNEt)_2](BF_4)_2$, $[Mo(bipy)(CNEt)_5](BF_4)_2$, and $[Mo(bipy)_2(CNEt)_3]$ $(BF_4)_2$ by simple substitution reactions (334).

Electrochemical investigations into reductions of the platinum(II) systems cis- and trans-dihalobis(isocyanide)platinum(II) (335) and M-Pt(CNR)$_2$-M (M = Mn(CO)$_5$, Fe(CO)$_3$NO, Co(CO)$_4$, CpM^1(CO)$_3$; M^1 = Cr, Mo, W; R = Bu^t, C$_6$H$_{11}$) (336) have shown a two-step reduction process. For PtX$_2$(CNR)$_2$ (X = Cl, Br, I; R = Bu^t, C$_6$H$_{11}$) the first reduction step is irreversible with the half-wave potential being critically dependent on the chemical nature of X. Potentiostatic reductive coulometry on PtX$_2$(CNR)$_2$ in the cavity of an ESR spectrometer produced a signal attributed to a monomeric paramagnetic Pt(I) complex. One halide ion is also liberated in this first step. An irreversible one-electron reduction step was also observed with the linear trimetallic compounds with a concomitant rupture of a Pt–metal bond to produce ·Pt(CNR)$_2$M radicals. These radical species in both systems are short-lived, and probably undergo dimerization or reaction with the solvent. The second reductive step is followed by decomposition of the products in both series. For the compounds cis-PtCl$_2$(CNR)(PMePh$_2$) (R = aryl) (337) and trans-[PtX(CNR)L$_2$]ClO$_4$ (X = Cl, Br; R = aryl; L = tertiary phosphine) (337a), electrochemical studies indicated that a stable platinum(I) dimer containing bridging isocyanide ligands was one of the main reduction products. This proposal was based upon strong ν(CN) bands in the region 1600–1660 cm^{-1} and terminal ν(Pt–Cl) bands in the 280–300 cm^{-1} range in the IR solution spectra. No products were isolated from these reductions.

Chemical oxidations of trans-M(CNR)$_2$(dpe)$_2$ (M = Mo, W) with iodine or Ag$^+$ gave the paramagnetic cations trans-[M(CNR)$_2$(dpe)$_2$]$^+$ (R = Bu^t, C$_6$H$_4$Me-p) and [MX(CNC$_6$H$_4$Me-p)$_2$(dpe)$_2$]$^+$ with chlorine or bromine (338). Attempts at single-electron oxidations on Mo(CNPh)$_6$ with NOPF$_6$ gave only the dark-purple nitrosyl [Mo(NO)(CNPh)$_5$]PF$_6$ (14). Nitrosyls of this type together with those of chromium have resulted from displacement reactions of NO and Cp with isocyanides in [Cr(NO)$_2$(NCMe)$_4$](PF$_6$)$_2$ (339) and [CpMo(NO)I$_2$]$_2$ (340, 340a), respectively, which gave [M(NO) (CNR)$_5$]PF$_6$ (M = Cr, R = Me, Bu^t, C$_6$H$_4$Cl-p; M = Mo, R = Me, Et, Pr^i, Bu^t, C$_6$H$_{11}$) and [Mo(NO)(CNR)$_4$I]. Further chemical oxidations of [Cr(NO)(CNR)$_5$]PF$_6$ produced paramagnetic [Cr(NO)(CNR)$_5$](PF$_6$)$_2$ and evidence for the existence of [Cr(NO)(CNR)$_5$]$^{3+}$ (339).

A nitromethane solution of NS$^+$PF$_6^-$ generated from N$_3$S$_3$Cl$_3$ and AgPF$_6$ (1:3) reacted with η^6-C$_6$H$_6$Cr(CO)$_3$ in acetonitrile solution to give the salt [Cr(NS)(NCMe)$_5$](PF$_6$)$_2$. In the presence of CNBu^t and zinc powder this thionitrosyl complex is reduced to [Cr(NS)(CNBu^t)$_5$]PF$_6$ whereas subsequent oxidation of this latter complex with either NOPF$_6$ or AgPF$_6$ formed [Cr(NS)(CNBu^t)$_5$](PF$_6$)$_2$ (341). The salt [Ni$_2$X(DMB)$_4$](PF$_6$)$_3$ (X = Cl, Br;

DMB = p-1,8-diisocyanomethane) reacted with iodine in the presence of [Ph$_4$As]I$_3$ to give [Ni$_2$X(DMB)$_4$](I$_3$)$_4$, in which an electron was removed from the $a_{2u}[d_z2(-)]$ Ni–X–Ni antibonding orbital (342).

The reductive coupling of adjacent isocyanide ligands in [MoX (CNBu^t)$_6$]$^+$ (X = Br, I) by zinc has produced the [Mo(CNBu^t)$_4$-(Bu^{t}HNCCNHBu^t)]$^+$ cation containing the novel (N,N^1-dialkyldiamino) acetylene molecule (343). Sodium amalgam reductions of the substituted halides CpMoX(CO)$_{3-x}$(CNMe)$_x$ (x = 1, 2) and MnBr(CO)$_{5-x}$(CNMe)$_x$ (x = 1–3) gave the anions [CpMo(CO)$_{3-x}$(CNMe)$_x$]$^-$ (160, 161) and [Mn(CO)$_{5-x}$(CNMe)$_x$]$^-$ (344), which were precursors to the products [CpMo(CO)$_{3-x}$(CNMe)$_x$]$_2$Hg, [Mn(CO)$_{5-x}$(CNMe)$_x$]$_2$Hg, CpMoR(CO)$_{3-x}$-(CNMe)$_x$, and MnR(CO)$_{5-x}$(CNMe)$_x$ (R = H, CH$_2$CN, HgI, GePh$_3$, SnMe$_3$, PbPh$_3$). Similar reductions with Cp$_2$Fe$_2$(CO)$_2$(CNMe)$_2$ produced CpFeH(CO)CNMe, which underwent a base-catalyzed reaction with MeCN and IVB halides to give CpFeR(CO)CNMe (R = CH$_2$CN, GeMe$_2$Cl, SnMe$_3$) and [CpFe(CO)CNMe]$_2$Hg (345).

C. Photochemical Redox Reactions

The blue cations [Ir(CNMe)$_4$]$^+$ react with L (L = CO, CNMe) in aqueous solution in light to give the orange-red five-coordinate cations [IrL(CNMe)$_4$]$^+$, isolated as the BPh$_4$ salt. The association does not take place in the dark and was originally proposed to be an example of the photochemical formation of a metal–ligand bond (photoassociation) (346). Later work, however, has provided evidence that in solution these blue cations are oligomers of the type [Ir(CNMe)$_4$]$_n^{n+}$, and the photochemical reaction is in fact a rapid photoinduced cleavage of these oligomers to give monomeric orange [Ir(CNMe)$_4$]$^+$. It is this monomer which reacts with CNMe or CO (= L) to give [IrL(CNMe)$_4$]$^+$ (347, 348). Analogous oligomers have been characterized in corresponding rhodium systems [Rh(CNPh)$_4$]$_n^{n+}$ (n = 2, 3) from characteristic $1a_{2u} \rightarrow 2a_{1g}$ and $2a_{1g} \rightarrow 2a_{2u}$ transitions at 568 and 727 nm, respectively, observed in the electronic absorption spectra. They were interpreted in terms of interactions expected between the occupied $a_{1g}(d_z2)$ and unoccupied $a_{2u}[p_z, \pi^*CNPh]$ monomer orbitals. Similar spectra were observed for the [Rh(CNR)$_4$]$_n^{n+}$ (R = Pr^i, Bu^t, vinyl, C$_6$H$_{11}$ (131, 132, 349) and [M(CNR)$_4$]$^{2+}$ (M = Pd, Pt; R = Me, Et) (349) cations.

This work has been extended by Gray and co-workers (37, 38, 135, 287, 288, 350–358) to the synthesis of dimeric rhodium(I) complexes containing bridging diisocyanopropane. The salts isolated, e.g., [Rh$_2$(L$_2$)$_4$]X$_2$ [L$_2$ = CN(CH$_2$)$_3$NC; X = Cl, BPh$_4$] (abbreviated Rh$_2^{2+}$), from reactions of [CODRhCl]$_2$ with the chelating diisocyanide have revealed some exceptionally interesting spectroscopic and redox properties. As with the CNPh dimer, the electronic absorption spectrum of the diisocyanide dimer contains the characteristic intense ($\epsilon = 14{,}500$) low-lying absorption band at 553 nm corresponding to the $1a_{2u} \rightarrow 2a_{1g}$ transition. Comparison of this transition for a range of chelating diisocyanides has shown that the rotameric configuration of the ligands has an important influence on the position of the lowest allowed electronic transition. For example, the structure of $[Rh_2(2,5\text{-dimethyl-}2,5\text{-diisocyanohexane})_4](PF_6)_2$ is partially staggered and a blue shift of the $1a_{2u} \rightarrow 2a_{1g}$ transition to 515 nm was observed (287).

A study of concentration effects of Rh_2^{2+} in methanol has shown that as the concentration increases low energy bands appear at 778, 990, 1140, and 1735 nm, the first three of which correspond to $2a_{2u} \rightarrow 3a_{1g}$, $3a_{2u} \rightarrow 4a_{1g}$, and $4a_{2u} \rightarrow 5a_{1g}$ transitions in tetrameric, hexameric, and octameric rhodium(I) oligomers. The band at 1735 nm appeared only at high concentrations and was assigned to an oligomer higher than octameric (359). Polarized single-crystal spectroscopic studies of the lowest triplet $\leftarrow$ singlet system have been completed (350), and the resonance Raman spectra of the $^1A_{1g}$ and $E_u(^3A_{2u})$ electronic states of Rh_2^{2+} have been recorded (351). Investigations of the redox chemistry of the relatively long-lived emission triplet $^3A_{2u}$ of Rh_2^{2+} has found that the Rh_2^{2+} and Rh_2^{3+} redox levels are produced by electron transfer reactions from the Rh_2^{2+} excited states (354), and flash kinetic studies on $[Rh_2(L_2)_4]^{2+}$ {L = $CNC_6H_4Me\text{-}p$, $L_2 = CN(CH_2)_3NC$} have given the lifetimes of these excited states (360). When Rh_2^{2+} is dissolved in 12 M HCl and irradiated with visible light (550 nm), H_2 and $Rh_2Cl_2^{2+}$ are produced (135). The reaction occurs in two distinct steps (356) [Eqs. (4) and (5)]:

$$Rh_2^{2+} + HCl \xrightarrow{\Delta} \tfrac{1}{2}Rh_4Cl_2^{4+} + \tfrac{1}{2}H_2 \tag{4}$$

$$\tfrac{1}{2}Rh_4Cl_2^{4+} + HCl \xrightarrow[550\ nm]{h\nu} Rh_2Cl_2^{2+} + \tfrac{1}{2}H_2 \tag{5}$$

and the proposed photoactive species $Rh_4Cl_2^{4+}$ was thought to be a linear mixed valence species, e.g., $[ClRh(I)Rh(II)Rh(II)Rh(I)Cl]^{4+}$.

The complex $[Rh_2Cl_2\{CN(CH_2)_3NC\}_4]^{2+}$ in Eq. (5) was also formed through a two-center oxidative addition reaction on Rh_2^{2+} with chlorine and characterized crystallographically (see Section V,D and E) (352, 358).

In 1 M H_2SO_4 in air Rh_2^{2+} slowly reacts according to Eq. (6) (356):

$$Rh_2^{2+} + H^+ + \tfrac{1}{4}O_2 \rightarrow \tfrac{1}{2}H_2O_2 + \tfrac{1}{2}Rh_4^{6+} \tag{6}$$

Flash photolysis on this species has generated a transient species which decays back to Rh_4^{6+} following second-order kinetics, suggesting that this oligomer is actually the dimer $\{Rh_2^{3+}\}_2$ (*353*). The dimer $\{Rh_2^{3+}\}_2$ also appears to be the product of the irradiation of the complex $Rh_2Cl_2^{2+}$ in 6 M HCl (*356*) at $-60°C$. These latter two intermediates were earlier assigned as $HRh_2Cl_2^{2+}$ (*135*) and $Rh_4H_2^{4+}$ (*352*), respectively.

Irradiation ($\lambda > 520$ nm) of $\{Rh_2^{3+}\}_2$ (= Rh_4^{6+}) in 12 M HCl solution produced H_2 and $Rh_2Cl_2^{2+}$. In 6 M HCl no dihydrogen is liberated because the photogenerated Rh_2^{2+} is trapped by Rh_4^{6+} to give Rh_6^{8+} (*356*).

The salt $H_3[Rh_4Cl\{CN(CH_2)_3NC\}_8]$ $[CoCl_4]_4$ has recently been obtained by addition of $CoCl_2·6H_2O$ to the photoactive solution of Rh_2^{2+} in 12 M HCl, and characterized by an X-ray structural analysis (see Section V,E). It is proposed that the photoactive species in 12 M HCl differs from the above cation only by the binding of an additional Cl^- ion (*357*).

It appears that additional polynuclear species are formed as Rh_4^{6+} is progressively reduced to Rh_2^{2+}, and a study of the binding of $[Rh_2(L_2)_4]^{2+}$ (L_2 = 2,5-dimethyl-2,5-diisocyanohexane) to Rh_4^{6+} has identified the hexanuclear cation $Rh_2(L_2)_4Rh_4^{8+}$, the octanuclear cation $Rh_2(L_2)_4Rh_4Rh_2(L_2)_4^{10+}$, and the dodecanuclear cation $\{Rh_2(L_2)_4Rh_4^{8+}\}_2$. An oxidation of the octanuclear cation gave $Rh_2(L_2)_4Rh_6^{6+}$ (*355*).

D. Oxidative-Addition Reactions

Cleavage of $[Cp_2Fe_2(CO)_3]_2CN(CH_2)_nNC$ (n = 2,6) with iodine produced $[CpFe(CO)I]_2CN(CH_2)_nNC$ and $CpFe(CO)_2I$ (*168*). Reactions of $Cp_2Fe_2(CO)_2(CNMe)_2$ with HgX_2 (X = Cl, Br, I) has given $CpFe(CO)_2HgX$ and $CpFe(CNMe)_2X$ as the sole products in quantitative yields. The proposed intermediate $Cp_2Fe_2(\mu\text{-}CO)\{\mu\text{-}CN(Me)HgX_2\}(CO)CNMe$ was not observed (*361*).

The electron-rich nature of the $M(CNBu^t)_5$ complexes is demonstrated by the ease with which they undergo oxidative addition. The compounds $M(CNBu^t)_5$ are protonated by nido-2,3-Me_2-2,3-$C_2B_4H_6$, $MnH(CO)_5$, $OsH_2(CO)_4$ and HBF_4 to give $[MH(CNBu^t)_5]X$ {M = Ru, X = nido-2,3-Me_2-2,3-$C_2B_4H_5$ (*362*); M = Fe, X = $Mn(CO)_5$, $OsH(CO)_4$, BF_4 (*363*)}. For $Fe(CNBu^t)_5$ some oxidative additions were followed by insertions and these are discussed in the relevant section. Noninserted products obtained were $[FeY(CNBu^t)_5]X$ (Y = Me, X = I; Y = C_6F_5, X = Br; Y = C_3H_5, X = Cl, Br) and $FeXY(CNBu^t)_4$ (X = CH_2Ph, Y = Br; X = $CF(CF_3)_2$, Y = I) (*364*).

The reactions of the cobalt dimers $Co_2(CNR)_8$ with halogens (= X_2), HY (Y = BF_4, ClO_4) in the presence of isocyanide, or with NO formed $CoX_2(CNXylyl)_4$, $[Co(CNXylyl)_5]Y$ (*25*, *123*) and $Co(NO)(CNBu^t)_3$ (*19*), respectively.

A novel series of dioxygen complexes $RhX(O_2)(CNR)L_2$ (X = halide; R = Bu^t, C_6H_{11}, C_6H_4Me-p; L = PPh_3, $AsPh_3$) and $[Rh(O_2)(CNBu^t)_2$-$(PPh_3)_2]Cl$ were prepared from interaction of the square-planar complexes in air. The CNBu^t derivatives bind dioxygen irreversibly, whereas with CNC_6H_4Me-p the binding is reversible (365).

Oxidative additions on the cations $[M(CNR)_4]^+$ (M = Rh, Ir) or on the substituted products derived therefrom have been performed with a variety of reagents including proton acids, halogens, alkyl, aryl, allyl, or acyl halides, $HgCl_2$, $SnCl_4$, $NOBF_4$, $[p$-$MeC_6H_4N_2]PF_6$, disulfides, and $[Me_3O]BF_4$. In most cases the product is derived from the well-established two-electron oxidative addition (125, 128, 314, 316, 317, 366), but in a few cases the resultant complex is derived from rare single-electron transannular oxidative additions. For example, treatment of $[Rh(CNBu^t)_3$-$(PPh_3)_2]PF_6$ or $[Rh(CNC_6H_4Cl$-$p)_2(PMePh_2)_2]PF_6$ with C_3F_7I or MeI, respectively, has afforded $[RhI(CNBu^t)_3(PPh_3)]_n(PF_6)_n$ and $[RhI(CNC_6H_4Cl$-$p)_2(PMePh_2)_2]_n(PF_6)_n$ (314). The diamagnetism of these complexes suggest n = 2 with a rhodium–rhodium bond. Unfortunately the fate of the C_3F_7 or Me residues was not investigated, and the mechanism was not discussed. These reactions may represent examples of metal–carbon homolytic bond fission. Other dimeric rhodium(II) compounds were formed from interacting $[Rh(CNR)_4]^+$ with $[RhX_2(CNR)_4]^+$, or from oxidation of $[Rh(CNR)_4]^+$ (R = alkyl) with half a mole of bromine or iodine (367). Corresponding complexes were obtained by treating the dimers (51) with iodine or trifluoromethyl disulfide which formed $[Rh_2X_2(CNR)_4(dpm)_2]^{2+}\{X = I$ (320), SCF_3 (321)}, or by reacting $[Rh_2\{CN(CH_2)_3NC\}_4]^{2+}$ with halogens or methyl iodide to give $[Rh_2XY\{CN(CH_2)_3NC\}_4]^{2+}$ (X = Y = Cl, Br, I; X = I, Y =Me) (353, 356, 358–359).

Further examples of unusual oxidation states stabilized by isocyanides are given by the reactions of $IrH_3(AsPh_3)_2(CNC_6H_4Me$-$p)$ with chelating ligands. Carboxylic acids, pentane-2,4-dione, and quadridentate Schiff bases have given the iridium(II) and iridium(IV) monomeric paramagnetic complexes $Ir(AsPh_3)(CNC_6H_4Me$-$p)L_2$ [L = O_2CR; R = Me, aryl (368), Schiff base (369)] and $Ir(AsPh_3)(CNC_6H_4Me$-$p)L_2$ (L = OC_6H_4O-o; OC_6H_4NH-o, $O_2CC_6H_4O$-o, $O_2CC_6H_4NH$-o, $O_2CCH(O)Me$), respectively, together with the iridium(III) compounds $IrH_2(AsPh_3)_2(CNC_6H_4Me$-$p)L$ (L = 1,1,1-trifluoropentane-2,4-dionato; $O_2CC_6H_4OMe$-p) (369).

The kinetics of the oxidative addition of iodine to $Rh(dtc)(CNC_6H_2Me_3$-$2,4,6)_2$ (dtc = S_2CNMe_2) have shown the reaction to proceed via a charge transfer complex $Rh(dtc)(CNR)_2 \cdot I_2$. This rearranges in two consecutive zero-order reactions with respect to iodine concentration, to give first the cis adduct, which then isomerizes to $trans$-$RhI_2(dtc)(CNR)_2$. This mechanism was further supported from kinetic data on the $I_2 + Rh(acac)(CNC_6H_2Me_3$-$2,4,6)_2$ system. The complexes $RhI(I_3)(dtc)(CNC_6H_2Me_3$-$2,4,6)_2$ and

RhI(I$_3$)(acac)(CNC$_6$H$_2$Me$_3$-2,4,6)$_2$ were isolated from reactions with excess iodine (*370*).

The compounds MO$_2$(CNBu^t)$_2$ (M = Ni, Pd) react with CNR, CO, CO$_2$, NO, N$_2$O$_4$, SO$_2$, C$_6$H$_5$NO, RCOX, and Ph$_3$CBF$_4$ to give products which form as a result of the following processes: (i) atom transfer redox reactions, (ii) atom transfer oxidation reactions, (iii) oxidative substitution reactions, and (iv) metal assisted peroxidation reactions. In the first two processes the O–O bond is cleaved, whereas in the remaining two it is retained. Examples of these reaction pathways are, for (i), the formation of Bu^{t}NCO and CO$_2$ from CNBu^t and CO interactions, for (ii), the formation of *cis*- and *trans*-M(NO$_3$)$_2$(CNBu^t)$_2$ [M = Ni(*trans*) and Pd(*cis*)] from N$_2$O$_4$, for (iii), the isolation of [Ni(NO)(CNBu^t)$_3$]NO$_3$ from NO reactions, and for (iv), the formation of benzoyl peroxide from benzoyl chloride (*233*). Addition of SO$_2$ to Pd$_3$(CNBu^t)$_6$ gave the compound Pd$_3$(SO$_2$)$_2$(CNBu^t)$_5$ in which an SO$_2$ molecule bridges two palladium atoms (**57**) (*371*).

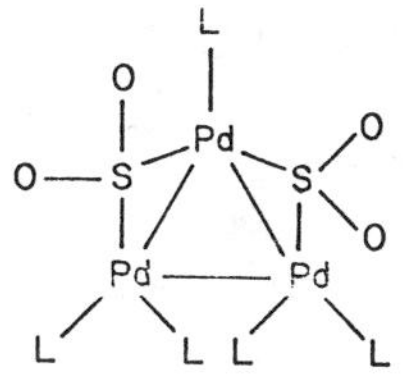

L = CNBu^t

(57)

The compound Pt(CNBu^t)$_2$(PPh$_3$)$_2$ undergoes oxidative additions with I$_2$, MeI, and CF$_3$I to give the well-defined stable five-coordinate salts [PtI(CNBu^t)$_2$(PPh$_3$)$_2$]I, and products formulated as either [PtR(CNBu^t)$_2$(PPh$_3$)$_2$]I (R = Me, CF$_3$) or [Pt{C(R)=NBu^t}(CNBu^t)(PPh$_3$)$_2$]I. With Ph$_3$SnCl a neutral compound analyzing for [Pt(CNBu^t)$_2$(PPh$_3$)$_2$]·SnPh$_3$Cl was isolated (*85*).

Treatment of Pt$_3$(CNBu^t)$_6$ with diphenylcyclopropenone has produced the yellow diplatinum complex (**58**), in which cleavage of the carbon–car-

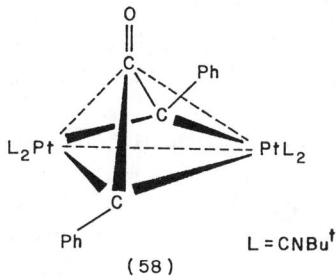

L = CNBu^t

(58)

bon double bond has occurred (47). It is interesting that platinum phosphine complexes have given the platinocyclobutenones, $L_2Pt\{C(=O)-C(Ph)=C(Ph)\}$, with diphenylcyclopropenone, and it is probable that the rearrangement of the preformed η^2-complex $Pt\{\eta^2-PhC=C(Ph)C(O)\}L_2$ (L = PPh$_3$ (372), CNBu^t, COD/2) is slowed down by the presence of the weaker σ donors sufficiently to permit attack by a second PtL_2 group on the bent σ orbitals of the C_3 ring. The protonated species $[Pt_2\{\mu-(PhC)_2C(OH)\}(CNBu^t)_2]^+$ was also observed (47).

With fluorocarbons $Pt_3(CNBu^t)_6$ gave the range of products $Pt\{OC(CF_3)_2OC(CF_3)_2\}(CNBu^t)_2$, $Pt\{\eta^2-CF(CF_3)CF(R)\}(CNBu^t)_2$ (R = F, CF$_3$), $Pt\{\eta^2-C(CN)_2C(CF_3)_2\}(CNBu^t)_2$, $Pt\{\eta^2-CF_2CF_2\}(CNBu^t)_2$, and $Pt\{\eta^2-CFClCF_2\}(CNBu^t)$, the latter of which then undergoes isomerization first to cis-$PtCl(CFCF_2)(CNBu^t)_2$ and then to the trans isomer (373).

A carbon–carbon double bond is cleaved in the reaction of hexakis(trifluoromethyl)benzene with $Pt_3(CNBu^t)_6$ to give 59 (374) (see Section V,E).

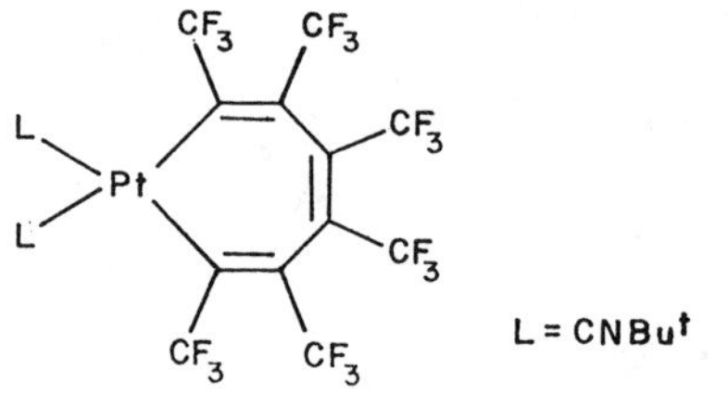

(59)

A variety of unusual products are formed from reactions of $Pt_3(CNBu^t)_6$ with acetylenes. With three equivalents of PhC≡CPh, Pt(PhC≡CPh)(CNBu^t)$_2$ is formed (48) which converts to the platinocyclopent-2,4-diene (60) (R = Ph) (Scheme 3) with excess acetylene (49). The symmetrical acetylene MeO$_2$CC≡CCO$_2$Me and the isocyanide trimer gave 61 which converted to 60 (R = CO$_2$Me) in refluxing toluene. The unsymmetrical acetylene MeC≡CCO$_2$Me, and RC≡CC≡CR (R = Me, Ph) produced the diplatinacyclobutene ring compound 62 (R = CO$_2$Me, R = Me; R = C≡CR^1; R^1 = Me, Ph) (49). The silane Pt(SiPh$_3$)$_2$(CNBu^t)$_2$ formed from $Pt_3(CNBu^t)_6$ and SiPh$_3$H (375).

Oxidative addition of X$_2$ (X = Br, I) to the dimers $[Pd_2(CNMe)_6]^{2+}$ gave the mixture $[Pd(CNMe)_4]^{2+}$, $[PdX(CNMe)_3]^+$, and $PdX_2(CNMe)_2$ (X = Br, I) (34).

A dealkylation occurred during oxidative addition reactions of chlorine to the cations $[PtCl(CNBu^t)(PMe_2Ph)_2]^+$ and $[Pt(CNBu^t)_2(PMe_2Ph)_2]^{2+}$ which produced $PtCl_3(CN)(PMe_2Ph)_2$, $[PtCl_2(CN)(CNBu^t)(PMe_2Ph)_2]^+$, and $PtCl_2(CN)_2(PMe_2Ph)_2$, respectively. Other oxidative additions were

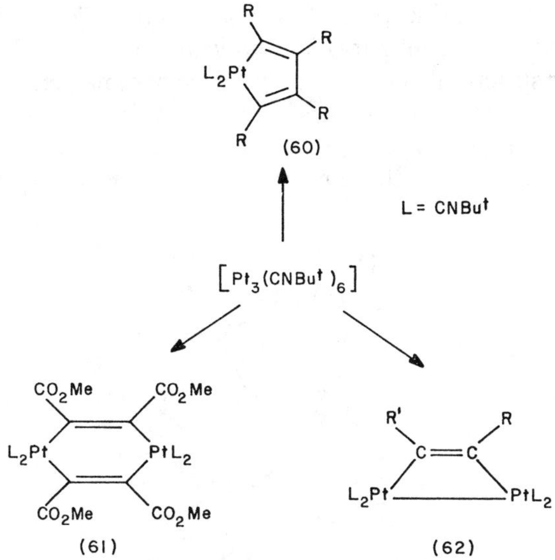

SCHEME 3

straightforward, giving $PtCl_4(CNR)_2$, $PtCl_4(CNR)(PMe_2Ph)$, $[PtCl_3-(CNR)(PMe_2Ph)_2]^+$, and $[PtCl_2(CNR)_2(PMe_2Ph)_2]^{2+}$ (R =Me, Bu^t, C_6H_{11}, C_6H_4Me-p) from the corresponding neutral, mono, and dicationic precursors (376).

E. X-Ray Structures

Crystal structure determinations of the cations $[MoX(CNBu^t)_4$ $(Bu^tHNCCNHBu^t)]^+$ (X = Br, I) (63) containing the (N,N-dialkylamino)

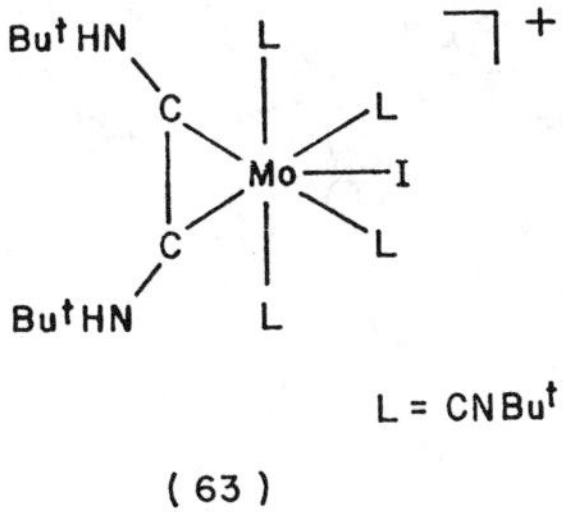

acetylene have been completed. The cations are very similar and resemble the capped trigonal-prismatic $[MoX(CNBu^t)_6]^+$ complexes (377). The structure of the hydride [RuH(CNBu^t)$_5$] [*nido*-2,3-Me$_2$-2,3-C$_2$B$_4$H$_5$] has been reported in a preliminary communication (*362*).

The X-ray structural determinations of the paramagnetic and diamagnetic modifications of CoI$_2$(CNPh)$_4$ have shown that crystals of the former contain monomeric CoI$_2$(CNPh)$_4$ units whereas those of the latter contain dimeric cationic units with a linear Co–I–Co bridge (**64**) and discrete I$^-$

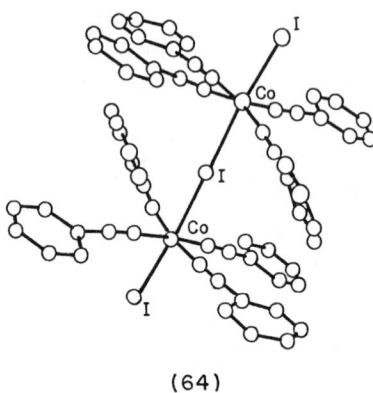

(64)

anions (*378*). The chemical interconversions between these two cobalt modifications has been studied (*379*). The crystal structures of [Rh$_2$I$_2$(CNC$_6$H$_4$Me-*p*)$_8$](PF$_6$)$_2$ (**65**) (*380*) and [Rh$_2$Cl$_2${CN(CH$_2$)$_3$NC}$_4$]-

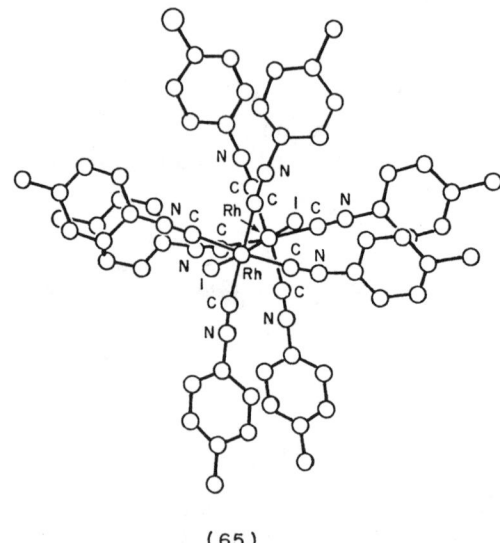

(65)

$Cl_2 \cdot 8H_2O$ **(66)**[7] (*352, 358*) have established the existence of rhodium – rho-

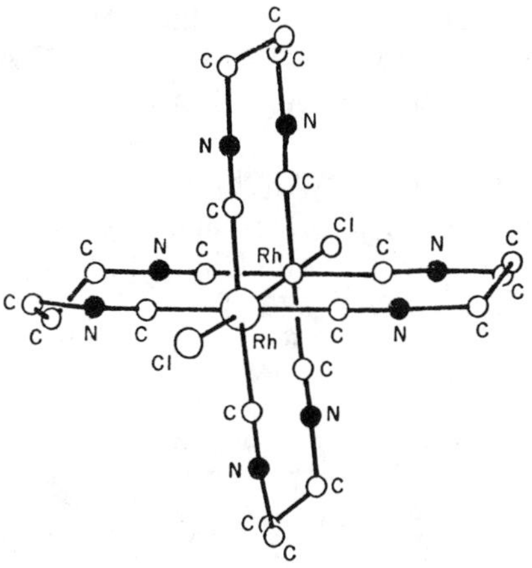

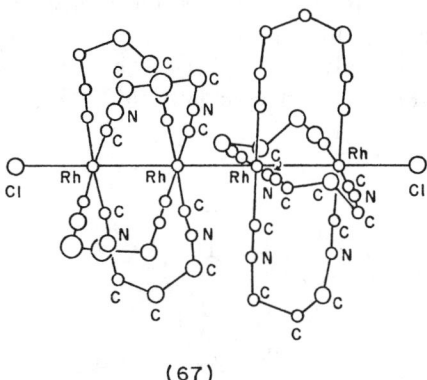

(66)

dium bonds (2.785 and 2.84 Å, respectively) with pseudo-octahedral geometry about each rhodium being completed by isocyanide and halide ligands. The salt $H_3[Rh_4Cl\{CN(CH_2)_3NC\}_8][CoCl_4]_4 \cdot nH_2O$ has been structurally characterized **(67)**.[8] The cation is made up of two binuclear

(67)

$Rh_2\{CN(CH_2)_3NC\}_4^{3+}$ units linked by an Rh – Rh bond. The Cl^- ion bridges the $Rh_4\{CN(CH_2)_3NC\}_8^{6+}$ units to form an infinite chain of repeat unit

[7] Reprinted with permission from *Inorg. Chem.* **18**, 2673 (1979). Copyright 1979 American Chemical Society.

[8] Reprinted with permission from *J. Am. Chem. Soc.* **102**, 3966 (1980). Copyright 1980 American Chemical Society.

(Rh_4Cl). The two binuclear $Rh_2\{CN(CH_2)_3NC\}_4^{3+}$ units have approximately eclipsed ligand systems but are nearly perfectly staggered with respect to each other. The Rh–Rh distance between the binuclear units is 2.775 Å and within the binuclear units Rh–Rh is 2.932 and 2.923 Å. Bond orders of $1\frac{1}{2}$ are proposed for these bonds, giving effectively a Rh(I)Rh(II)Rh(II)Rh(I) chain (*357*).

The structures of the dimers $Pt_2(C_3Ph_2O)\,(CNBu')_4$ (**58**) (*47*) and $[Pd_2(\mu\text{-}CNMe)(CNMe)_2(dpm)_2](PF_6)_2$ (*327a*) and the two trimers $Pd_3(SO_2)_2$-$(CNBu')_5$ (**57**) (*371*) and $[Pd_3(CNMe)_6(PPh_3)_2](PF_6)_2$ (**68**)[9] (*144*) have been

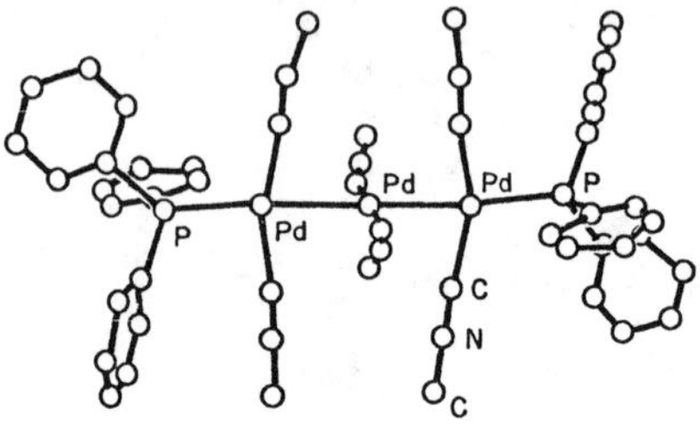

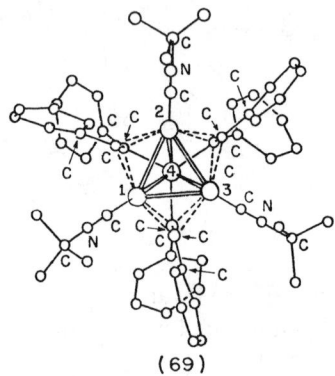

(68)

resolved. For the latter, the Pd atoms are collinear with the two phosphorus atoms, forming a five-atom chain. Interesting aspects of the structure are the short Pd–Pd bonds of 2.592(5) Å and the displacement of the equatorial isocyanide ligands toward the center of the molecule (*144*). The metal–metal force constants in this and related trimetallic cations has been estimated at ca. 0.7 mdyn Å⁻¹ (*381*). The solid-state structure of $Ni_4(CNBu')_4(PhC_2Ph)_3$ has been elucidated (**69**).[10] The unique apical nickel,

(69)

[9] Reprinted with permission from *J. Am. Chem. Soc.* **98**, 7431 (1976). Copyright 1976 American Chemical Society.

[10] Reprinted with permission from *Inorg. Chem.* **19**, 1553 (1980). Copyright 1980 American Chemical Society.

Ni(4), lies on the pseudo-threefold axis and is bonded to a terminal isocyanide. Complete details of this structure have not yet appeared but the acetylene bond order appears to be considerably reduced and may account for the catalytic activity of these complexes in converting acetylenes to exclusively *cis*-olefins (*21*).

VI

REACTIONS OF ISOCYANIDE LIGANDS

A. *Insertions and Related Reactions*

1. *Into Metal–Carbon Bonds*

a. Metal-Alkyl and -Aryls. Insertion reactions of isonitriles into metal–alkyl or metal–aryl bonds are now well established, occurring with metal–alkyl or –aryl groups from group IVA to IB and, recently, with uranium and thorium carbon bonds. In many cases the intermediate alkyl metal isocyanide complex could be isolated, and the ease of insertion (on the limited evidence available) can be related to the size of ligands in the coordination sphere, the bulk of the isonitrile (*382*), the nucleophilicity of the halide and migrating carbon, and the electron-withdrawing groups on the isocyanide and phosphine (*383, 384*). In comparison with carbonyl–metal–alkyl systems, isonitrile insertions are more facile, occurring in minutes (*43*) rather than hours for comparable carbonyl systems (*385*), giving iminoacyl groups which are thermodynamically more stable than their acyl counterparts, so much so that no reverse migration of an alkyl or aryl group from an iminoacyl ligand has yet been observed. In contrast to carbonyl systems, multiple insertions of isocyanides readily occur, especially with the more nucleophilic metals, giving usually heterometallacycles containing four- or five-membered rings. As a further contrast, $\mu^2-\eta^2$ and $\mu^3-\eta^2$ bonding modes as well as η^1 and η^2 modes have been observed in iminoacyl systems. In some instances unusual products have resulted from metal alkyl isonitrile interactions, as well as carbene complexes from electrophilic additions to the basic iminoacyl nitrogen atom.

Insertions to give iminoacyl ligands have now been observed to occur in titanium (*386–389*), zirconium (*388*), tantalum (*390*), and rhenium (*388*) alkyl or aryl bonds. For titanium, reaction of Cp_2TiRR^1 ($R = R^1 = Me$, CH_2Ph; $R = Me$, $R^1 = Ph$, C_6F_5) with isonitriles at $-30°C$ gave unstable compounds of formula $Cp_2TiR(\eta^2\text{-}CR^1 = NR^2)$ [$R = R^1 = Me$, $R^2 = Bu^t$ (*388*), C_6H_{11} (*387*); $R = Me$, $R^1 = C_6F_5$, Ph, $R^2 = C_6H_{11}$ (*389*)]. In the alkyl–aryl or alkyl–fluoroaryl compounds, preferential insertion of the isonitrile into the Ti–C_6F_5 or Ti–Ph bond occurred, whereas with carbon monoxide, insertion into the Ti–Me bond was the preferred pathway. The complex $Cp_2TiMe\{\eta^2\text{-C}(C_6F_5)NC_6H_{11}\}$ represents the first example of an isonitrile insertion into a fluoroaryl bond, although fluoroaryl iminoacyl complexes of iron have previously been synthesized from LiC_6F_5 additions to $[CpFe(CNMe)_3]^+$ (6). With Cp_2TiR {R = alkyl (391), aryl (386)} the intermediate $Cp_2TiR(CNR^1)$ was isolated, and shown to slowly rearrange to $Cp_2Ti(\eta^2\text{-CR}{=}NR^1)$ (R^1 = xylyl; R = alkyl or aryl). Oxidations of these η^2-iminoacyls with X_2 (X = I, SPh) gave the products $Cp_2Ti(\eta^2\text{-CR}{=}NR^1)X$ (386, 391).

Air-sensitive, uncharacterizable oils were isolated from reactions of $NbMe_5$ and $TaMe_5$ with isonitriles (388), and the thermally unstable solids $MeTaCl_2(CMe{=}NR)_2$ and $TaCl_4(CMe{=}NR)$ resulted from interactions of Me_xTaCl_{5-x} (x = 1–3) with CNR (R = C_6H_4Me-p, CH_2Ph, C_6H_{11}) (390). In contrast, Me_3TaCl_2 and $CNBu^t$ gave air-stable orange crystals of $TaCl_2\{CMe{=}NBu^t\}_3$ (388). The tetraneopentylzirconium complex $Zr(CH_2CMe_3)_4$ and $CNBu^t$ produced white air-sensitive crystals of $Zr(CH_2CMe_3)_3(CNBu^t)\{C(CH_2CMe_3){=}NBu^t\}$, whereas the paramagnetic rhenium alkyl $ReMe_6$ yielded air-sensitive $Re(CNBu^t)_2\{CMe{=}NBu^t\}_3$. The fate of the three missing methyl groups in the latter reaction was not ascertained (388).

Other iminoacyl complexes which have been prepared by insertions into metal–alkyl or –aryl bonds are $CpFe(CMe{=}NBu^t)(CO)(CNBu^t)$ (392), $Ru(CR{=}NR)Cl(CO)(PPh_3)_2$ (R = C_6H_4Me-p) (393), $CpNi(CR{=}NC_6H_4Me\text{-}p)(CNC_6H_4Me\text{-}p)$ (R = $CpMn(CO)_3$, Ph, $C_6H_2Me_3$-2,4,6) (394), and $[MX(CR{=}NR^1)L_x]_n$ (M = Pd, Pt; x = 1, n = 2; x = 2, n = 0) (382–384, 395). The crystal structure of trans-$PtI(\eta^1\text{-CMe}{=}NC_6H_4Cl\text{-}p)(PEt_3)_2$ has shown the complex to be square planar (396), and not five-coordinate through η^2-iminoacyl coordination as previously proposed (397).

Intramolecular insertions into palladium carbon bonds of cyclometallated complexes have been observed. The compound $Pd(o\text{-}C_6H_4CH_2NMe_2)Cl(CNR)$ (R = Bu^t, C_6H_4Me-p) produced $[Pd\{C({=}NR)\text{-}o\text{-}C_6H_4CH_2NMe_2)Cl]_2$, which was converted with Grignard reagents or $LiAlH_4$ to N,N-dimethyl(o-aminomethyl-, or o-iminoacyl)benzylamine (398). The enyl compounds $[M(C_{10}H_{12}OMe)ClL]$ (M = Pd (399), Pt (400); L = PPh_3, CNR; R = Me, C_6H_{11}) react with isonitriles to give different products. With platinum, insertion into the Pt–C σ bond occurs to give 70, whereas with palladium product 71 results.

Iminoacyls have also been obtained by the addition of alkylating agents to metal–isocyanide compounds. The anions $[CpM(CO)_2(CNR)]^-$ (M = Mo,

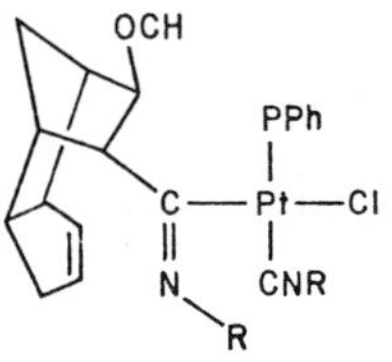

(70) 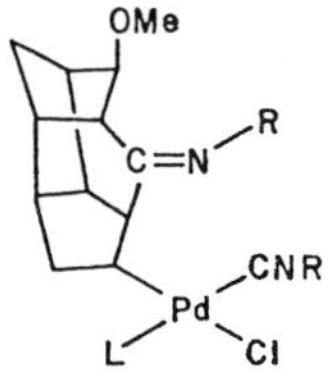   (71)

R = Me, Ph; M = W, R = Me) react with methyl iodide to give the compounds $CpMo(CO)_2(\eta^2\text{-}CMe\text{=}NR)$ (R = Me, Ph) and $CpW(CO)_2$ $(CNMe)Me$ (43). The dihapto arrangement (72) (M = $CpMo(CO)_2$) for the

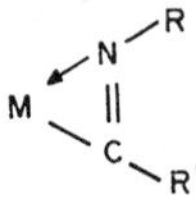

(72)

aminoacyl ligand was shown by an X-ray structural determination of $CpMo(CO)_2\{\eta^2\text{-}CMe\text{=}NPh\}$ (401). Both molybdenum complexes react with TCNE, $P(OMe)_3$, and PPh_3 to produce the η^1-iminoacyl complexes $CpMo(CO)_2L(CMe\text{=}NR)$ [L = TCNE, R = Me; L = TCNE,$P(OMe)_3$, PPh_3, R = Ph] (43) and one of these [L = $P(OMe)_3$, R = Ph] was resolved structurally (401). Alkylations with RX or $[Me_3O]BF_4$ of $Fe(CNR)_5$ and $CpM(CNR)(PMe_3)$ (M = Co, Rh) produced the compounds $[FeI(CR\text{=}NXylyl)(CNXylyl)_4]$ (R = Me, Et), $[Fe(CR\text{=}NBu^t)(CNBu^t)_4]X$ (R = Me, Et, Pr^i; X = I, BF_4) (364), $CpMI(CMe\text{=}NR)(PMe_3)$ {M = Co, R = Ph (98); M = Rh, R = Me (402)}, and $[CpCo(CMe\text{=}NBu^t)PMe_3]I$ (98).

Other routes to iminoacyl complexes have been reported. The reaction of the anions $[CpW(CO)_3]^-$ and $[Co(CO)_3PPh_3]^-$, or $Pd(PPh_3)_4$, with imidoyl chlorides has given the neutral compounds $CpW(CPh\text{=}NPh)(CO)_3$ (403), $Co(CR^1\text{=}NR)(CO)_3(PPh_3)$ (404), and $PdCl(CR^1\text{=}NR)(PPh_3)_2$ (403) for a range of alkyl and aryl iminoacyls. Treatment of $Os_3(CO)_{12}$ with $PhCH\text{=}NMe$ in refluxing octane under carbon monoxide atmosphere for 15 hr gave the three products $HOs_3(\mu\text{-}\eta^2\text{-}CPh\text{=}NMe)(CO)_{10}$ (73), $H_2Os_3(\mu\text{-}\eta^2\text{-}CPh\text{=}NMe)_2(CO)_8$ (74), and the ortho-metallated complex cis-Os-$(o\text{-}C_6H_4CH\text{=}NMe)_2(CO)_2$ in yields of 18, 5, and 1%, respectively. Similar reactions of $Os_3(CO)_{12}$ with Me_3N and $PhCH_2NMe_2$ gave the above three products in low yield together with the formimidoyl compounds (75)

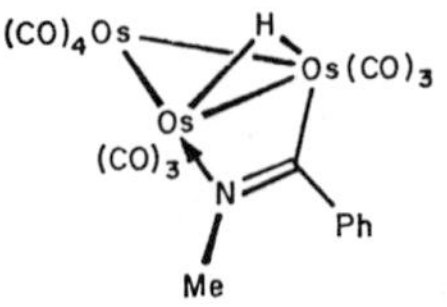

(73)

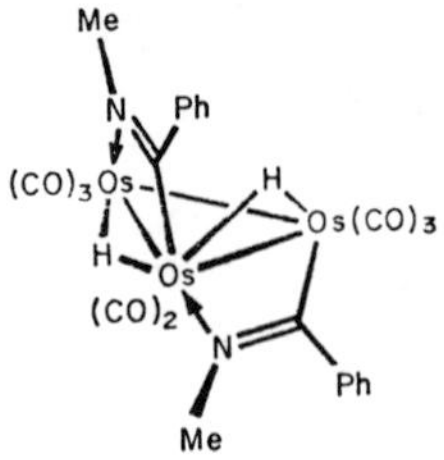

(74)

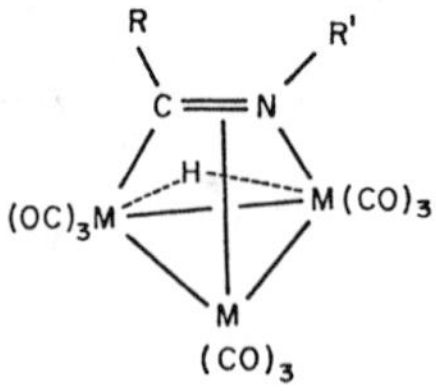

(75)

(M = Os; R = H; R^1 = Me), the methyl analog of 75 (M = Os; R = Me; R^1 = Ph), and the bridging carbyne (76) (M = Os; R^1 = R^2 = Me) (405).

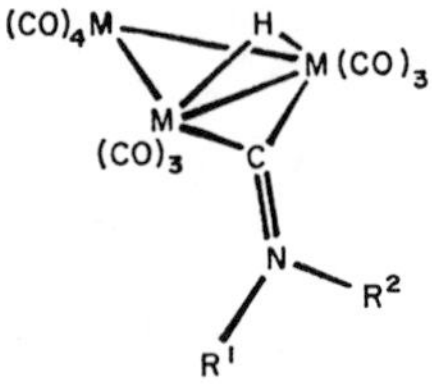

(76)

The iminoacyl analog of the μ^3-formimidoyl complex (75) has been obtained (406) from reactions of mixtures of [Fe$_2$(CO)$_8$]$^{2-}$ and [W(CO)$_5$I]$^-$, or from [HFe$_3$(CO)$_{11}$]$^-$, or from mixtures of Fe(CO)$_5$ and NaI, in refluxing acetonitrile over extended periods. The complex formed, [Fe$_3$(μ^3-CMe=NH)(CO)$_9$]$^-$, was readily protonated to give HFe$_3$(μ^3-CMe=NH)(CO)$_9$ which was shown to have structure 75 (M = Fe; R = Me; R^1 = H) by an X-ray analysis (407). The iminoacyl group in these complexes may be considered to be derived from insertions into a hypothetical Fe–Me group by CNH.

An unusual rearrangement occurred during the reaction of CNBu^t with the thorium and uranium metallacycles[(Me$_3$Si)$_2$N]$_2$MCH$_2$Si(Me)$_2$NSiMe$_3$ to give 77. The five-membered ring is not fluxional in the ^{1}H NMR of these

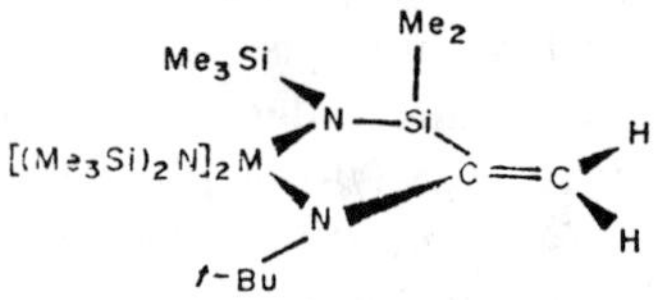

M = Th or U

(77)

compounds, and to explain the chemical equivalence of the $SiMe_2$ methyls it was proposed that either rapid inversion at the nitrogen is occurring or the five-membered ring is planar (*408*).

An equilibrium has been shown to exist between the enamine-ketimine form of the iminoacyl group (Scheme 4) by the addition of D_2O, which effected deuteration of the methyl group, and by the addition of $MeO_2C\equiv CCO_2Me$, which gave **78** (R = C_6H_4Me-*p*). This cyclized product was characterized structurally. The intermediate **79** in this reaction was characterized by ^{1}H NMR and by additions of HBF_4 which formed **80** (*409*).

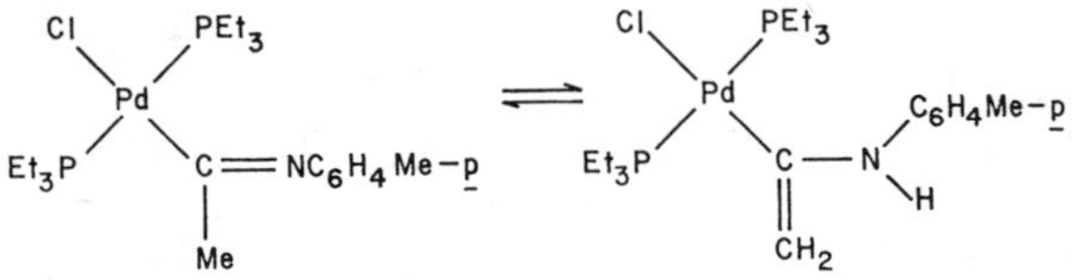

SCHEME 4

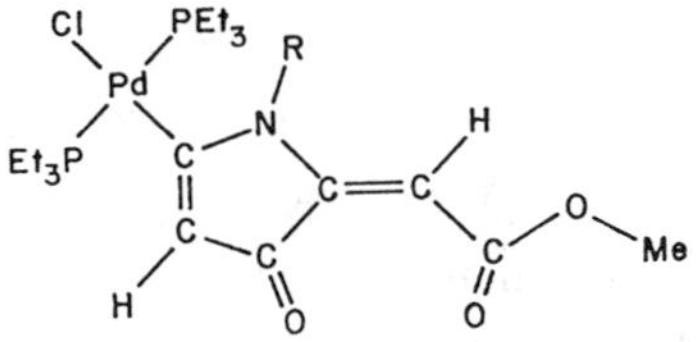

(78)

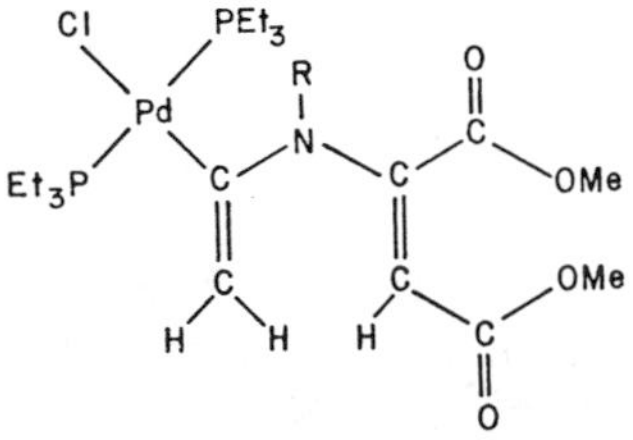

(79)

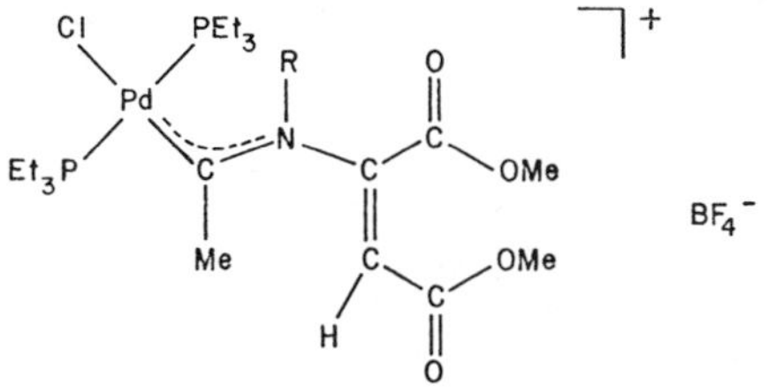

(80)

The strongly basic character of the iminoacyl nitrogen has been demonstrated by protonations with NH_4PF_6 (*410*) and HBF_4 (*411*) and alkylations with methyl iodide (*384*) to give metal–carbene complexes. Further examples include the formation of the secondary carbene $[Ru\{C(C_8H_{13})\text{-}NHBu^t\}(CNBu^t)_5](PF_6)_2$ from the reaction of $[(\eta^1,\eta^2\text{-}C_8H_{13})Ru$

(CNBu^t)$_4$]PF$_6$ and CNBu^t in ethanol, and the isolation of the salts [Ru(Me-CNHR)(CNR)$_5$](BPh$_4$)$_2$ (R = Bu^t, xylyl) from reactions of [CODRu-MeCl(NCMe)]$_2$ and CNR in refluxing ethanol in the presence of NaBPh$_4$ (36). The structures of two of these ruthenium carbenes were elucidated (e.g., **81** and **82**) (279). The reaction of PtI(Me)(PEt$_3$)$_2$ with CNMe has

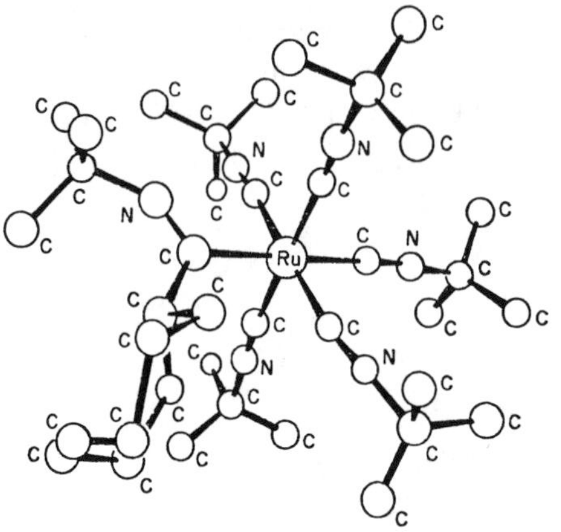

(81) 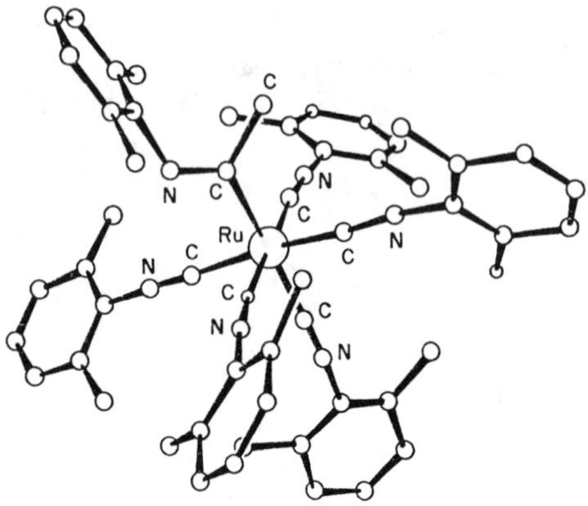   (82)

produced [PtI(η^1-MeCNMe$_2$)(PEt$_3$)$_2$]I, which was apparently formed by reaction of a preformed PtI(η^1-MeCNMe)(PEt$_3$)$_2$ with methyl iodide generated from PtCN(Me)(PEt$_3$)$_2$ by a dealkylation step (384).

Methyl iodide additions to the anions [CpMo(CO)$_{3-x}$(CNMe)$_x$]$^-$ (x = 1, 2) in THF gave the products CpMoI(CO)$_2$(MeCNHMe) (43) and CpMoI(CO){C(NMe$_2$)C(Me)NMe} (**83**) (43). In the latter complex the

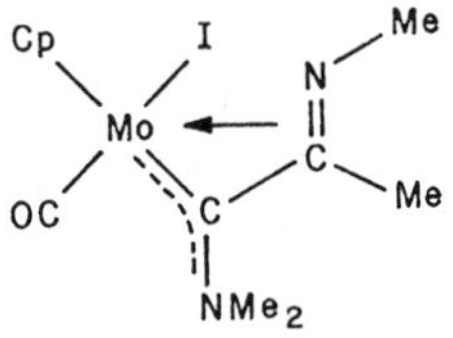

(83)

iminodimethylaminocarbene results from two consecutive methylation steps. Interesting examples of the reactivity of the iminoacyl nitrogen atom come from the isolation of acetimidoylcobalt heterocycles from [3+2]-cycloadditions of acetone or acetonitrile to the cobalt iminoacyl [CpCo(PMe$_3$){η^2-MeCNMe}]I. The products formed were **84** and **85**, the

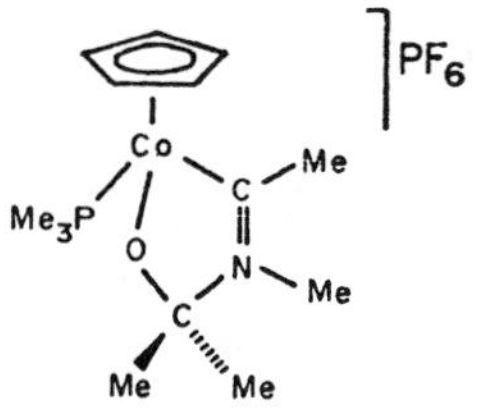

(84) 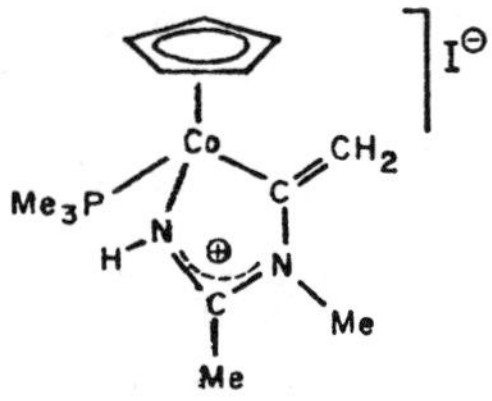   (85)

latter of which was structurally characterized (98, 402). The acetimi-doylruthenium compound (86) has recently been characterized from reac-

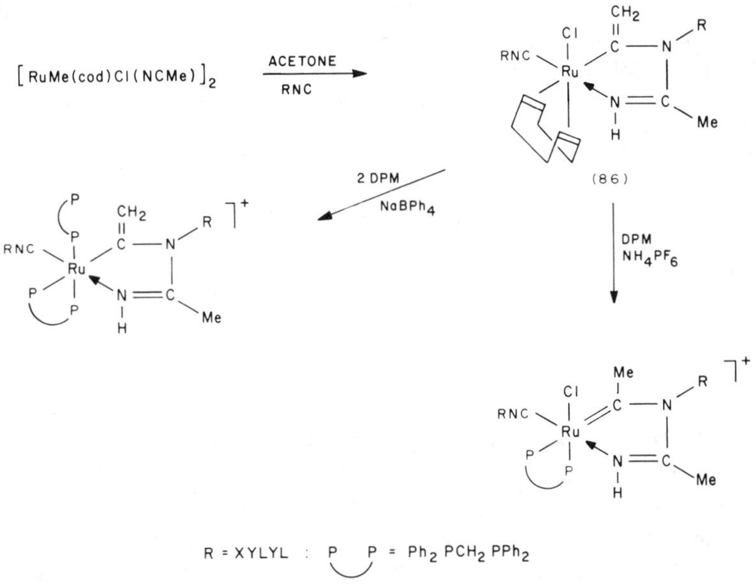

R = XYLYL :   P⌣P = Ph₂PCH₂PPh₂

SCHEME 5

tions of the methyl complex [Ru(Me)Cl(cod)NCMe]₂ with xylyl isocyanide in acetone solution. Compound 86 reacted further with dpm in acetone solution to form the two products given in Scheme 5 (18).

A complicated insertion and rearrangement process was observed from reactions of CNBuᵗ with WMe₆. At −78°C in diethylether air-stable crystals of the five-coordinate tungsten(VI) compound W{N(Buᵗ)CMe₂}(Me) (NBuᵗ){N(Buᵗ)CMeCMe₂} were obtained, and this complex was shown to have the structure 87 from an X-ray diffraction study. A mechanism for

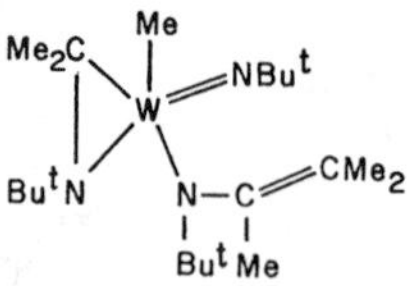

(87)

the formation of the unusual dialkylamido group $N(Bu^t)CMe{=}CMe_2$ was proposed. Compound **87** reacts with refluxing methanol to give $W(OMe)_6$, and with HCl at $-20°C$ to form $[Me_2C{=}C(Me)NH_2Bu^t][W\{NH(Bu^t)CMe_2\}(Me)(NBu^t)Cl_3]$, the anion of which was characterized crystallographically as **88** (*388*).

Three routes to η^1-bonded polyimino ligands have been utilized. Multiple insertions into metal–alkyl or –aryl bonds or into the intermediate mono-imino complex has given $PdI\{\eta^1\text{-}(C{=}NC_6H_{11})_2Me\}L_2$ ($L{=}$ PMe_2Ph, $PMePh_2$) from $[PdMeIL_x]_n$ ($x = 2$, $n = 1$; $x = 1$, $n = 2$) or $PdI(MeCNC_6H_{11})L_2$ and CNC_6H_{11} (*382*), and $[Pd\{\eta^1\text{-}(C{=}NBu^t)_3CH{=}$

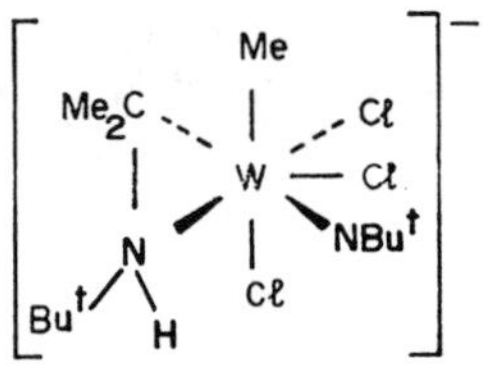

(88)

$CHCO_2Me\}(CNBu^t)(dpe)]Br$ from $PdBr(CH{=}CHCO_2Me)dpe$ and $CNBu^t$ (*395*). The second route involves the displacement of the N bond in an η^2-polyimino ligand. For example, $CpFe\overline{(CO)\{C({=}NC_6H_{11})C(Me){=}NC_6}}$ $\overline{H_{11}\}}$ and $CNBu^t$ gave $CpFe(CO)\{\eta^1\text{-}C({=}NC_6H_{11})C(Me){=}NC_6H_{11}\}$- $(CNBu^t)$ (*412*). Interestingly, this complex rapidly converted into the cyclic carbene (**89**) on heating at 70°C. The third route has involved reactions of

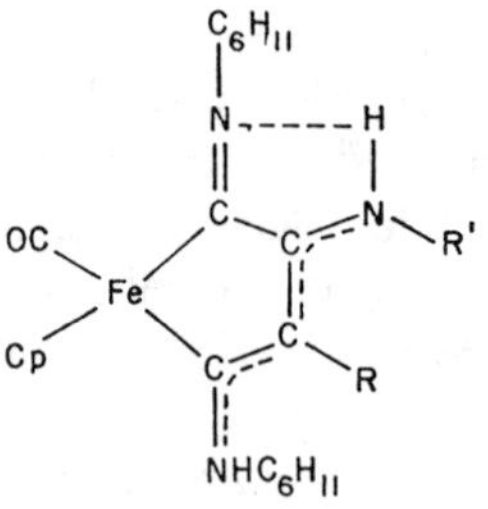

(89)

cis-PdCl$_2$(CNR)$_2$ (R = Ph, C$_6$H$_4$X-p; X = Me, OMe) with HgR$_2^1$ (R^1 = Me, Ph) and PPh$_3$, which gave $trans$-PdCl{C(=NR)C(R^1)=NR}(PPh$_3$)$_2$ (413). The chloro-bridged intermediate [PdCl(R^{1}CNR)PPh$_3$]$_2$ was reported earlier (414). Different substituents on the nitrogen atoms were obtained from either HgMe$_2$ alkylations on PdCl$_2$(CNR)(CNR^1), or by a circular route involving first the conversion of PdCl{η^1-C(=NR)C(Me)=NR^1}(PPh$_3$)$_2$ to PdCl{η^1-C(=NR)C(=O)Me}(PPh$_3$)$_2$ with HClO$_4$ and Et$_3$N (415) and then condensation of the carbonyl group with primary aliphatic amines (416).

The nitrogen atoms in these iminoalkyl groups have pronounced nucleophilic character and can act as nitrogen donor ligands towards a series of metal atoms forming complexes of type **90** [M = Fe, Co, Ni, Cu, Zn;

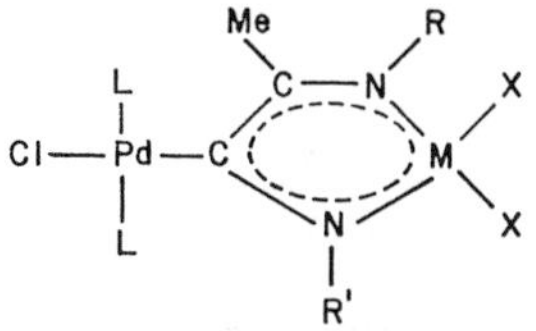

(90)

X = Cl, Br $(416-418)$; M = Pd, Pt; X = Cl (419)]. In contrast, reactions of the derivatives Pd{η^1-C(=NR)C(Me)=NR}XL$_n$ [X = S$_2$CNMe$_2$, L$_n$ = PPh$_3$; X = Cl, L$_n$ = dpe; X = Cl, L$_n$ = (PPh$_3$)$_2$] with platinum (419) and rhodium (420) complexes has affected ligand migrations to give the products shown in Scheme 6.

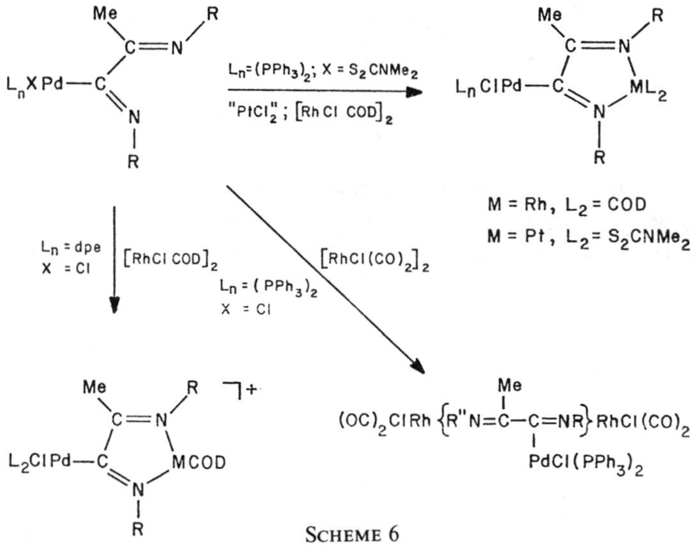

SCHEME 6

Complexes containing η^2-iminoacyl ligands obtained from multiple insertions usually contains the ligand bonded as a four- or five-membered ring depending on the number of insertions. Thus, photochemical activation of CpFe(CO)(CNC$_6$H$_{11}$)R (R = Me, COMe) produced **91** (M = CpFe(CO); R = C$_6$H$_{11}$) (*412*), whereas addition of MeI to Fe(CNBu^t)$_5$ in THF gave **91**

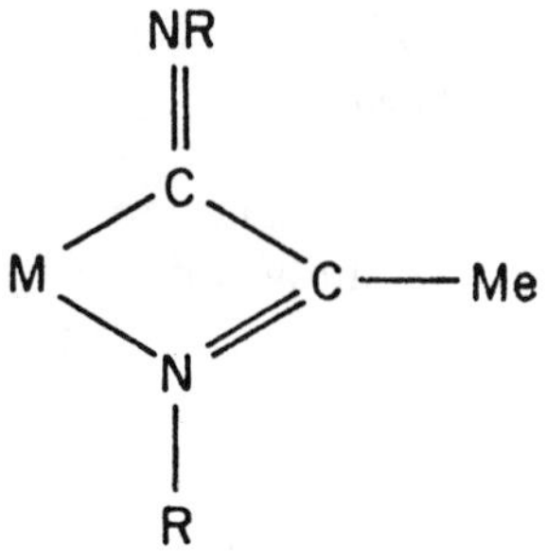

(91)

{M = FeI(CNBu^t)$_3$} (*364*). An unusual rearrangement has accompanied the addition of MeI to the anion [Mn(CO)$_4$CNMe]$^-$. The product was shown to be **92** from X-ray diffraction studies (*44*).

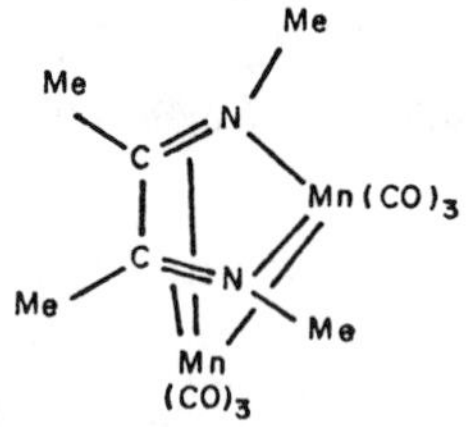

(92)

Dihapto iminoacyls bonded through five-membered rings have been synthesized from CNR^1 (R^1 = C$_6$H$_{11}$) insertions into the metal–methyl and –aryl bonds in *trans*-PdIMe(PMePh$_2$)$_2$ (*382*) and CpFe(CO)-(CNC$_6$H$_{11}$)C$_6$H$_4$X-*p* (X = H, Cl) (*392*) which gave **93** [R = Me, M =

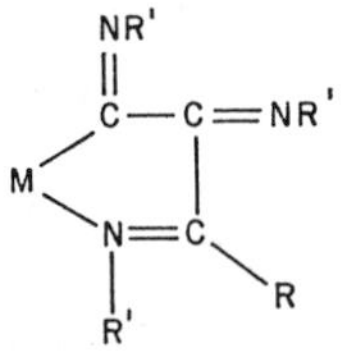

(93)

PdI(PMePh$_2$); R = C$_6$H$_4$X-p, R' = C$_6$H$_{11}$, M = CpFe(CO)]. The latter complex catalyzed the polymerization of isonitriles.

Facile insertions into nickel carbon σ bonds were observed during oxidation reactions of Ni(CNBu^t)$_4$ with alkyl chloroformates and alkyl iodides. The compounds formed 93 [M = NiX(CNBu^t); X = Cl, I; R = alkyl, COR; R' = Bu^t], further converted to 94 with excess CNBu^t (421).

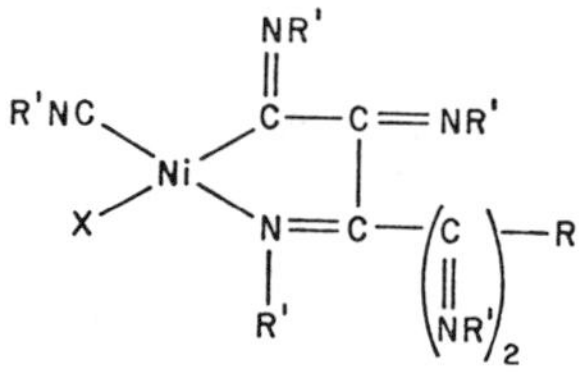

(94)

b. *Metal–Allyl, –Acetylene, and –Vinyls.* Insertions of isocyanides into the metal–carbon bond in η^3-allylpalladium chloride to give [Pd{C(=NR)CH=CH—R^1}(CNR)Cl]$_2$ (R = C$_6$H$_{11}$, Ph, C$_6$H$_4$NO$_2$-p; R^1 = H, Me) (422, 423) have been observed to occur via the η^1-allyl intermediate [Pt(σ-allyl)(CNR)Cl]$_2$ (424). These reactions have been used as synthetic routes to ketenimines.

Two types of products have resulted from the insertions into metal π-acetylene complexes, depending upon the number of coordinated acetylenes and the functional groups on the acetylene. For example, CpMCl(η^2-PhC$_2$Ph)$_2$ and CpMCl(η^2-PhC$_2$Ph)(η^2-CF$_3$C$_2$CF$_3$) react with an excess of CNBu^t to give the salts [CpM(CNBu^t)$_3$(η^2-RC$_2$R)]Cl (M = Mo, W; R = Ph, CF$_3$). With CpMCl(η^2-CF$_3$C$_2$CF$_3$)$_2$ and excess CNBu^t, however, cyclopentadienimines are formed by a mechanism given in Scheme 7. The crystal structure of the molybdenum complex has been elucidated (425). The cobalt complex CpCo(η^2-PhC$_2$Ph)PPh$_3$ produces the metallocycles (95) with excess isocyanide, which are intermediates in the formation of diiminocyclobutenes and triiminocyclopentenes (Scheme 8) (426). An iron product Fe{1-4-η-C(=NBu^t)—C(Ph)=C(Ph)—C(=NBu^t))(CNBu^t)$_3$, containing an identical metallocycle as in 95, was isolated from PhC$_2$Ph additions to Fe(CNBu^t)$_5$ (364). The asymmetry in this metallocycle shown in the X-ray structural determination is not reflected in the NMR spectra, and this is ascribed to a rotational process similar to that observed in (1,3-diene)Fe(CO)$_3$.

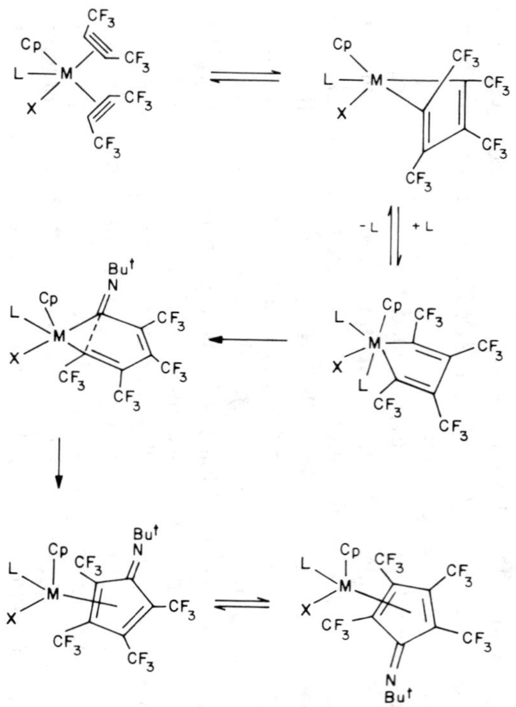

SCHEME 8[11]

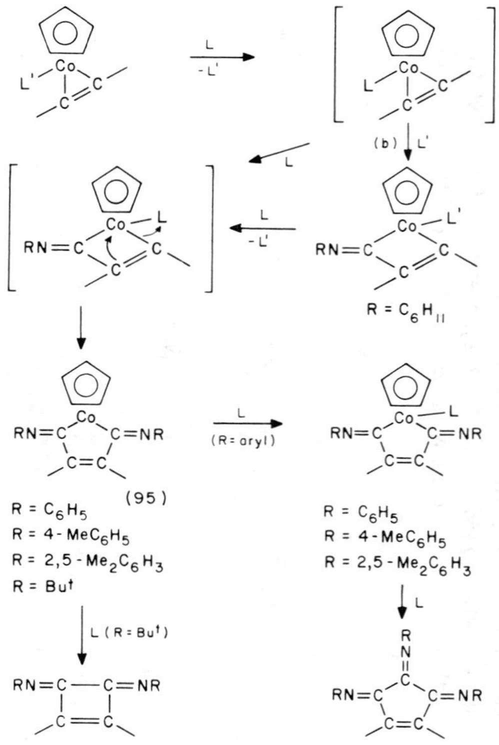

SCHEME 7

Nucleophilic attack at a carbocationic carbon, similar to that observed in metal–carbynes, has been observed from $CNBu^t$ additions to the μ_2-η^2-acetylide in the complex $Fe_2(CO)_6(C\equiv CPh)PPh_2$ (427). The ligand dipole, indicated from the X-ray structural determination of the product **96**, makes it susceptible to additions of primary amines, generating the carbene **97**.

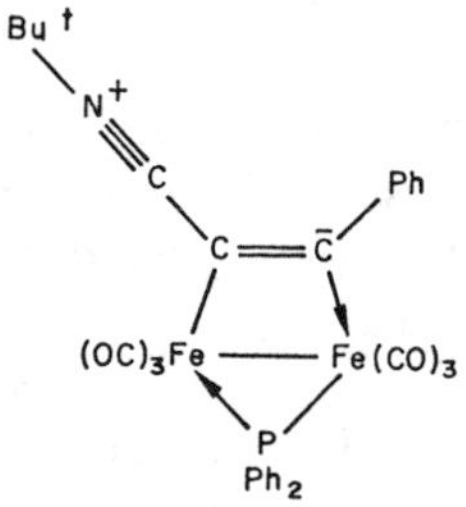

(96) 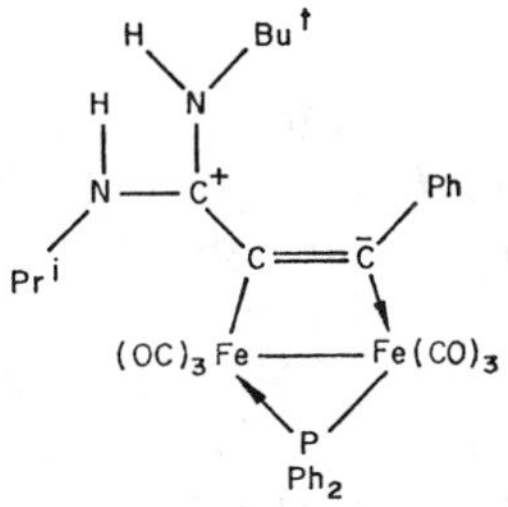   (97)

Protonation reactions in the cobalt metallocycles (**95**) with, $HX(X = Cl, \ BF_4, \ O_2CCF_3)$ gave $CpClCo\{C(NHR)C(Ph)=C(Ph)C(=NR)\}$, $[CpCo\{C(NHR)C(Ph)=C(Ph)C(=NR)\}]BF_4$, and $[CpCo\{C(NHR)C(Ph)=C(Ph)C(NHR)\}](O_2CCF_3)_2$, respectively (411).

Aromatic isonitriles insert into platinum–vinyl but not platinum–acetylide bonds, to form $trans$-$Pt\{C(CH=CH_2)NR\}ClL_2$ (L = tertiary phosphine) (428). The intramolecular insertion of $trans$-$MBr(cis$-$CH=CHCO_2Me)(CNBu^t)_2$ gives $trans$-$[MBr\{C(trans$-$CH=CHCO_2Me)=NBu^t\}(CNBu^t)]_2$ whereas insertion of CNC_6H_4Me-p into $trans$-$MBr(cis$-$CH=CHCO_2Me)(PPh_3)_2$ yields $trans$-$MBr\{C(cis$-$CH=CHCO_2Me)=NC_6H_4Me$-$p\}(PPh_3)_2$ (395).

c. Metal–Carbenes and –Carbynes. Treatment of the carbene $(OC)_5Cr=C(OMe)Me$ with CNC_6H_{11} produced the aziridene complex (**98**). This was characterized by peroxide oxidations and HCl reductions,

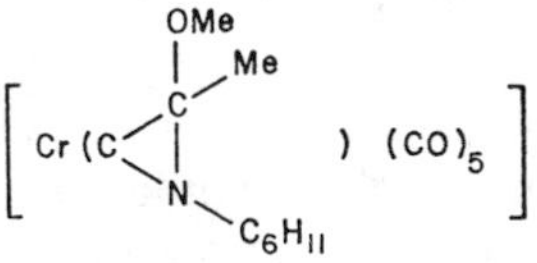

(98)

to give the derivatives $Cr\{C(NHC_6H_{11})C(OMe)=CH_2\}(CO)_5$ and $Cr\{C(NHC_6H_{11})C(Me)=O\}(CO)_5$ (429, 430). Complex **98** dissolved in methanol to produce $Cr\{C(NHC_6H_{11})C(OMe)_2Me\}(CO)_5$ (430), which is analogous to the manganese derivative, $[Mn\{C(NHR)C(OMe)_2Me\}$

---

[11] Scheme 8 reprinted with permission from *J. Am. Chem. Soc.* **97**, 3546 (1975). Copyright 1975 American Chemical Society.

$(CO)_3\{P(OMe)_3\}_2]PF_6$ (**99**) (R = Me, aryl), obtained from [Mn{C(OMe)-

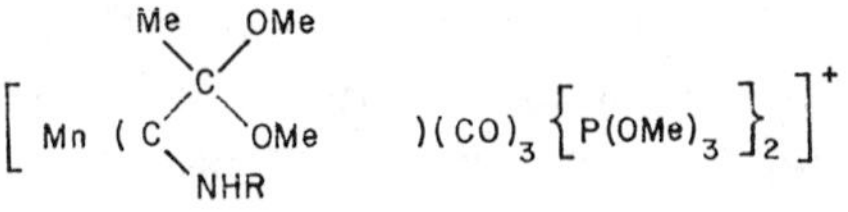

(99)

Me}$(CO)_3\{P(OMe)_3\}_2]^+$ and CNR in benzene (*431*). The reaction of $CpFe(CO)(CNC_6H_{11})CH_2R$ with CNC_6H_{11} in refluxing THF has produced the cyclic carbene **89** (R = H, aryl; $R^1 = C_6H_{11}$). Analogous compounds were prepared from $CpFe(CO)(CNR^1)R$ (R = H, Ph, $R^1 = Bu^t$, C_6H_{11}, CH_2Ph) and CNC_6H_{11}, or from $CpF\overline{e\{C(N\!=\!C_6H_{11})CH\!=\!NC_6H_{11}}})\text{-}$ (CO) and $CNBu^t$ (*392*). The products result from an insertion of two isocyanide molecules into a metal–carbon σ bond and an unusual third isocyanide insertion into the C–H bond of an alkyl group. The structure of the complex $CpF\overline{e\{C(\!=\!NC_6H_{11})C}(NHC_6H_{11})\!=\!CHC(NHBu^t)\}(CO)$ has been elucidated by X-ray diffraction methods (*432*). This has prompted a reformulation of the complex $CpF\overline{e(CO)\{C(\!=\!NC_6H_{11})C(\!=\!NC_6H_{11})}C\text{-}$ $(CH_2C_6H_4X\text{-}p)\!=\!NC_6H_{11}\}$ reported earlier (*433*), as a derivative of **89** with the hydrogen-bonded proton removed (*434*).

Insertions of CNR (R = Bu^t, C_6H_{11}, Ph) into the iron–carbene bond in the (η^3-vinylcarbene)irontricarbonyl complex (**100**) formed the η^4-vinylketenimine (**101**) which converted to **102** on alkylation with $MeOSO_2F$. The structural elucidation of **101** (R = C_6H_{11}) was also reported (*435*).

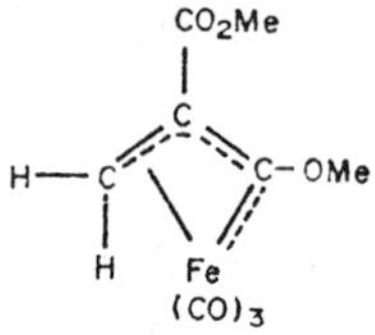

(100) 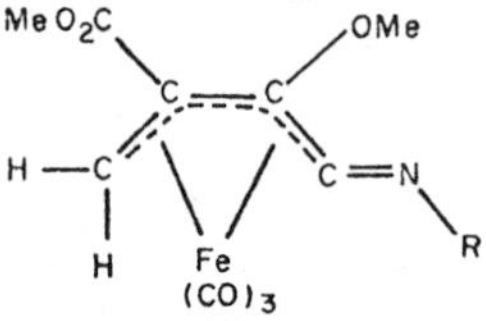   (101) 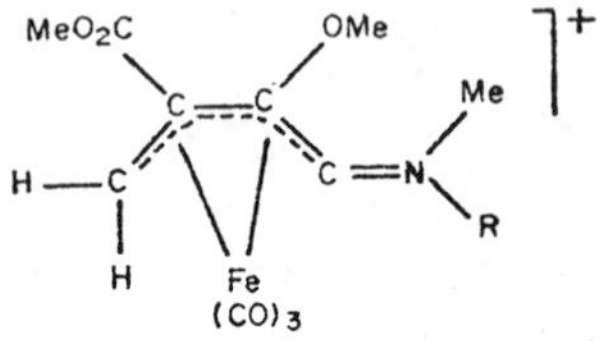   (102)

Two cyclic carbene complexes have been prepared by either insertions in the metal–carbene bond in complexes containing the $:C(R)NHR^1$ group or by protonation of the product derived from multiple insertions into metal–acylimino systems. Thus, the ruthenium carbene **81** undergoes insertions with $CNBu^t$ in refluxing acetone to give **103** and **104** (*36*), whereas protonations of the imino complexes $PdX\{C(NC_6H_4OMe\text{-}p)\text{-}$ $(MeC\!=\!NC_6H_4OMe\text{-}p)\}(PPh_3)_2$ with methanolic $HClO_4$ gave a salt designated **105** because of a fast exchange of the proton between the two nitrogen atoms (*413*).

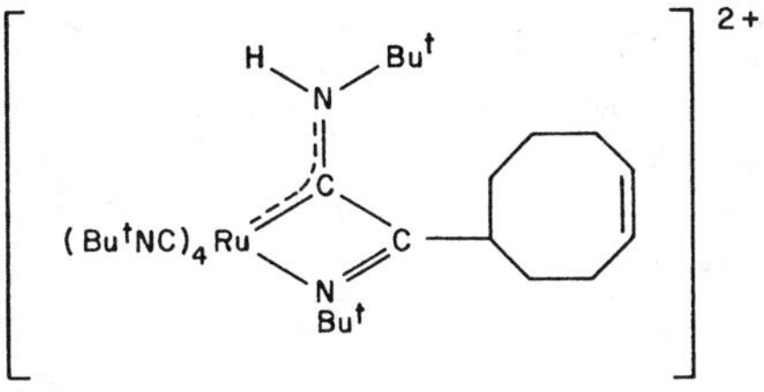

(103)

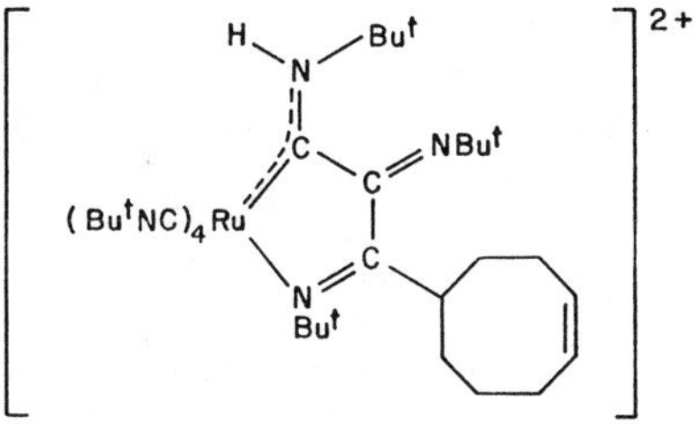

(104)

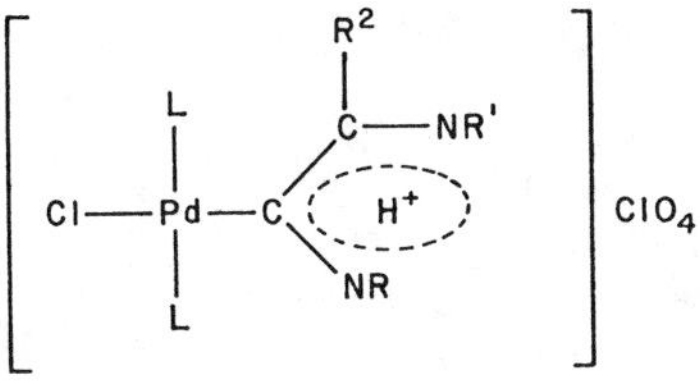

(105)

The intermediate $Ni(CNBu^t)_3\{N_2C(CN)_2\}$, formed by adding diazodi-cyanomethane to $Ni(CNBu^t)_4$, rearranges, probably via a carbene insertion, to give the ketenimine complex $Ni\{Bu^tN=C=C(CN)_2\}(CNBu^t)_2$. An X-ray structural determination of the product has been reported (*436*).

The addition of isocyanides CNR (R = Me, Bu^t, C_6H_{11}) to the carbon atom in the metal carbynes $[CpMCR^1(CO)_2]BCl_4$ (M = Mn, R^1 = Ph; M = Re, R^1 = Me) gave the ketenimyl salts $[CpM=C(R)-C\equiv CR^1(CO)_2]BCl_4$. The rhenium compound is extremely labile and was only characterized by IR techniques (*437*). With the alkylidyne complex $CpM(CCH_2Bu^t)\{P(OMe)_3\}_2$ (M = Mo, W), the reaction with xylyl isocya-nide produced **106** which was proposed to occur via a novel cheletropic reaction involving CNR attack on the metal–carbon triple bond. The metallacyclopropenimine then rearranges with loss of $P(OMe)_3$, or alterna-tively opens up to form an η^1-bonded iminoketenyl. Migration of this iminoketenyl onto a coordinated isocyanide then forms **106** (Scheme 9) (*438*). An X-ray diffraction study on **106** was reported.

2. Into Metal–Hydride Bonds

Very few reports on the formation of formimidoyl complexes have appeared. They have been prepared by various routes, with the majority of

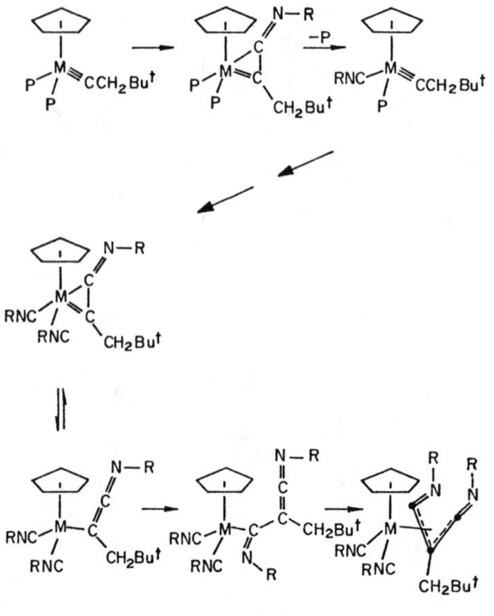

(106)

SCHEME 9

pathways involving either insertions of isocyanides into metal–hydrides or additions of proton donors to metal–isocyanide complexes.

Insertion into the Pt–H bond in *trans*-[PtH(CNC$_6$H$_4$Me-*p*)L$_2$]Cl (L = PEt$_3$, PMe$_2$Ph) occurred in nonpolar solvents to give the formimidoyl complexes *trans*-PtCl(CH=NC$_6$H$_4$Me-*p*)L$_2$ (*439, 440*). Syn and anti isomers of the formimidoyl ligand were observed during ^{1}H NMR studies, which interconverted via rotation through a polarized transition state (*440*). Heating the dioxygen adduct Ru(O$_2$)(CO)(CNC$_6$H$_4$Me-*p*)(PPh$_3$)$_2$ in alcohols produced the carboxylato–formimidoyl complex Ru(O$_2$CR)-(CH=NC$_6$H$_4$Me-*p*)(CO)(PPh$_3$)$_2$ (R = Me, Et, Ph) (*441*), of which the acetate has been structurally characterized (*442*). No mechanism for this interesting alcohol–acetate conversion was proposed.

The cluster Pt$_3$(CNBu^t)$_6$ reacted with triorganosilanes or germanes to give the platinum dimers [Pt(CHNBu^t)(ER$_3$)(CNBu^t)]$_2$ (E = Si, R$_3$ = Me$_3$, Et$_3$, Me$_2$Ph, MePh$_2$, Ph$_3$, (OEt)$_3$; E = Ge, R = Me$_3$). The X-ray crystal structure determination has shown the platinum atoms in [Pt(CHCNBu^t)(Si-MePh$_2$)(CNBu^t)]$_2$ to be bridged by the formimidoyl ligands to give a six-membered ring in a boat conformation (*375*).

Examples of formimidoyl ligands bridging three metal atoms have been observed in reactions involving Ru$_3$(CO)$_{11}$CNBu^t with dihydrogen (*443, 444*), or the reaction of the hydrides [HFe$_3$(CO)$_{11}$]$^-$ (*445*) and H$_2$Os$_3$(CO)$_{10}$

(*446–448*) with isocyanides. Relevant products isolated were (**75**) (M = Fe, Ru, Os; R = H; R^1 =Me, Bu^t, Ph) and $HRu_3(CO)_8(CNBu^t)(CH=NBu^t)$. The osmium intermediates $H(\mu\text{-}H)Os_3(CO)_{10}CNR$ (R = Me, Ph) were isolated and the products of their further conversions were found to be solvent dependent. In strong donor solvents the carbene complexes **76** are formed by a mechanism which involved deprotonation and formation of a transient anion $[HOs_3(CO)_{10}CNR]^-$. In the anionic form the isocyanide ligand moves into a bridging position, developing lone-pair character on the nitrogen atom which is then protonated. In weak donor solvents preferential hydride transfer to the isocyanide carbon atom occurs via an intramolecular process to give $(\mu\text{-}H)Os_3(\mu\text{-}\eta^2\text{-}CH=NR)(CO)_{10}$. The formimidoyl ligand in this complex bridges the edge of the cluster, but rearranges when the complex is further heated to give **75** (M = Os; R = Me, Ph; R^1 = H) through carbonyl loss. The compound **75** (M = Os; R = Me; R^1 =H) was also formed as the product of the interaction of $Os_3(CO)_{12}$ with trimethylamine (*405*). The reaction of **75** (M = Os; R = Ph; R^1 = H) with $P(OMe)_3$ formed the edge-bridging formimidoyl complex $(\mu\text{-}H)Os_3(\mu\text{-}\eta^2\text{-}CHNPh)(CO)_9P(OMe)_3$. The crystal structure determinations of this latter complex (**107**)[12] and **75** (M = Os; R = Ph; R^1 = H) have been elucidated (*446*). A biproduct of these osmium reactions from refluxing *n*-butylether was the benzyne complex $H_3Os_3(\mu\text{-}\eta^2\text{-}C_6H_4)(\mu\text{-}\eta^2\text{-}CHNPh)(CO)_8$ (**108**), in which the benzyne

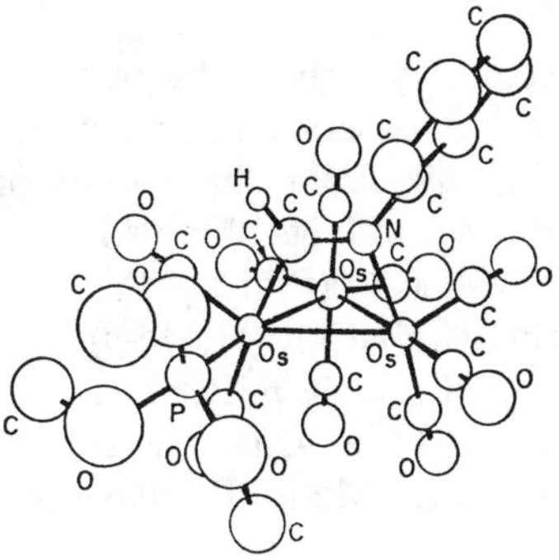

(107)

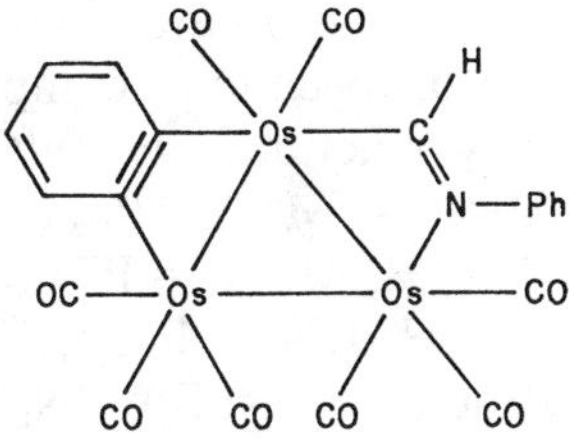

(108)

ligand presumably formed by abstraction of a phenyl group from CNPh. No mechanistic details are yet available on this reaction (*448*).

Complete hydrogenation of isocyanides has not yet been achieved with these osmium compounds although it occurs readily with the nickel cluster $Ni_4(CNBu^t)_7$ (*449*). In the osmium case, however, the two different hydride

12 Reprinted with permission from *J. Am. Chem. Soc.* **101**, 2581 (1979). Copyright 1979 American Chemical Society.

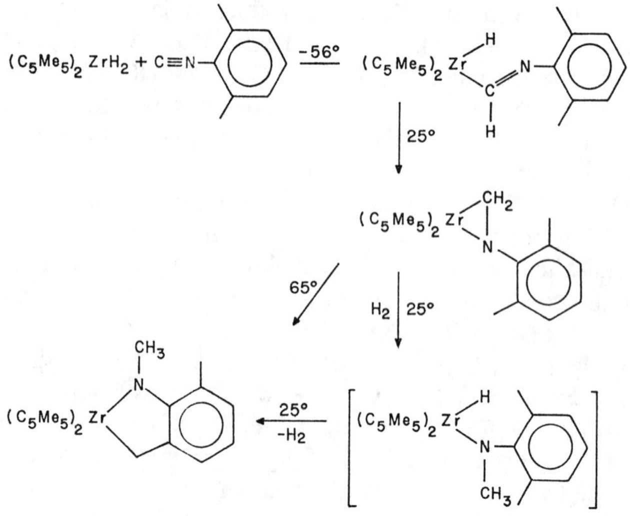

SCHEME 10

transfer pathways, e.g., to the nitrogen atom or the carbon atom, could lead to fundamentally different reduction products. Complexes containing stepwise reduction products of isonitriles have been obtained, however, from reactions of Cp_2ZrH_2 with isocyanides. The reaction products obtained with xylyl isocyanide are given in Scheme 10[13] (39).

Conversions of formimidoyl ligands into secondary carbenes by protonations or alkylations of the imino nitrogen have been observed in a few cases. In general, protonation–deprotonation studies have found the basicity of this nitrogen to be comparable with weak amines. Products obtained from these reactions are $[Ru\{CHN(Y)C_6H_4Me\text{-}p\}(O_2CR)(CO)(PPh_3)_2]^+$ (Y = H, Me; R = Me, Et, Ph) (441, 450) and $trans\text{-}[PtX(CHN(Y)C_6H_4Me\text{-}p)L_2]^{n+}$ (Y = H, Me, L = PEt_3, X = Cl, n = 1; X = $CNC_6H_4Me\text{-}p$, PPh_3, n = 2; Y = H, L = PMe_2Ph, X = $CNC_6H_4Me\text{-}p$, n = 2) (439, 451). The crystal structure of $RuI_2\{CHN(Me)C_6H_4Me\text{-}p\}(CO)(CNC_6H_4Me\text{-}p)PPh_3$ has been reported (452).

3. Into Metal–Nitrogen, –Oxygen, and –Sulfur Bonds

As an extension of earlier work (453), the insertions of isocyanides into the metal–nitrogen bond in the azido complexes $Co(N_3)(chelate)L$ (chelate = dianion of $N,N^1\text{-}o$-phenylenebis(salicylidenimine); L = PPh_3, py or free coordination site) has been investigated. The tetrazole complexes (109)

[13] Reprinted with permission from J. Am. Chem. Soc. 101, 6451 (1979). Copyright 1979 American Chemical Society.

{M = Co(chelate)L;   R = C$_6$H$_{11}$,   (CH$_2$)$_n$OH   (n = 2,3),   (CH$_2$)$_2$OCO$_2$ (CH$_2$)$_2$NC; L = PPh$_3$, py} result from CNR reactions in chloroform, whereas the oxazolidinylidene compound (110) [M = Co(chelate)N$_3$; R =

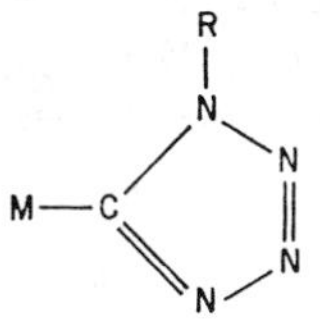

(109)

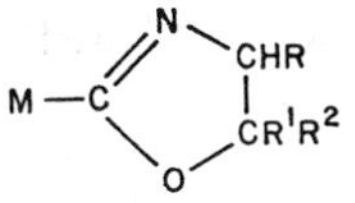

(110)

R^1 = R^2 = H] is the preferred product of CN(CH$_2$)$_2$OH reactions in THF. Similarly CNNPr$^i_2$ produced 109 [M = Co(chelate)PPh$_3$; R = NPr$^i_2$] from CHCl$_3$ and Co(CN)(chelate){N(Pr^i)=CMe$_2$} from acetone (454). The crystal structure of the gold tetrakis(1-isopropyltetrazol-5-ato) complex [AsPh$_4$][Au(CN$_4$Pr^i)$_4$] has been reported (455).

Other oxazolidinylidene compounds have been synthesized by three routes. Intramolecular condensations of palladium and gold compounds containing the hydroxyisocyanide CN(CH$_2$)$_n$OH (n = 2,3) have produced 111 (M = Pd, n = 2, 3, x = 4, a = 2; M = AuCl$_3$, AuCl, n = 3, x = 1, a = 0) (456). Additions of 1,3-dipolarophiles to [PtCl(CNCH$_2$CO$_2$Et)-(PPh$_3$)$_2$]BF$_4$ gave a range of products in high yield, including 110 (M = PtCl(PPh$_3$)$_2$; R = CO$_2$Et; R^1 = H; R^2 = alkyl, aryl), 112 (M = PtCl(PPh$_3$)$_2$;

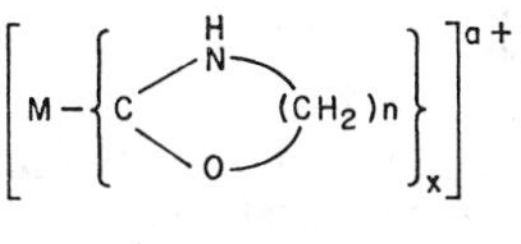

(111)

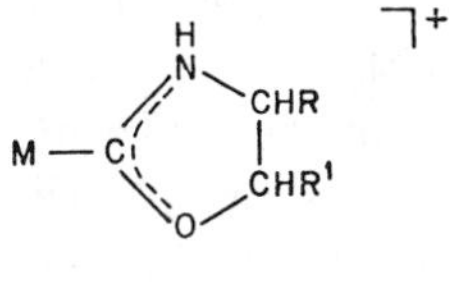

(112)

R = CO$_2$Et; R^1 = H), pyrrolines, pyrrols, and imidazoles (457). The reaction of aldehydes and ketones with OsHCl(CO)(CNR)L$_2$ and [OsCl(CO)$_2$(CNR)L$_2$]$^+$ (L = PPh$_3$; R = tosylmethyl) in the presence of NaOMe gave the four product types 113 (M = OsHCl(CO)L$_2$), 110 (M =

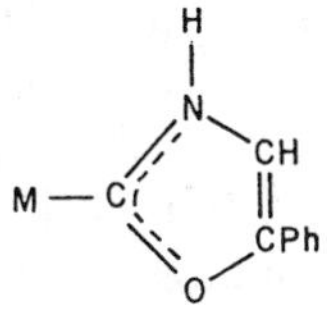

(113)

$OsCl(CO)_2L_2$, R = SO_2Me; R^1 = H; R^2 = Me); 112 (M = $OsHCl(CO)L_2$;
R = OMe; R^1 = Me), and $OsHCl\{COCMe_2CH(OMe)NH\}(CO)(PPh_3)_2$
from PhCHO, CH_3CHO, and acetone, respectively (458).
    Nitrilimines or nitrilylides were found to produce the cyclic carbene
compounds 114 (R = R^1 = C_6H_4Me-p; R = C_6H_4Me-p, R^1 = H; L = PPh_3,
$PEtPh_2$; X = Cl, Br, I) and 115 (R = $C_6H_4NO_2$-p, R^1 = R^2 = C_6H_4Me-p;

(114)   (115)

L = PPh_3, $PEtPh_2$; X = Cl, Br, I) on reacting with cis-$PdX_2L(CNC_6H_4$
Me-p) (459–461).
    The insertion of CNMe into the diethylamine–palladium bond in
chloro(3-diethylaminopropionyl)(diethylamine)palladium(II)   produced
the carbene $ClPd\{C(=O)(CH_2)_2NEt_2\}\{C(NHMe)NEt_2\}$ (462). Treatment of
the ylide $[Ph_3PN(SiMe_3)PdCl_2]_2$ with CNC_6H_{11} gave, after hydrolysis, the
carbene $Pd\{C(NHC_6H_{11})N=PPh_3\}(CNC_6H_{11})Cl_2$ (463). The reaction of
Cu_2O,   CNBu^t,   and   carboxamides   gave $Cu\{N(Bu^t)CH=N—$
$C(R)=O\}(CNBu^t)_2$ which were assumed to be formed by isonitrile insertion
into the copper–nitrogen bond of a copper(I) amide intermediate (464).
Deprotonation of the carbenes $[AuCl\{C(NHR)NHR^1\}]_n$ (n = 1, 2) has given
the insoluble species $[Au\{C(NHR)=NR^1\}]_x$ (R, R^1 = alkyl, aryl) (465)
related to the trimers $[Au\{C(OR)=NR^1\}]_3$ (466).
    The insertion of CNR (R = Me, C_6H_4OMe-p, xylyl) into the platinum–
oxygen bonds in $Pt(OH)R^1L_2$ (L_2 = $Ph_2PCH=CHPPh_2$; R^1 = CF_3,
CH_2CN) to give the carboximido derivative $Pt(CONHR)(R^1)L_2$ represents
the only example known of this type of insertions (467).
    Carboxyimido complexes have, in one other earlier instance, been pre-
pared from OH^- reactions on coordinated isocyanides (6), but recently the
reaction of $ReCl_3(CNMe)dpe$ with moist methanol in the presence of
$NaClO_4$ has given $[ReCl(CONHMe)(dpe)_2]ClO_4$ (468). The rhenium carboximido group was readily deprotonated by bases to give the first example
of an η^2-bonded isocyanate ligand, e.g., $Re\{\eta^2$-$C(=O)NMe\}Cl(dpe)_2$. The
osmium   complex $Os(\eta^2$-$SCNR)L(CO)(PPh_3)_2$ containing   an η^2-isothio-
cyanato group has been isolated as the product of SH^- attack on the
isocyanide in $[OsClL(CO)(CNR)(PPh_3)_2]^+$ [R = C_6H_4Me-p; L = CO (469),
L = CS (470)]. In this case η^2-bonding to the osmium is proposed to occur
via carbon and sulfur in contrast to the bonding to carbon and nitrogen in
the rhenium case. Alkylation of the isothiocyanato group produced [Os{η^2-

SCNRMe}L(CO)(PPh$_3$)$_2$]$^+$ which converted to the monohapto dithioester Os{C(S)NRMe}H(CS)(CO)(PPh$_3$)$_2$ with NaBH$_4$. A further rearrangement and alkylation produced **116** and **117** {M = OsH(CO)(PPh$_3$)$_2$}, respectively

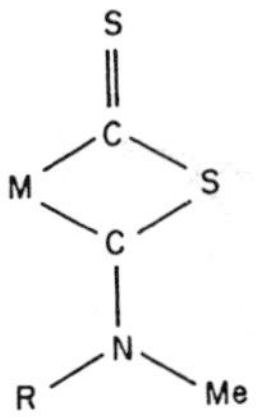

(116) 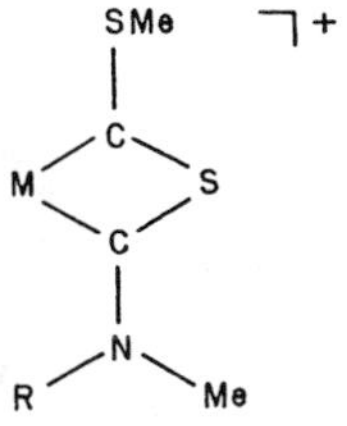 (117)

(470). The crystal structure of **116** has been completed (471). The dimeric molybdenum complexes CpMo{SC(NR)S}$_2$MoCp (R = Me, Bu^t, C$_6$H$_{11}$, CH$_2$Ph), containing bridging dithiocarbonimidate ligands, were synthesized from [CpMoS(CH$_2$)$_3$S]$_2$ and excess isonitrile (472).

There is one example to date of the formation of an imidoyl complex, viz., Pt{η^1-C(OMe)NR}R^{1}L$_2$ (R = Me, aryl; L$_2$ = Ph$_2$PCH=CHPPh$_2$; R^1 = CF$_3$, CH$_2$CN), by the insertion of an isocyanide into a metal–alkoxy bond (467). All other reactions have utilized the attack of alkoxide ions on bonded isocyanide ligands. Minghetti and Bonati (473, 474) first used alcoholic KOH to synthesize Pt{C(OR)NR^1}$_2$(CNR^1)PPh$_3$ from PtCl$_2$(PPh$_3$)$_2$–CNR^1 mixtures and later extended this work to the synthesis of the silver and gold imidoyl trimers [M{C(OR)NR^1}]$_3$ (M = Ag, Au; R = alkyl; R^1 = alkyl, C$_6$H$_4$Me-p) from simple silver(I) and gold(I) chloroisocyanide compounds (474–476). The crystal structure determination of [Au{C(OEt)NC$_6$H$_4$Me-p}]$_3$ has been reported (466). These trimers are cleaved by donor ligands to give the monomeric compounds Au{C(OR)NR^1}L (L = PPh$_3$, CNR) (477) and by HCl to give the carbenes AuCl{C(OMe)NHMe} (475). They also readily transfer the imidoyl group to form HgY$_2$, [AuY]$_3$ and PtClY(PPh$_3$)$_2$ {Y = C(OEt)NC$_6$H$_4$Me-p} from suitable precursors (476).

Typical platinum complexes that have recently been prepared from alcoholic KOH solutions and platinum isocyanide precursors are *trans*-Pt{C(OMe)NR}(CHNR)(PEt$_3$)$_2$ (R = C$_6$H$_4$Me-p) (440, 451) and Pt{C(OR)NC$_6$H$_4$Me-p}$_2$(CNC$_6$H$_4$Me-p)PPh$_3$ (R = Me, Et) (478).

The compounds Pt{C(OMe)NR}$_2$dpe, Pt{C(OMe)NR}$_2$(CNR)PPh$_3$ (478), Au{C(OMe)NR}PPh$_3$, and Hg{C(OEt)NR}$_2$ (R = C$_6$H$_4$Me-p) (479, 480) have been shown to act as monodentate or bidentate ligands to a series of metal complexes.

A linear tetramer has recently been isolated from reactions on xylyl isocyanide with (CODRhCl)$_2$ in methanol solution (*32*). An X-ray structural determination (*279*) has shown it to be **118** containing four rhodium

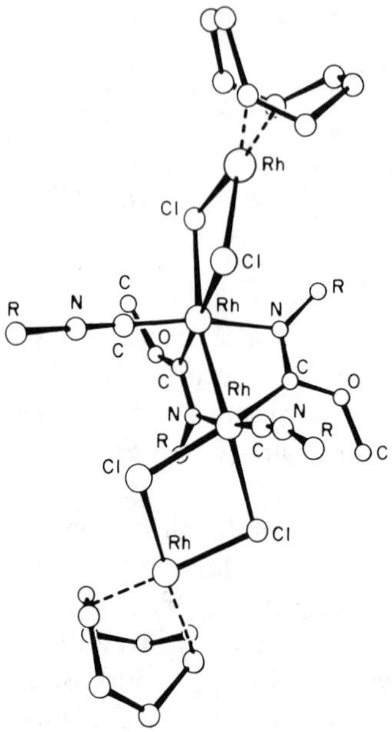

(118)

atoms bonded in two pairs by chloride bridges. The two pairs are linked by two imidoyl bridges and a rhodium–rhodium bond of 2.57 Å. A formal oxidation state of rhodium(II) is proposed for the two central rhodium atoms.

Insertions into the metal–oxygen bond in mercuric acetate with aromatic isocyanides has given the compound **119** (R = Ph, xylyl), which readily hydrolyzed to give organic acetamide derivatives (*480*).

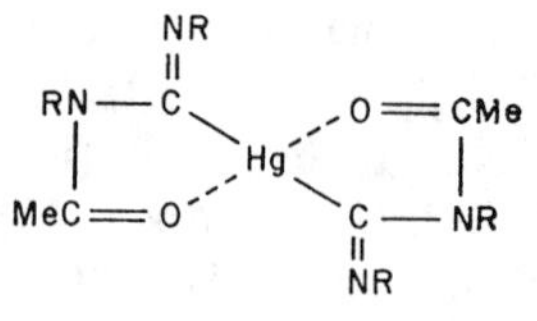

(119)

4. *Into Metal–Halogen Bonds*

Methyl and *t*-butyl isocyanide are proposed to insert into metal–halogen bonds in the compounds MCl_4 (M = Ti, Zr, Hf), MCl_3 (M = Ti, V) and $NbCl_4(THF)_2$ to give the chloroimino bridged compounds $[MCl_2X\{\mu\text{-}\eta^2\text{-}C(Cl)NBu^t\}(CNBu^t)]_2$ (M = Ti, Hf,Nb; X = Cl; M = V; X = $CNBu^t$), which for vanadium is proposed to have structure **120** (*204, 481*). Similar

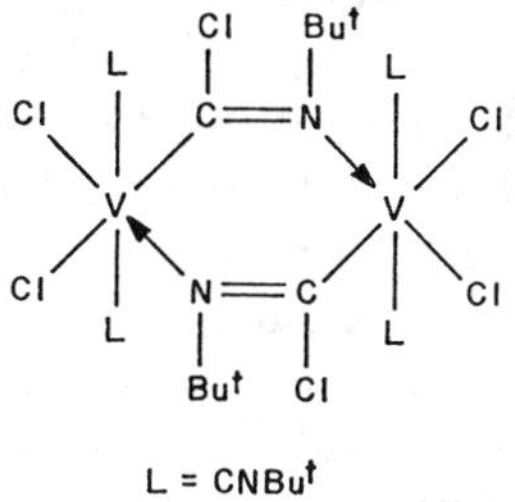

L = CNBu^t

(120)

reactions with the pentahalides MX_5 (M = Nb, Ta; X = Cl, Br) and CNR (R = Me, Bu^t) and $TaOBr_3$ gave $MX_{4-n}\{C(X)NR\}_n(CNR)$ (n = 1, 2, M = Nb, X = Cl, R = Me; n = 3, M = Ta, X = Br, R = Me) and $TaBr_2O\{C(Br)NMe\}CNMe$ (*482, 483*). No inserted products were observed for chromium (*481*) molybdenum and tungsten halides (*204, 481*).

The formation of chloroimino ligands has been proposed solely on weak $\nu(C=N)$ vibrations in the 1600–1750 cm^{-1} IR region. A repeat of the vanadium reaction by other workers (*109*) found the product of their reaction to be $VCl_3(CNBu^t)_3$, so some caution should be applied to the formulation of the proposed vanadium dimer at least. Substitution reactions on these compounds with a series of mono- and bidentate tertiary phosphorus ligands and metathetical replacements with lithium quinolin-8-olate and sodium diethyl dithiocarbamate have given a range of neutral and cationic products (*204, 483*).

B. *Nucleophilic Addition by Alcohols and Amines*

These reactions involve the addition across the $C≡N$ bond in isonitriles by the O–H or N–H bonds of alcohols or amines to form carbenes. The general reactions may be represented as Eqs. (7) and (8):

$$M—C≡N—R + HOR^1 \rightarrow M=C(OR^1)—NHR \qquad (7)$$

$$M—C≡N—R + HNR^1R^2 \rightarrow M=C(NR^1R^2)—NHR \qquad (8)$$

and the carbene products are related to the imidoyl $M—C(OR)=NR$ and imido $M—C(NR^1R^2)=NR$ precursors by deprotonations. Some comments about these reactions can be made. As expected the addition is facilitated by electron-withdrawing groups on the isocyanide and electron-releasing groups on the amine, and alcohol additions tend to require more forcing conditions than amines. The carbene products have been shown to be better σ donors than their isocyanide precursors and poorer π acceptors (*484, 485*). The C 1*s* binding energies show that the carbenoid carbon atoms are less positively charged than the carbon atom of the isocyanide (*484*). Geometrical isomers have been separated in these complexes, and NMR studies have demonstrated rapid rotation about the metal–carbon bond but restricted rotation about the carbon–nitrogen bond. In some cases restricted rotation has been observed about the metal–carbon bond, but this has been ascribed to steric rather than electronic influences (*3, 486*).

For alcohol additions, mono- and dicarbene complexes have been synthesized, for palladium, platinum (*487–489*), and gold (*475, 490–493*), from methanol, ethanol, or ethanethiol (*494*) additions to the isocyanide precursor. The stability of the gold(I) carbenes can be gauged from reactions with halogens which gave the oxidative addition product without loss of the carbene ligand (*492*).

For amino additions, two general reactions occur which generate either monocarbenes or chelating dicarbenes (*495–497*). A variety of types of chelating dicarbenes have been synthesized depending upon whether the attacking amine is mono- or bifunctional. Primary amines produce four-membered metallocycle rings of general formula **121** (*495*) whereas with bifunctional amines five-membered metallocycle rings of the type **122** are produced (*496, 497*).

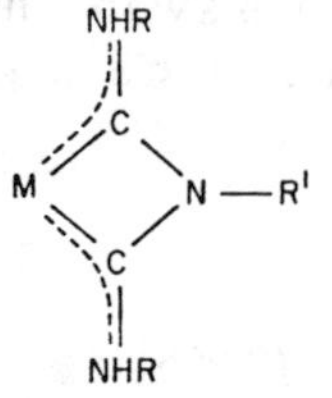

(121) 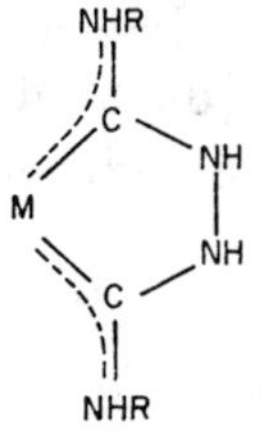   (122)

With monodentate carbenes, complexes containing one, two, or three of these ligands in the coordination sphere are known, and quantitative data on the reaction of amines to a coordinated isocyanide ligand have shown the addition to proceed by a stepwise mechanism through an intermediate

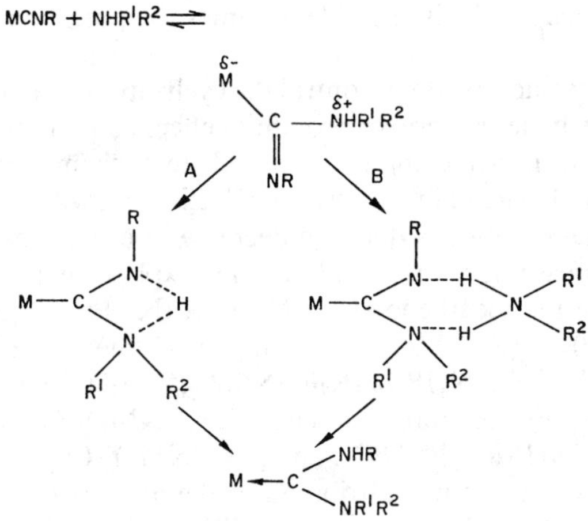

$$MCNR + NHR^1R^2 \rightleftharpoons$$

$$\underset{\substack{\delta- \\ M}}{} \diagdown C \overset{\delta+}{\text{—}} NHR^1R^2$$

SCHEME 11

produced by amine attack on the isocyanide carbon atom. This then undergoes proton transfer to give the final product (Scheme 11) (498–500). Proton transfer to give the final carbene can occur intramolecularly via a four-center transition state (Path A) or intermolecularly with a second amine acting as a catalyst in a six-membered transition state (Path B) (498). Solvents which had a decreasing ability to form hydrogen bonds with the attacking amine increased the rate as did electron-withdrawing substituents on the isocyanide, and the electronegativity of the halide (501–504). With ortho-substituted anilines, steric strain in the intermediate process produces an overall decrease in the rate such that reversible attack by amine on the isocyanide carbon becomes comparable in rate with subsequent stages leading to carbene formation (505). Variations in tertiary phosphorus ligands affected the rate through a balance of steric and electronic factors, i.e., amine attack requires low steric hindrance and a high π-acceptor property of the phosphorus donor ligand (501). Steric factors were cited as the reason why biscarbenes in cis-$PdCl_2(CNAr)_2$ were formed with primary amines but only monocarbenes resulted with secondary amines (499). In a study on the reaction of $MeNHNH_2$ with cis-$PdCl_2(CNAr)_2$ in dichloromethane, preliminary attack of the solvent on the carbene carbon is proposed (506).

In a competing reaction between bonded carbonyls and isocyanides with amines, amine attack on the carbonyl appears to be much more rapid than on the isocyanide. However, as the electron density on the metal, and hence on the carbonyl carbon increases the tendency to react with amines decreases and attack on the isocyanide becomes the preferred pathway (507, 508).

There are a few factors which control the cyclization of a monocarbene to a chelating dicarbene. A steric or proximity effect could prevent the mono-carbene ligand from achieving the required stereochemistry for attack on another isocyanide ligand in cis position. Cyclization will be less favored in low-valent metal complexes due to a decrease in electropositive character through an increase in π backbonding. This explains why $[Rh(CNBu^t)_4]^+$ forms only the monocarbene $[Rh(CNBu^t)_3\{C(NHPr^n)NHBu^t\}]^+$ whereas $[Rh(R)I(CNBu^t)_4]^+$ (R = Me, Et) rapidly gave the chelating dicarbene $[Rh\{C(NHBu^t)N(Pr^n)C(NHBu^t)\}(R)I(CNBu^t)_2]^+$ (509). In a comparison of the formation of the monocarbenes $[Fe(CNMe)_5\{C(NHMe)NH_2\}]^{2+}$, $[Ru(CNMe)_5\{C(NHMe)_2\}]^{2+}$ (496), and $[Os(CNMe)_3\{C(NHMe)_2\}_3]^{2+}$ (510) the absence of chelation was attributed to the amino-substituted carbene ligand being less sterically demanding, and that the greater radius of ruthenium and osmium decreases the proximity of the coordinated ligands.

These carbenes exhibit some unusual properties and also undergo some interesting reactions. For example, heating $[Ru(CNMe)_5\{C(NHEt)-NHMe\}]^{2+}$ in DMSO for 6 hr resulted in an isomerization to $[Ru(CNMe)_4-(CNEt)\{C(NHMe)_2\}]^{2+}$ via a chelating dicarbene similar to 121 (495). The carbene ligand in $CoMe(dimethylglyoximato)\{C(NHMe)_2\}$ undergoes an isomerization, which is thermally irreversible, to give the formamidine complex $CoMe(dimethylglyoximato)\{N(Me)CHNHMe\}$ (511). Deprotonation reactions in mono- and dicarbene complexes readily occur to give compounds containing the neutral amidino ligand—$C(NHR)=NR^1$. For example the yellow isomer of Chugaeuv's salt $[Pt\{C(NHMe)NHNHC(NHMe)\}(CNMe)_2]^{2+}$ converts to the red isomer $[Pt\{C(NHMe)=NNHC(NHMe)\}(CNMe)_2]^+$ with base (497). Displacement of carbenes in gold complexes by PPh$_3$ in alcohol, or cyanide ions in DMSO, gave the formamidines $CH(=NR)NR^1R^2$ (R = Bu^t, R^1 = R^2 = CH$_2$Ph; R = R^2 = Me, R^1 = H; R = R^1 = R^2 = Me), and these reactions have established the carbenes as intermediates in the α-addition reactions of protic nucleophiles to isocyanides (512, 512a).

C. Formation of Metal–Carbynes

The compound $HFe_3(CO)_{10}(CNMe_2)$ (76) (M = Fe; R^1 = R^2 = Me), characterized (513) in 1970 from reactions of DMF with $Fe_3(CO)_{12}$, has recently been prepared by new routes. In one reaction, 76 (M = Fe; R^1 = R^2 = Me, Et) was one of two products resulting from additions of chlorofor-miminium chlorides to $Fe_3(CO)_{12}$ (514). Another route has involved addition of CNR^1 (R^1 = Me, Et, Pr^i) to the anion [HFe$_3$(CO)$_{11}$]$^-$ followed by alkylation and protonation. This gave the two products **76** (M = Fe; R^1 = R^2 = Et; R^1 = Et, R^2 = Me, Pr^i; R^1 = Me, R^2 = H) and HFe$_3$(μ-CNEt$_2$)(CO)$_9$CNEt (445). The ruthenium analog **76** (M = Ru; R^1 = R^2 = Me) was prepared from Me$_2$NCH$_2$SnMe$_3$ and Ru$_3$(CO)$_{12}$ (515), whereas, corresponding osmium derivatives **76** {M = Os; R^1 = H, R^2 = Bu^t (446); R^1 = R^2 = Me (405)} resulted from reactions of H$_2$Os$_3$(CO)$_{10}$ with CNBu^t, or by treating Os$_3$(CO)$_{12}$ with Me$_3$N or PhCH$_2$NMe$_2$ in high boiling solvents, respectively.

X-Ray crystallographic structural determinations of **76** [M = Ru, R^1 = R^2 = Me (515); M = Os, R^1 = H, R^2 = Bu^t (516)] have shown an iminium-type structure (**123B**) for the CNR$_2$ ligand, but variable-temperature ^{1}H NMR has indicated a significant contribution of the carbyne structure (**123A**) to the resonance hybrid (445). Large energy barriers to rotation in the μ-CNR^{1}R^2 ligands have been observed in the iron (445) and osmium (405) complexes.

(123)

Other bridging iminyl groups have been synthesized from protonations and alkylations of cyclopentadienyl iron carbonyl isocyanide complexes. Protonations of [Cp$_2$Fe$_2$(CO)$_3$]$_2$CN(CH$_2$)$_n$NC, and (dienyl)$_2$Fe$_2$(CO)$_{4-x}$-(CNR)$_x$ occur at a rate dependent upon the nature of the dienyl ligand, the R group, and the degree of isocyanide substitution. Thus, indenyl salts are less readily formed than their cyclopentadiene counterparts, smaller R groups react faster and gave more stable products than bulky R groups, and an increase in the number of isocyanides in the coordination sphere increases the rate of protonation or alkylation (517). Representative products formed in these reactions are [{Cp$_2$Fe$_2$(CO)$_3$}$_2$·CNH-(CH$_2$)$_n$HNC](PF$_6$)$_2$ (168), cis-[(dienyl)$_2$Fe$_2$(CO)$_2$(μ-CO)(μ-CNHR)]PF$_6$, and cis-[(dienyl)$_2$Fe$_2$(CO)$_2$(μ-CNHR)$_2$](PF$_6$)$_2$ (518). The latter two products are also the hydrolysis products of acetyl or benzoyl chloride reactions on Cp$_2$Fe$_2$(CO)$_{3-x}$(CNR)$_x$ (x = 1, 2; R = Me, Et, CH$_2$Ph) via the unstable intermediate [Cp$_2$Fe$_2$(CO)$_{3-x}${CN(R)COR}$_x$]Cl (x = 1, 2) (519). In ethanol or phenolic solvents the ^{1}H NMR of (dienyl)$_2$Fe$_2$(CO)$_{4-x}$(CNR)$_x$ (dienyl = MeC$_5$H$_4$, Cp; R = alkyl; x = 1, 2) contains resonances attributable to species hydrogen bonded to solvent protons through the isocyanide nitrogen atom. In the more acidic solvents the "free" species, hydrogen-bonded species, and the protonated species [(dienyl)$_2$Fe$_2$(CO)$_3${CN(R)H}]$^+$ are shown to be in equilibrium (520). The crystal structure of cis-[Cp$_2$Fe$_2$(CO)$_2$(μ-CO){μ-CN(H)Me}]BF$_4$ has been elucidated and the C–N bond distance of 1.28 Å is evidence for the iminyl-type bonding (123B) of this group (518). Addition of alkyl iodides to (dienyl)$_2$Fe$_2$(CO)$_3$CNR (dienyl = C$_5$H$_4$Me, Cp; R = alkyl) and Cp$_2$Fe$_2$-(CO)(CNMe)$_2$CS gave [(dienyl)$_2$Fe$_2$(CO)$_3$(μ-CNR^{l}R)]I (191, 517, 521) and [Cp$_2$Fe$_2$(μ-CO)(μ-CNMe$_2$)(CNMe)CSMe]I$_2$ (522). The crystal structure of cis-(C$_5$H$_4$Me)$_2$Fe$_2$(μ-CO)(μ-CNMe$_2$)(CO)$_2$]I was resolved (517). In the completely substituted product Cp$_2$Fe$_2$(CNXylyl)$_4$, protonations with HPF$_6$ have given the diprotonated species Cp$_2$Fe$_2$(CNXylyl)$_2$(μ-CN(H)Xylyl)$_2$](PF$_6$)$_2$, the reactions being extremely rapid when compared with mono- and disubstituted species. In polar solvents this salt rapidly converts to [CpFe(CNXylyl)$_3$]PF$_6$. Alkylations on Cp$_2$Fe$_2$(CNXylyl)$_4$ with RI, (R = Me, Et) effected alkylation of only one isonitrile to give [Cp$_2$Fe$_2$(CNXylyl)$_3$(μ-CN(R)Xylyl)]I (R = Me, Et). These reactions exemplify the greater nucleophilic character of the bridging nitrogen atom through an increased contribution of the resonance form B (Section II) in the bonding mode, when compared with monodentate isocyanide ligands (32).

Other bridging carbynes that have been prepared are [Fe$_2$(CNEt)$_6$(μ-CNEt){μ-CN(R)Et}$_2$]I$_2$ (R = Me, Et) from alkyl halide additions to Fe$_3$(CNEt)$_9$ (17), and [W$_2$(CNMe)$_4$(μ-CNHMe)$_2$(PMe$_2$Ph)$_4$](BF$_4$)$_2$ from reactions of mer-W(CNMe)$_3$(PMe$_2$Ph)$_3$ with HX (X = HSO$_4$, Cl) (523). In the iron example the bridging carbynes were proposed to have character (123A) from their low contact-carbon atom shifts in the ^{13}C NMR spectrum.

Protonation of terminal isocyanide ligands has been effected in the electron rich complexes trans-ReCl(CNR)(dpe)$_2$ (R = Me, Bu^t) (524, 525) and M(CNR)$_2$(dpe)$_2$ (M = Mo, W; R = Me, Bu^t, Ph, C$_6$H$_4$Me-p) (57, 280). This is due to the considerably induced π character in the metal–carbon bond through electron release from the metal. This causes a bending in the MCNR angle to 156(1)° as observed in the crystal structure determination of trans-Mo(CNMe)$_2$(dpe)$_2$ (280). Complexes isolated from these protonation reactions with HX (X = BF$_4$, HSO$_4$, SFO$_3$) include trans-[ReCl(CNHR)(dpe)$_2$]BF$_4$ trans-[M(CNHR)(CNR)(dpe)$_2$]X, and trans-[M(CNHR)$_2$(dpe)$_2$]X$_2$. Under carefully controlled conditions, the intermediate hydrides [MH(CNR)$_2$(dpe)$_2$]X and [MH(CNHR)-(CNR)(dpe)$_2$]X$_2$ were formed (526). Irradiation of trans-Mo(CNMe)$_2$(PMe$_2$Ph)$_4$ and mer-W(CNMe)$_3$(PMe$_2$Ph)$_3$ in alcohol solutions containing H$_2$SO$_4$ or HCl produced low yields of methylamine, ammonia, and hydrocarbons (mainly methane). The dimeric carbyne [W$_2$(CNMe)$_4$(μ-CNHMe)$_2$(PMe$_2$Ph)$_4$]$^{2+}$ was implicated as an intermediate in these conversions (523).

Alkylations on trans-$M(CNMe)_2(dpe)_2$ were also readily effected with $MeFSO_3$, Me_2SO_4, or $[Et_3O]BF_4$ in benzene to give the monoalkylated salts trans-$[M(CNRMe)(CNMe)(dpe)_2]X$ (M = Mo, W; R = Me, Et; X = FSO_3, $MeSO_4$, BF_4), which isomerized to the cis isomer in CH_2Cl_2 solution. No dialkylations were observed with these complexes (58).

X-Ray structural determination on trans-$[ReCl(CNHMe)(dpe)_2]BF_4$ (525) has found the C(carbyne)–N bond length to be 1.35 Å (i.e., bonding mode 123A). In contrast, the bridging $>CNR^1R^2$ ligands in 76 (M = Os; R^1 = H; R^2 = Bu^t) (516), 76 (M = Ru; R^1 = R^2 = Me) (515), and $[Cp_2Fe_2(CO)_2(\mu$-CO)$(\mu$-CNHMe)]BF_4$ (518) have C(carbyne)–N bond lengths of 1.298, 1.28, and 1.28 Å, respectively, demonstrating an iminyl-type resonance mode (123B) for these bridging ligands.

D. Organic Syntheses

An in-depth survey of the numerous organic compounds that have been synthesized from isocyanide–metal interactions is beyond the scope of this review. However, the versatility of the isocyanide molecule in organic synthesis can be gauged from its involvement in the following series of products: amides (462, 527), amines (527, 528), bicyclic ring systems (529, 530), C_{11}–C_{15} rings (531, 532), dianionic acids (533, 534), diketones (535), ethylene derivatives (529, 535), formamides (534), formamidines (536), formimidates (536), formylaminoacrylic esters (534), guanidines (537), heterocycles (534, 538–540), indazole derivatives (541, 542) and heterobicyclic ring systems (543), isocyanates and diisocyanates (544, 545), macrocycles (528), Michael additions (546, 547), saturated and unsaturated cyclic olefin derivatives (529, 543, 548–554), serines (534), ureas (555, 556), and urethanes (555).

References

1. S. J. Lippard, Prog. Inorg. Chem. 21, 91 (1976).
2. Y. Yamamoto, Coord. Chem. Rev. 32, 193 (1980).
3. F. A. Cotton and C. M. Lukehart, Prog. Inorg. Chem. 16, 487 (1972).
4. D. J. Cardin, M. J. Doyle, and M. F. Lappert, Chem. Soc. Rev. 2, 99 (1973).
5. C. P. Casey, Org. Chem. (N.Y.) 33, Part 1, 195 (1976).
6. P. M. Treichel, Adv. Organomet. Chem. 11, 21 (1973).
7. F. Bonati and G. Minghetti, Inorg. Chim. Acta 9, 95 (1974).
8. W. A. LaRue, A. Thu Liu, and J. San Filippo, Jr. Inorg. Chem., 19, 315 (1980).
9. P. Brant, F. A. Cotton, J. C. Sekutowski, T. E. Wood, and R. A. Walton, J. Am. Chem. Soc. 101, 6588 (1979).
10. G. S. Girolami and R. A. Andersen, J. Organomet. Chem. 182, C43 (1979).
11. G. S. Girolami and R. A. Andersen, Inorg. Chem. 20, 2040 (1981).

12. T. E. Wood, J. C. Deaton, J. Corning, R. E. Wild, and R. A. Walton, *Inorg. Chem.* **19,** 2614 (1980).
13. W. S. Mialki, R. E. Wild, and R. A. Walton, *Inorg. Chem.* **20,** 1380 (1981).
14. D. D. Klendworth, W. W. Welters, III, and R. A. Walton, *Organometallics* **1,** 336 (1982).
15. W. S. Mialki, T. E. Wood, and R. A. Walton, *Inorg. Chem.* **21,** 480 (1982).
16. J. M. Bassett, D. E. Berry, G. K. Barker, M. Green, J. A. K. Howard, and F. G. A. Stone, *J. Chem. Soc., Dalton Trans.* p. 1003 (1979).
17. J. M. Bassett, G. K. Barker, M. Green, J. A. K. Howard, F. G. A. Stone, and W. C. Wolsey, *J. Chem. Soc., Dalton Trans.* p. 219 (1981).
18. H. E. Swanepoel, unpublished results.
19. W. E. Carroll, M. Green, A. M. R. Galas, M. Murray, T. W. Turney, A. J. Welch, and P. Woodward, *J. Chem. Soc., Dalton Trans.* p. 80 (1980).
20. Y. Yamamoto, K. Aoki, and H. Yamazaki, *Chem. Lett.* p. 391 (1979).
21. E. L. Muetterties, E. Band, A. Kokorin, W. R. Pretzer, and M. G. Thomas, *Inorg. Chem.* **19,** 1552 (1980).
22. M. G. Thomas, W. R. Pretzer, B. F. Beier, F. J. Hirsekorn, and E. L. Muetterties, *J. Am. Chem. Soc.* **99,** 743 (1977), and references therein.
23. M. Green, J. A. K. Howard, M. Murray, J. L. Spencer, and F. G. A. Stone, *J. Chem. Soc. Dalton Trans.* p. 1509 (1977).
24. L. Malatesta and F. Bonati, "Isocyanide Complexes of Metals." Wiley (Interscience), New York, 1969.
25. Y. Yamamoto and H. Yamazaki, *Inorg. Chem.* **17,** 3111 (1978).
26. M. O. Albers, N. J. Coville, T. V. Ashworth, E. Singleton, and H. E. Swanepoel, *J. Chem. Soc., Chem. Commun.* p. 489 (1980).
27. M. O. Albers, N. J. Coville, T. V. Ashworth, E. Singleton, and H. E. Swanepoel, *J. Organomet. Chem.* **199,** 55 (1980).
28. N. J. Coville, M. O. Albers, T. V. Ashworth, and E. Singleton, *J. Chem. Soc., Chem. Commun.* p. 408 (1981).
29. M. O. Albers, N. J. Coville, P. ten Doeschate, and E. Singleton, *S. Afr. J. Chem.* **34,** 82 (1981).
30. M. O. Albers, N. J. Coville, C. P. Nicolaides, R. A. Webber, T. V. Ashworth, and E. Singleton, *J. Organomet. Chem.* **217,** 247 (1981).
31. M. O. Albers, N. J. Coville, and E. Singleton, *J. Chem. Soc., Chem. Commun.* p. 96 (1982).
32. E. Singleton, unpublished results.
33. S. Otsuka, Y. Tatsuno, and K. Ataka, *J. Am. Chem. Soc.* **93,** 6705 (1971).
34. J. R. Boehm, D. J. Doonan, and A. L. Balch, *J. Am. Chem. Soc.* **98,** 4845 (1976).
35. M. F. Rettig, E. A. Kirk, and P. M. Maitlis, *J. Organomet. Chem.* **111,** 113 (1976).
36. H. E. Swanepoel, Ph.D. Thesis, Rand Afrikaans University (1980).
37. H. B. Gray, V. M. Miskowski, S. J. Milder, T. P. Smith, A. W. Maverick, J. D. Buhr, W. L. Gladfelter, I. S. Sigal, and K. R. Mann, *Fundam. Res. Homogeneous Catal.* **3,** 819 (1979).
38. H. B. Gray, K. R. Mann, N. S. Lewis, J. A. Thich, and R. M. Richman, *Adv. Chem. Ser.* **168,** 44 (1978).
39. P. T. Wolczanski and J. E. Bercaw, *J. Am. Chem. Soc.* **101,** 6450 (1979).
40. S. Z. Goldberg, R. Eisenberg, J. S. Miller, and A. J. Epstein, *J. Am. Chem. Soc.* **98,** 5173 (1976).
41. T. Iinuma and T. Tanaka, *Inorg. Chim. Acta* **49,** 79 (1981).
42. H. Isci and W. R. Mason, *Inorg. Chem.* **13,** 1175 (1974).
43. R. D. Adams and D. F. Chodosh, *J. Am. Chem. Soc.* **99,** 6544 (1977).
44. R. D. Adams, *J. Am. Chem. Soc.* **102,** 7476 (1980).

45. N. J. Coville, M. O. Albers, and E. Singleton, *J. Chem. Soc., Dalton Trans.* p. 1389 (1982).
46. A. L. Balch, *J. Am. Chem. Soc.* **98**, 285 (1976).
47. W. E. Carroll, M. Green, J. A. K. Howard, M. Pfeffer, and F. G. A. Stone, *J. Chem. Soc., Dalton Trans.* p. 1472 (1978).
48. N. M. Boag, M. Green, D. M. Grove, J. A. K. Howard, J. L. Spencer, and F. G. A. Stone, *J. Chem. Soc., Dalton Trans.* p. 2170 (1980).
49. N. M. Boag, G. H. M. Dias, M. Green, J. L. Spencer, F. G. A. Stone, and J. Vicente, *J. Chem. Soc., Dalton Trans.* p. 1981 (1981).
50. A. Efraty, I. Feinstein, F. Frolow, and A. Goldman, *J. Chem. Soc., Chem. Commun.* p. 864 (1980).
51. A. Efraty, I. Feinstein, F. Frolow, and L. Wackerle, *J. Am. Chem. Soc.* **102**, 6341 (1980).
52. A. Efraty, I. Feinstein, L. Wackerle, and F. Frolow, *Angew. Chem., Int. Ed. Engl.* **19**, 633 (1980).
53. P. Fantucci, L. Naldini, F. Cariati, V. Valenti, and C. Bussetto, *J. Organomet. Chem.* **64**, 109 (1974).
54. A. C. Sarapu and R. F. Fenske, *Inorg. Chem.* **14**, 247 (1975).
55. J. Chatt, C. M. Elson, A. J. L. Pombeiro, R. L. Richards, and G. H. D. Royston, *J. Chem. Soc., Dalton Trans.* p. 165 (1978).
56. A. J. L. Pombeiro, J. Chatt, and R. L. Richards, *J. Organomet. Chem.* **190**, 297 (1980).
57. J. Chatt, A. J. L. Pombeiro, and R. L. Richards, *J. Chem. Soc., Dalton Trans.* p. 492 (1980).
58. J. Chatt, A. J. L. Pombeiro, and R. L. Richards, *J. Organomet. Chem.* **184**, 357 (1980).
59. W. P. Fehlhammer, F. Degel, and H. Stolzenberg, *Angew. Chem., Int. Ed. Engl.* **20**, 214 (1981).
60. V. W. Day, R. O. Day, J. S. Kristhoff, F. J. Hirsekorn, and E. L. Muetterties, *J. Am. Chem. Soc.* **97**, 2571 (1975).
61. L. S. Benner, M. M. Olmstead, and A. L. Balch, *J. Organomet. Chem.* **159**, 289 (1978).
62. A. V. Rivera, G. M. Sheldrick, and M. B. Hursthouse, *Acta Crystallogr. Sect. B* **B34**, 1985 (1978).
63. M. I. Bruce, J. G. Matisons, J. R. Rodgers, and R. C. Wallis, *J. Chem. Soc., Chem. Commun.* p. 1070 (1981).
64. C. J. Commons and B. F. Hoskin, *Aust. J. Chem.* **28**, 1663 (1975).
65. R. D. Adams and F. A. Cotton, *in* "Dynamic Nuclear Magnetic Resonance Spectroscopy" (L. M. Jackman and F. A. Cotton, eds.), p. 489. Academic Press, New York, 1975.
66. G. F. Stuntz and J. R. Shapley, *J. Organomet. Chem.* **213**, 389 (1981).
67. J. A. Dineen and P. L. Pauson, *J. Organomet. Chem.* **71**, 91 (1974).
68. M. I. Bruce and R. C. Wallis, *Aust. J. Chem.* **34**, 209 (1981).
69. W. P. Fehlhammer and F. Degel, *Angew. Chem., Int. Ed. Engl.* **18**, 75 (1979).
70. E. O. Fischer and R. J. J. Schneider, *J. Organomet. Chem.* **12**, P27 (1968).
71. M. L. H. Green and W. E. Lindsell, *J. Chem. Soc. A* p. 2150 (1969).
72. M. Novotny, D. F. Lewis, and S. J. Lippard, *J. Am. Chem. Soc.* **94**, 6961 (1972).
73. R. V. Parish and P. G. Simms, *J. Chem. Soc., Dalton Trans.* p. 2389 (1972).
74. M. Novotny and S. J. Lippard, *J. Chem. Soc., Chem. Commun.* p. 202 (1973).
75. M. Schaal and W. Beck, *Angew. Chem., Int. Ed. Engl.* **11**, 527 (1972).
76. M. Höfler and W. Kemp, *Chem. Ber.* **112**, 1934 (1979).
77. H. Behrens, M. Moll, W. Popp, H. J. Seibold, E. Sepp, and P. Würstl, *J. Organomet. Chem.* **192**, 389 (1980).
78. G. Simonneaux, P. le Maux, G. Jaouen, and R. Dabard, *Inorg. Chem.* **18**, 3167 (1979).
79. P. le Maux, G. Simmonneaux, P. Caillet, and G. Jaouen, *J. Organomet. Chem.* **177**, C1 (1979).

80. P. le Maux, G. Simonneaux, G. Jaouen, L. Quahab, and P. Batail, *J. Am. Chem. Soc.* **100**, 4312 (1978).
81. G. Höfle, *Z. Naturforsch., B: Anorg. Chem., Org. Chem.* **30**, 982 (1975).
82. B. V. Johnson, D. P. Sturtzel, and J. E. Shade, *Inorg. Chim. Acta* **32**, 243 (1979).
83. D. L. Reger, *Inorg. Chem.* **14**, 660 (1975).
84. C. Eaborn, N. Farrell, J. L. Murphy, and A. Pidcock, *J. Chem. Soc., Dalton Trans.* p. 58 (1976).
85. G. A. Larkin, R. Mason, and M. G. H. Wallbridge, *J. Chem. Soc., Dalton Trans.* p. 2305 (1975).
86. B. Weinberger and W. P. Fehlhammer, *Angew. Chem., Int. Ed. Engl.* **19**, 480 (1980).
87. H. Alper and R. A. Partis, *J. Organomet. Chem.* **35**, C40 (1972).
88. J. Kiji, A. Matsumura, S. Okazaki, T. Haishi, and J. Furukawa, *J. Chem. Soc., Chem. Commun.* p. 751 (1975).
89. W. Jetz and R. J. Angelici, *J. Organomet. Chem.* **35**, C37 (1972).
90. W. P. Fehlhammer and A. Mayr, *Angew. Chem., Int. Ed. Engl.* **14**, 757 (1975).
91. W. P. Fehlhammer, A. Mayr, and M. Ritter, *Angew. Chem., Int. Ed. Engl.* **16**, 641 (1977).
92. W. P. Fehlhammer, G. Christian, and A. Mayr, *J. Organomet. Chem.* **199**, 87 (1980).
93. W. P. Fehlhammer, A. Mayr, and G. Christian, *J. Organomet. Chem.* **209**, 57 (1981).
94. F. B. McCormick and R. J. Angelici, *Inorg. Chem.* **18**, 1231 (1979).
95. F. B. McCormick and R. J. Angelici, *Inorg. Chem.* **20**, 1118 (1981).
96. D. Mansuy, M. Lange, J. C. Chottard, and J. F. Bartoli, *Tetrahedron Lett.* p. 3027 (1978).
97. W. P. Fehlhammer and A. Mayr, *J. Organomet. Chem.* **191**, 153 (1980).
98. H. Werner, S. Lotz, and B. Heiser, *J. Organomet. Chem.* **209**, 197 (1981).
99. R. O. Harris, J. Powell, A. Walker, and P. V. Yaneff, *J. Organomet. Chem.* **141**, 217 (1977).
100. W. M. Greaves and R. J. Angelici, *Inorg. Chem.* **20**, 2983 (1981).
101. B. D. Dombek and R. J. Angelici, *Inorg. Chem.* **15**, 2403 (1976).
102. B. D. Dombek and R. J. Angelici, *J. Am. Chem. Soc.* **95**, 7516 (1973).
103. A. M. English, K. R. Plowman, I. M. Baibich, J. P. Hickey, and I. S. Butler, *J. Organomet. Chem.* **205**, 177 (1981).
104. F. Faraone, P. Piraino, V. Marsala, and S. Sergi, *J. Chem. Soc., Dalton Trans.* p. 859 (1977).
105. L. Bussetto and A. Palazzi, *Inorg. Chim. Acta* **19**, 233 (1976).
106. R. O. Harris and P. Yaneff, *J. Organomet. Chem.* **134**, C40 (1977).
107. J. Furukawa, A. Matsumura, Y. Matsuoka, and J. Kiju, *Bull. Chem. Soc. Jpn.* **49**, 829 (1976).
108. W. P. Fehlhammer, A. Mayr, and B. Olgemoeller, *Angew. Chem., Int. Ed. Engl.* **14**, 369 (1975).
109. L. D. Silverman, J. C. Dewan, C. M. Giandomenico, and S. J. Lippard, *Inorg. Chem.* **19**, 3379 (1980).
110. L. D. Silverman, P. W. R. Corfield, and S. J. Lippard, *Inorg. Chem.* **20**, 3106 (1981).
111. J. Müller and W. Holzinger, *Z. Naturforsch., B* **33**, 1309 (1978).
112. E. P. Kündig and P. L. Tims, *J. Chem. Soc., Chem. Commun.* p. 912 (1977).
113. E. P. Kündig and P. L. Tims, *J. Chem. Soc., Dalton Trans.* p. 991 (1980).
114. P. M. Treichel and G. J. Essenmacher, *Inorg. Chem.* **15**, 146 (1976).
115. E. B. Dreyer, C. T. Lam, and S. J. Lippard, *Inorg. Chem.* **18**, 1904 (1979).
116. C. T. Lam, M. Novotny, D. L. Lewis, and S. J. Lippard, *Inorg. Chem.* **17**, 2127 (1978).
117. M. Novotny and S. J. Lippard, *Inorg. Chem.* **13**, 828 (1974).
118. F. A. Cotton, P. E. Fanwick, and P. A. McArdle, *Inorg. Chim. Acta* **35**, 289 (1979).
119. C. A. L. Becker, *Inorg. Chim. Acta* **27**, L105 (1978).

120. C. A. L. Becker, *J. Inorg. Nucl. Chem.* **37**, 703 (1975).
121. C. A. L. Becker, *Inorg. Nucl. Chem. Lett.* **11**, 295 (1975).
122. C. A. L. Becker, *J. Inorg. Nucl. Chem.* **35**, 1875 (1973).
123. Y. Yamamoto and H. Yamazaki, *J. Organomet. Chem.* **137**, C31 (1977).
124. M. O. Albers and N. J. Coville, personal communication.
125. P. R. Branson and M. Green, *J. Chem. Soc., Dalton Trans.* p. 1303 (1972).
126. Y. Yamamoto, K. Aoki, and H. Yamazaki, *Inorg. Chem.* **18**, 1681 (1979).
127. F. Faraone, B. Pietropaolo, and E. Rotondo, *J. Chem. Soc., Dalton Trans.* p. 2262 (1974).
128. J. W. Dart, M. K. Lloyd, R. Mason, and J. A. McCleverty, *J. Chem. Soc., Dalton Trans.* p. 2039 (1973).
129. E. L. Muetterties, *Inorg. Chem.* **13**, 495 (1974).
130. A. J. Deeming, *J. Organomet. Chem.* **175**, 105 (1979).
131. K. R. Mann, J. G. Gordon, II, and H. B. Gray, *J. Am. Chem. Soc.* **97**, 3553 (1975).
132. K. R. Mann, N. S. Lewis, R. M. Williams, H. B. Gray, and J. G. Gordon, II, *Inorg. Chem.* **17**, 828 (1978).
133. K. Kawakami, M. Okajima, and T. Tanaka, *Bull. Chem. Soc. Jpn.* **51**, 2327 (1978).
134. P. V. Yaneff and J. Powell, *J. Organomet. Chem.* **179**, 101 (1979).
135. K. R. Mann, N. S. Lewis, V. M. Miskowski, D. K. Erwin, G. S. Hammond, and H. B. Gray, *J. Am. Chem. Soc.* **99**, 5525 (1977).
136. Y. Ohtani, S.-Y. Miya, Y. Yamamoto, and H. Yamazaki, *Inorg. Chim. Acta* **53**, L53 (1981).
137. P. W. Jolly and G. Wilke, "The Organic Chemistry of Nickel," Vol. 2. Academic Press, New York, 1974.
138. J. L. Spencer, *Inorg. Synth.* **19**, 213 (1979).
139. M. G. Thomas, E. L. Muetterties, R. O. Day, and V. W. Day, *J. Am. Chem. Soc.* **98**, 4645 (1976).
140. M. Green, J. A. K. Howard, J. L. Spencer, and F. G. A. Stone, *J. Chem. Soc., Chem. Commun.* p. 3 (1975).
141. R. W. Stephany and W. Drenth, *Recueil* **91**, 1453 (1972).
142. T. J. Weaver and C. A. L. Becker, *J. Inorg. Nucl. Chem.* **35**, 3739 (1973).
143. M. F. Rettig and P. M. Maitlis, *Inorg. Synth.* **17**, 134 (1977).
144. A. L. Balch, J. R. Boehm, H. Hope, and M. M. Olmstead, *J. Am. Chem. Soc.* **98**, 7431 (1976).
145. G. L. Geoffroy and M. S. Wrighton, "Organometallic Photochemistry." Academic Press, New York, 1979.
146. B. H. Byers and T. L. Brown, *J. Am. Chem. Soc.* **97**, 947 (1975).
147. B. H. Byers and T. L. Brown, *J. Am. Chem. Soc.* **99**, 2527 (1977).
148. J. M. Saveant, *Acc. Chem. Res.* **13**, 323 (1980).
149. P. M. Treichel and H. J. Mueh, *Inorg. Chim. Acta* **22**, 265 (1977).
150. Y. S. Varshavsky, E. P. Shestakova, N. V. Kiseleva, T. G. Cherkasova, N. A. Buzina, L. S. Bresler, and V. A. Karmer, *J. Organomet. Chem.* **170**, 81 (1979).
151. B. F. G. Johnson, J. Lewis, and D. Pippard, *J. Chem. Soc., Dalton Trans.* p. 407 (1981).
152. J. Chatt, G. J. Leigh, and N. Thankarajan, *J. Organomet. Chem.* **29**, 105 (1971).
153. S. A. Al-Jibori and B. L. Shaw, *J. Organomet. Chem.* **192**, 83 (1980).
154. R. B. King and M. S. Saran, *Inorg. Chem.* **13**, 74 (1974).
155. P. M. Treichel and D. B. Shaw, *J. Organomet. Chem.* **139**, 21 (1977).
156. M. Bigorgne, *J. Organomet. Chem.* **1**, 101 (1963).
157. R. B. King and P. R. Heckley, *J. Coord. Chem.* **7**, 193 (1978).
158. J. A. Connor, E. M. Jones, G. K. McEwen, M. K. Lloyd, and J. A. McCleverty, *J. Chem. Soc., Dalton Trans.* p. 1246 (1972).

159. M. O. Albers, Ph.D. Thesis, Witwatersrand University, Johannesburg (1982).
160. R. D. Adams, *Inorg. Chem.* **15**, 169 (1976).
161. R. D. Adams, *J. Organomet. Chem.* **88**, C38 (1975).
162. R. B. King, M. S. Saran, D. P. McDonald, and S. P. Diefenbach, *J. Am. Chem. Soc.* **101**, 1138 (1979).
163. R. B. King and M. S. Saran, *Inorg. Chem.* **13**, 364 (1974).
164. N. J. Coville, *J. Organomet. Chem.* **190**, C84 (1980).
165. W. Hieber and D. von Pigenot, *Chem. Ber.* **89**, 193 (1956).
166. S. M. Grant and A. R. Manning, *J. Chem. Soc., Dalton Trans.* p. 1789 (1979).
167. S. T. Wilson, N. J. Coville, J. R. Shapley, and J. A. Osborn, *J. Am. Chem. Soc.* **96**, 4038 (1974).
168. J. A. S. Howell and A. J. Rowan, *J. Chem. Soc., Dalton Trans.* p. 297 (1981).
169. H. Brunner and M. Vogel, *J. Organomet. Chem.* **35**, 169 (1972).
170. T. M. Chan, J. W. Connolly, C. D. Hoff, and F. Millich, *J. Organomet. Chem.* **152**, 287 (1978).
171. P. M. Treichel and D. C. Molzahn, *J. Organomet. Chem.* **179**, 275 (1979).
172. S. Grant, J. Newman, and A. R. Manning, *J. Organomet. Chem.* **96**, C11 (1975).
173. M. I. Bruce, D. Schultz, R. C. Wallis, and A. D. Redhouse, *J. Organomet. Chem.* **169**, C15 (1979).
174. M. J. Mays and P. D. Gavins, *J. Chem. Soc., Dalton Trans.* p. 911 (1980).
175. B. F. G. Johnson, J. Lewis, and D. Pippard, *J. Organomet. Chem.* **145**, C4 (1978).
176. M. Tachikwa and J. R. Shapley, *J. Organomet. Chem.* **124**, C19 (1977).
176a. R. D. Adams and N. M. Golembeski, *Inorg. Chem.* **18**, 1909 (1979).
177. J. Newman and A. R. Manning, *J. Chem. Soc., Dalton Trans.* p. 2549 (1974).
178. J. E. Ellis and K. L. Fjare, *J. Organomet. Chem.* **214**, C33 (1981).
179. P. M. Treichel, D. W. Firsich, and G. P. Essenmacher, *Inorg. Chem.* **18**, 2405 (1979).
180. H. J. Mueh, *Gov. Rep. Announce. Index (U.S.)* **77**, 94 (1977).
181. P. N. Prasad, *Spectrochim. Acta, Part A* **33A**, 335 (1977).
182. P. M. Treichel, G. E. Dirreen, and H. J. Mueh, *J. Organomet. Chem.* **44**, 339 (1972).
183. P. M. Treichel, K. P. Wagner, and H. J. Mueh, *J. Organomet. Chem.* **86**, C13 (1975).
184. P. M. Treichel and J. P. Williams, *J. Organomet. Chem.* **135**, 39 (1977).
185. R. D. Adams and F. A. Cotton, *Inorg. Chem.* **13**, 249 (1974).
186. R. D. Adams and F. A. Cotton, *J. Am. Chem. Soc.* **95**, 6589 (1973).
187. R. D. Adams, F. A. Cotton, and J. M. Troup, *Inorg. Chem.* **13**, 257 (1974).
188. R. D. Adams and F. A. Cotton *J. Am. Chem. Soc.* **94**, 6193 (1972).
189. J. A. S. Howell, T. W. Matheson, and M. J. Mays, *J. Chem. Soc., Chem. Commun.* p. 865 (1975).
190. J. A. S. Howell and P. Mathur, *J. Organomet. Chem.* **174**, 335 (1979).
191. J. A. S. Howell and A. J. Rowan, *J. Chem. Soc., Dalton Trans.* p. 503 (1980).
192. J. A. S. Howell, M. J. Mays, I. D. Hunt, and O. S. Mills, *J. Organomet. Chem.* **128**, C29 (1977).
193. M. J. Boylan, J. Bellerby, J. Newman, and A. R. Manning, *J. Organomet. Chem.* **47**, C33 (1973).
194. M. Ennis, R. Kumar, A. R. Manning, J. A. S. Howell, P. Mathur, A. J. Rowan, and F. S. Stephens, *J. Chem. Soc., Dalton Trans.* p. 1251 (1981).
195. J. Bellerby, M. J. Boylan, M. Ennis, and A. R. Manning, *J. Chem. Soc., Dalton Trans.* p. 1185 (1978).
196. W. P. Fehlhammer, A. Mayr, and G. Christian, *Angew. Chem., Int. Ed. Engl.* **17**, 866 (1978).
197. M. J. Mays and P. D. Gavens, *J. Organomet. Chem.* **177**, 443 (1979).
198. C. R. Eady, P. D. Gavens, B. F. G. Johnson, J. Lewis, M. C. Malatesta, M. J. Mays, A. G.

Orpen, A. V. Rivera, G. M. Sheldrick, and M. B. Hursthouse, *J. Organomet. Chem.* **149,** C43 (1978).

199. Y. Yamamoto, T. Mise, and H. Yamazaki, *Bull. Chem. Soc. Jpn.* **51,** 2743 (1978).
200. E. W. Powell and M. J. Mays, *J. Organomet. Chem.* **66,** 137 (1974).
201. G. Fachinetti and C. Floriani, *J. Chem. Soc., Chem. Commun.* p. 578 (1975).
202. G. Fachinetti, S. Del Nero, and C. Floriani, *J. Chem. Soc., Dalton Trans.* p. 1046 (1976).
203. A. H. Klazinga and J. H. Teuben, *J. Organomet. Chem.* **192,** 75 (1980).
204. M. Behnam-Dahkordy, B. Crociani, M. Nicolini, and R. L. Richards, *J. Organomet. Chem.* **181,** 69 (1979).
205. R. D. Adams, M. Brice, and F. A. Cotton, *J. Am. Chem. Soc.* **95,** 6594 (1973).
206. R. B. King and M. S. Saran, *Inorg. Chem.* **13,** 2453 (1974).
207. M. Kamata, K. Hirotsu, T. Higuchi, K. Tatsumi, R. Hoffmann, T. Yoshida, and S. Otsuka, *J. Am. Chem. Soc.* **103,** 5772 (1981).
208. A. J. L. Pombeiro, C. J. Pickett, R. L. Richards, and S. A. Sangokoya, *J. Organomet. Chem.* **202,** C15 (1980).
209. A. J. L. Pombeiro, *Rev. Port. Quim.* **21,** 90 (1979).
210. K. R. Mann, M. Cimolini, G. L. Geoffroy, G. S. Hammond, A. A. Orio, G. Albertin, and H. B. Gray, *Inorg. Chim. Acta* **16,** 97 (1976).
211. L. Verdonck, T. Tulin, and G. P. van der Kelen, *Spectrochim. Acta, Part A* **35A,** 867 (1979).
212. P. Romiti, M. Freni, and G. D'Alfonso, *J. Organomet. Chem.* **135,** 345 (1977).
213. M. Freni and P. Romiti, *J. Organomet. Chem.* **87,** 241 (1975).
214. P. M. Treichel, J. P. Williams, W. A. Freeman, and J. I. Gelder, *J. Organomet. Chem.* **170,** 247 (1979).
215. P. E. Ellis, Jr., R. D. Jones, and F. Basolo, *J. Chem. Soc., Chem. Commun.* p. 54 (1980).
216. T. G. Traylor, D. Campbell, S. Tsuchiya, M. Mitchell, and D. V. Stynes, *J. Am. Chem. Soc.* **102,** 5939 (1980).
217. P. I. Reisberg and J. S. Olson, *J. Biol. Chem.* **255,** 4159 (1980).
218. T. G. Traylor and D. V. Stynes, *J. Am. Chem. Soc.* **102,** 5938 (1980).
219. F. Stetzkowski, R. Banerjee, J.-Y. Lallemand, B. Cendrier, and D. Mansuy, *Biochimie* **62,** 795 (1980).
220. G. Mercati and F. Morazzoni, *Gazz. Chim. Ital.* **109,** 161 (1979).
221. D. Gussoni, G. Mercati, and F. Morazzoni, *Gazz. Chim. Ital.* **109,** 545 (1979).
222. A. Araneo, G. Mercati, F. Morazzoni, and T. Napoletano, *Inorg. Chem.* **16,** 1196 (1977).
223. T. Tsuihiji, T. Akiyama, and A. Sugimori, *Bull. Chem. Soc. Jpn.* **52,** 3451 (1979).
224. B. E. Prater, *J. Organomet. Chem.* **34,** 379 (1972).
225. F. Faraone and V. Marsala, *Inorg. Chim. Acta* **27,** L109 (1978).
226. L. Malatesta, G. Padoa, and A. Sonz, *Gazz. Chim. Ital.* **85,** 1111 (1955).
227. A-R. Al-Ohaly, R. A. Head, and J. F. Nixon, *J. Organomet. Chem.* **205,** 99 (1981).
228. T. V. Ashworth, A. A. Chalmers, E. Singleton, and H. E. Swanepoel, *J. Chem. Soc., Chem. Commun.* p. 214 (1982).
229. E. Singleton and H. E. Swanepoel, *Inorg. Chim. Acta* **57,** 217 (1982).
230. V. N. Shafranskii and T. A. Malkova, *Zh. Obshch. Khim.* **46,** 1197 (1976).
231. R. S. Dickson and J. A. Ibers, *J. Organomet. Chem.* **36,** 191 (1972).
232. R. S. Dickson and J. A. Ibers, *J. Am. Chem. Soc.* **94,** 2988 (1972).
233. S. Otsuka, A. Nakamura, Y. Tatsuno, and M. Miki, *J. Am. Chem. Soc.* **94,** 3761 (1972).
234. S. D. Ittel and J. A. Ibers, *J. Organomet. Chem.* **74,** 121 (1974).
235. M. Matsumoto and K. Nakatsu, *Acta Crystallogr., Sect. B* **B31,** 2711 (1975).
236. A. Nakamura, T. Yoshida, M. Cowie, S. Otsuka, and J. A. Ibers, *J. Am. Chem. Soc.* **99,** 2108 (1977).
237. S. D. Ittel, *Inorg. Chem.* **16,** 2589 (1977).

238. S. Otsuka, A. Nakamura, T. Koyama, and Y. Tatsuno, *Liebigs Ann. Chem.* 626 (1975).
239. S. Otsuka, A. Nakamura, T. Koyama, and Y. Tatsuno, *J. Chem. Soc., Chem. Commun.* p. 1105 (1972).
240. J. Clemens, M. Green, and F. G. A. Stone, *J. Chem. Soc., Dalton Trans.* p. 1620 (1973).
241. M. T. Chicote, M. Green, J. L. Spencer, F. G. A. Stone, and J. Vicente, *J. Chem. Soc., Dalton Trans.* p. 536 (1979).
241a. N. M. Boag, M. Green, J. A. K. Howard, J. L. Spencer, R. F. D. Stansfield, M. D. O. Thomas, F. G. A. Stone, and P. Woodward, *J. Chem. Soc., Dalton Trans.* p. 2182 (1980).
242. R. Ros, R. A. Michelin, G. Carturan, and U. Belluco, *J. Organomet. Chem.* **133**, 213 (1977).
243. G. Minghetti, A. L. Bandini, G. Banditelli, and F. Bonati, *J. Organomet. Chem.* **179**, C13 (1979).
244. W. J. Cherwinski, H. C. Clark, and L. E. Manzer, *Inorg. Chem.* **11**, 1511 (1972).
245. L. E. Manzer, *J. Chem. Soc., Dalton Trans.* p. 1535 (1974).
246. H. C. Clark and L. E. Manzer, *Inorg. Chem.* **11**, 503 (1972).
247. R. R. Cooke and J. L. Burmeister, *J. Organomet. Chem.* **63**, 471 (1973).
248. H. C. Clark and L. E. Manzer, *J. Organomet. Chem.* **59**, 411 (1973).
249. H. C. Clark and L. E. Manzer, *Inorg. Chem.* **11**, 2749 (1972).
250. H. C. Clark and L. E. Manzer, *Inorg. Chem.* **12**, 362 (1973).
251. G. C. Stocco, N. Bertazzi, and L. Pellerito, *Atti Accad. Sci., Lett. Arti Palermo, Parte 1* **33**, 155 (1973).
252. P. M. Treichel and W. J. Knebel, *J. Coord. Chem.* **2**, 67 (1972).
253. J.-P. Barbier and P. Braunstein, *J. Chem. Res., Synop.* p. 412 (1978).
254. D. L. Cronin, J. R. Wilkinson, and L. J. Todd, *J. Magn. Reson.* **17**, 353 (1975).
255. D. E. Axelson, C. E. Holloway, and A. J. Oliver, *Inorg. Nucl. Chem. Lett.* **9**, 885 (1973).
256. M. H. Chisholm, H. C. Clark, L. E. Manzer, J. B. Stothers, and J. E. H. Ward, *J. Am. Chem. Soc.* **95**, 8574 (1973).
257. H. C. Clark and J. E. H. Ward, *J. Am. Chem. Soc.* **96**, 1741 (1974).
258. B. Crociani and R. L. Richards, *J. Organomet. Chem.* **144**, 85 (1978).
259. R. W. Stephany, M. J. A. de Bie, and W. Drenth, *Org. Magn. Reson.* **6**, 45 (1974).
260. R. D. Adams and D. F. Chodosh, *J. Organomet. Chem.* **87**, C48 (1975).
261. J. Evans, B. F. G. Johnson, J. Lewis, J. R. Norton, and F. A. Cotton, *J. Chem. Soc., Chem. Commun.* p. 807 (1973).
262. L. E. Orgel, *J. Inorg. Nucl. Chem.* **14**, 136 (1960).
263. C. M. Giandomenico, J. C. Dewan, and S. J. Lippard, *J. Am. Chem. Soc.* **103**, 1407 (1981).
264. D. F. Lewis and S. J. Lippard, *Inorg. Chem.* **11**, 621 (1972).
265. D. L. Lewis and S. J. Lippard, *J. Am. Chem. Soc.* **97**, 2697 (1975).
266. P. Brant, W. S. Mialki, and R. A. Walton, *J. Am. Chem. Soc.* **101**, 5453 (1979).
267. C. T. Lam, D. L. Lewis, and S. J. Lippard, *Inorg. Chem.* **15**, 989, (1976).
268. G. A. Sim, J. G. Sime, D. I. Woodhouse, and G. R. Knox, *J. Organomet. Chem.* **74**, C7 (1974).
269. G. A. Sim, J. G. Sime, D. I. Woodhouse, and G. R. Knox, *Acta Crystallogr., Sect. B* **B35**, 2406 (1979).
270. M-S Ericsson, S. Jagner, and E. Ljungström, *Acta Chem. Scand., Ser. A* **A33**, 371 (1979).
271. M-S. Ericsson, S. Jagner, and E. Ljungström, *Acta Chem. Scand., Ser. A* **A34**, 535 (1980).
272. A. C. Sarapu and R. F. Fenske, *Inorg. Chem.* **11**, 3021 (1972).
273. D. Bright and O. S. Mills, *J. Chem. Soc., Dalton Trans.* p. 219 (1974).
274. M. J. Mays, D. W. Prest, and P. R. Raithby, *J. Chem. Soc., Dalton Trans.* p. 771 (1981).
275. I. L. C. Campbell and F. S. Stephens, *J. Chem. Soc., Dalton Trans.* p. 982 (1975).
276. F. A. Cotton and B. A. Frenz, *Inorg. Chem.* **13**, 253 (1974).

277. I. D. Hunt and O. S. Mills, *Acta Crystallogr., Sect. B* **B33**, 2432 (1977).
278. W. P. Fehlhammer, A. Mayr, and W. Kehr, *J. Organomet. Chem.* **197**, 327 (1980).
279. G. Kruger, personal communication.
280. J. Chatt, A. J. L. Pombeiro, R. L. Richards, G. H. D. Royston, K. W. Muir, and R. Walker, *J. Chem. Soc., Chem. Commun.* p. 708 (1975).
281. J. E. Field, personal communication.
282. A. G. Orpen and G. M. Sheldrick, *Acta Crystallogr., Sect. B* **B34**, 1989 (1978).
283. M. R. Churchill and F. J. Hollander, *Inorg. Chem.* **19**, 306 (1980).
284. F. A. Jurnak, D. R. Greig, and K. N. Raymond, *Inorg. Chem.* **14**, 2585 (1975).
285. L. D. Brown, D. R. Greig, and K. N. Raymond, *Inorg. Chem.* **14**, 645 (1975).
286. G. Albertin, E. Bordignon, A. Orio, G. Pelizzi, and P. Tarasconi, *Inorg. Chem.* **20**, 2862 (1981).
287. K. R. Mann, J. A. Thich, R. A. Bell, C. L. Coyle, and H. B. Gray, *Inorg. Chem.* **19**, 2462 (1980).
288. K. R. Mann, *Cryst. Struct. Commun.* **10**, 451 (1981).
289. M. R. Churchill and J. P. Hutchinson, *Inorg. Chem.* **18**, 2451 (1979), and references therein.
290. K. Tatsumi, T. Fueno, A. Nakamura, and S. Otsuka, *Bull. Chem. Soc. Jpn.* **49**, 2161 (1976).
291. K. Tatsumi, T. Fueno, A. Nakamura, and S. Otsuka, *Bull. Chem. Soc. Jpn.* **49**, 2170 (1976).
292. W. L. Gladfelter, M. W. Lynch, W. P. Schaefer, D. N. Hendrikson, and H. B. Gray, *Inorg. Chem.* **20**, 2390 (1981).
293. D. J. Doonan, A. L. Balch, S. Z. Goldberg, R. Eisenberg, and J. S. Miller, *Inorg. Chem.* **15**, 535 (1976), and references therein.
294. P. Braunstein, E. Keller, and H. Vahrenkamp, *J. Organomet. Chem.* **165**, 233 (1979).
295. N. A. Bailey, N. W. Walker, and J. A. W. Williams, *J. Organomet. Chem.* **37**, C49 (1972).
296. B. Jovanović, Lj. Manojlović-Muir, and K. W. Muir, *J. Chem. Soc., Dalton Trans.* p. 1178 (1972).
297. B. Jovanović and Lj. Manojlović-Muir, *J. Chem. Soc., Dalton Trans.* p. 1176 (1972).
298. K. R. Mann, H. B. Gray, and G. S. Hammond, *J. Am. Chem. Soc.* **99**, 306 (1977).
299. M-A. Haga, K. Kawakami, and T. Tanaka, *Inorg. Chim. Acta* **12**, 93 (1975).
300. K. Kawakami, K. Take-Uchi, and T. Tanaka, *Inorg. Chem.* **14**, 877 (1975).
301. K. Kawakami, T. Kaneshima, and T. Tanaka, *J. Organomet. Chem.* **34**, C21 (1972).
302. K. Sato, K. Kawakami, and T. Tanaka, *Inorg. Chem.* **18**, 1532 (1979).
303. R. B. King and M. S. Saran, *Inorg. Chem.* **11**, 2112 (1972).
304. C. A. L. Becker and B. L. Davis, *J. Inorg. Nucl. Chem.* **39**, 781 (1977).
305. S. Otsuka and M. Rossi, *Bull. Chem. Soc. Jpn.* **46**, 3411 (1973).
306. C. A. L. Becker, *J. Inorg. Nucl. Chem.* **42**, 27 (1980).
307. G. Albertin, E. Bordignon, U. Croatto, and A. A. Orio, *Gazz. Chim. Ital.* **104**, 1041 (1974).
308. E. Bordignon, U. Croatto, U. Mazzi, and A. A. Orio, *Inorg. Chem.* **13**, 935 (1974).
309. C. A. L. Becker, *J. Organomet. Chem.* **104**, 89 (1976).
310. C. A. L. Becker, *Inorg. Nucl. Chem. Lett.* **16**, 297 (1980).
311. C. A. L. Becker, *Inorg. Chim. Acta* **36**, L441 (1979).
312. J. W. Dart, M. K. Lloyd, R. Mason, J. A. McCleverty, and J. Williams, *J. Chem. Soc., Dalton Trans.* p. 1747 (1973).
313. R. Graziani, G. Albertin, E. Forsellini, and A. A. Orio, *Inorg. Chem.* **15**, 2422 (1976).
314. J. W. Dart, M. K. Lloyd, R. Mason, and J. A. McCleverty, *J. Chem. Soc., Dalton Trans.* p. 2046 (1973).
315. C. A. Reed and W. R. Roper, *J. Chem. Soc., Dalton Trans.* p. 1365 (1973).

316. K. Kawakami, M. Haga, and T. Tanaka, *J. Organomet. Chem.* **60**, 363 (1973).
317. W. M. Bedford and G. Rouschias, *J. Chem. Soc., Dalton Trans.* p. 2531 (1974).
318. K. Kawakami and M. Okajima, *J. Inorg. Nucl. Chem.* **41**, 1501 (1979).
319. Y. Yamamoto and H. Yamazaki, *J. Organomet. Chem.* **140**, C33 (1977).
320. J. T. Mague and S. H. de Vries, *Inorg. Chem.* **19**, 3743 (1980).
321. A. L. Balch, *J. Am. Chem. Soc.* **98**, 8049 (1976).
322. A. L. Balch, J. W. Labodie, and G. Delker, *Inorg. Chem.* **18**, 1224 (1979).
323. J. R. Boehm and A. L. Balch, *J. Organomet. Chem.* **112**, C20 (1976).
324. J. R. Boehm and A. L. Balch, *Inorg. Chem.* **16**, 778 (1977).
325. C-L Lee, C. T. Hunt, and A. L. Balch, *Inorg. Chem.* **20**, 2498 (1981), and references therein.
326. L. S. Benner and A. L. Balch, *J. Am. Chem. Soc.* **100**, 6099 (1978).
327. P. Brant, L. S. Benner, and A. L. Balch, *Inorg. Chem.* **18**, 3422 (1979).
327a. M. M. Olmstead, H. Hope, L. S. Benner, and A. L. Balch, *J. Am. Chem. Soc.* **99**, 5502 (1977).
328. P. M. Treichel and G. E. Dirreen, *J. Organomet. Chem.* **39**, C20 (1972).
329. M. K. Lloyd, J. A. McCleverty, D. G. Orchard, J. A. Connor, M. B. Hall, I. H. Hillier, E. M. Jones, and G. K. McEwen, *J. Chem. Soc., Dalton Trans.* p. 1743 (1973).
330. P. M. Treichel and H. J. Mueh, *Inorg. Chem.* **16**, 1167 (1977).
331. B. E. Bursten and R. F. Fenske, *Inorg. Chem.* **16**, 963 (1977).
332. G. J. Essenmacher and P. M. Treichel, *Inorg. Chem.* **16**, 800 (1977).
333. J. Hanzlik, G. Albertin, E. Bordignon, and A. A. Omo, *Inorg. Chim. Acta* **38**, 207 (1980).
334. J. A. Connor, E. J. James, C. Overton, and N. E. Murr, *J. Organomet. Chem.* **218**, C31 (1981).
335. P. Lemoine, A. Giraudeau, M. Gross, and P. Braunstein, *J. Organomet. Chem.* **202**, 447 (1980).
336. A. Giraudeau, P. Lemoine, M. Gross, and P. Braunstein, *J. Organomet. Chem.* **202**, 455 (1980).
337. P. Zanello, R. Seeber, A. Cinquantini, and B. Crociani, *Transition Met. Chem.* **5**, 226 (1980).
337a. P. Zanello, R. Seeber, B. Crociani, and M. Nicolini, *Transition Met. Chem.* **5**, 45 (1980).
338. A. J. L. Pombeiro and R. L. Richards, *J. Organomet. Chem.* **179**, 459 (1979).
339. M. K. Lloyd and J. A. McCleverty, *J. Organomet. Chem.* **61**, 261 (1973).
340. W. G. Kita, J. A. McCleverty, B. Patel, and J. Williams, *J. Organomet. Chem.* **74**, C9 (1974).
340a. J. A. McCleverty and J. Williams, *Transition Met. Chem.* **1**, 288 (1976).
341. M. Herberhold and L. Haumaier, *Z. Naturforsch., B: Anorg. Chem. Org. Chem.* **33**, 1277 (1980).
342. W. L. Gladfelter and H. B. Gray, *J. Am. Chem. Soc.* **102**, 5909 (1980).
343. C. T. Lam, P. W. R. Corfield, and S. J. Lippard, *J. Am. Chem. Soc.* **99**, 617 (1977).
344. R. D. Adams, *Inorg. Chem.* **15**, 174 (1976).
345. R. D. Adams, *J. Organomet. Chem.* **82**, C7 (1974).
346. W. M. Bedford and G. Rouschias, *J. Chem. Soc., Chem. Commun.* p. 1224 (1972).
347. G. L. Geoffroy, M. G. Bradley, and M. E. Keeney, *Inorg. Chem.* **17**, 777 (1978).
348. G. L. Geoffroy, M. G. Bradley, and M. E. Keeney, *Ann. N. Y. Acad. Sci.* **313**, 588 (1978).
349. H. Isci and W. R. Mason, *Inorg. Chem.* **14**, 913 (1975).
350. S. F. Rice and H. B. Gray, *J. Am. Chem. Soc.* **103**, 1593 (1981).
351. R. F. Dallinger, V. M. Miskowski, H. B. Gray, and W. H. Woodruff, *J. Am. Chem. Soc.* **103**, 1595 (1981).

352. K. R. Mann and H. B. Gray, *Adv. Chem. Ser.* **173**, 225 (1979).
353. V. M. Miskowski, I. S. Sigal, K. R. Mann, H. B. Gray, S. J. Milder, G. S. Hammond, and P. R. Ryason, *J. Am. Chem. Soc.* **101**, 4383 (1979).
354. S. J. Milder, R. A. Goldbeck, D. S. Kliger, and H. B. Gray, *J. Am. Chem. Soc.* **102**, 6761 (1980).
355. I. S. Sigal and H. B. Gray, *J. Am. Chem. Soc.* **103**, 2220 (1981).
356. I. S. Sigal, K. R. Mann, and H. B. Gray, *J. Am. Chem. Soc.* **102**, 7252 (1980).
357. K. R. Mann, M. J. DiPierro, and T. P. Gill, *J. Am. Chem. Soc.* **102**, 3965 (1980).
358. K. R. Mann, R. A. Bell, and H. B. Gray, *Inorg. Chem.* **18**, 2671 (1979).
359. N. S. Lewis, K. R. Mann, J. G. Gordon, II, and H. B. Gray, *J. Am. Chem. Soc.* **98**, 7461 (1976).
360. V. M. Miskowski, G. L. Nobinger, D. S. Kliger, G. S. Hammond, N. S. Lewis, K. R. Mann, and H. B. Gray, *J. Am. Chem. Soc.* **100**, 485 (1978).
361. R. Kumar and A. R. Manning, *J. Organomet. Chem.* **216**, C61 (1981).
362. G. K. Barker, M. Green, T. P. Onak, F. G. A. Stone, C. B. Ungermann, and A. J. Welch, *J. Chem. Soc., Chem. Commun.* p. 169 (1978).
363. J-M. Bassett, L. J. Farrugia, and F. G. A. Stone, *J. Chem. Soc., Dalton Trans.* p. 1789 (1980).
364. J-M. Bassett, M. Green, J. A. K. Howard, and F. G. A. Stone, *J. Chem. Soc., Dalton Trans.* p. 1779 (1980).
365. A. Nakamura, Y. Tatsuno, and S. Otsuka, *Inorg. Chem.* **11**, 2058 (1972).
366. R. V. Parish and P. G. Simms, *J. Chem. Soc., Dalton Trans.* p. 809 (1972).
367. A. L. Balch and M. M. Olmstead, *J. Am. Chem. Soc.* **98**, 2354 (1976).
368. A. Araneo, F. Morazzoni, and T. Napoletano, *J. Chem. Soc., Dalton Trans.* p. 2039 (1975).
369. G. Mercati and F. Morazzoni, *J. Chem. Soc., Dalton Trans.* p. 569 (1979).
370. R. Kuwae, T. Tanaka, and K. Kawakami, *Bull. Chem. Soc. Jpn.* **52**, 437 (1979).
371. S. Otsuka, Y. Tatsuno, M. Miki, T. Aoki, M. Matsumoto, H. Yoshioka, and K. Nakatsu, *J. Chem. Soc., Chem. Commun.* p. 445 (1973).
372. W. Wong, S. J. Singer, W. D. Pitts, S. F. Watkins, and W. H. Baddley, *J. Chem. Soc., Chem. Commun.* p. 672 (1972).
373. J. Forniés, M. Green, A. Laguna, M. Murray, J. L. Spencer, and F. G. A. Stone, *J. Chem. Soc., Dalton Trans.* p. 1515 (1977).
374. J. Browning, M. Green, A. Laguna, L. E. Smart, J. L. Spencer, and F. G. A. Stone, *J. Chem. Soc., Chem. Commun.* p. 723 (1975).
375. M. Ciriano, M. Green, D. Gregson, J. A. K. Howard, J. L. Spencer, F. G. A. Stone, and P. Woodward, *J. Chem. Soc., Dalton Trans.* p. 1294 (1979).
376. B. Crociani, M. Nicolini, and R. L. Richards, *Inorg. Chim. Acta* **12**, 53 (1975).
377. P. W. R. Corfield, L. M. Baltusis, and S. J. Lippard, *Inorg. Chem.* **20**, 922 (1981).
378. D. Baumann, H. Endres, H. J. Keller, B. Nuber, and J. Weiss, *Acta Crystallogr., Sect. B* **B31**, 40 (1975).
379. D. Baumann, H. J. Keller, D. Nöthe, H. H. Rupp, and G. Uhlmann, *Z. Naturforsch., B: Anorg. Chem., Org. Chem.* **31**, 912 (1976).
380. M. M. Olmstead and A. L. Balch, *J. Organomet. Chem.* **148**, C15 (1978).
381. R. J. H. Clark and C. Sourisseau, *Nouv. J. Chim.* **4**, 287 (1980).
382. Y. Yamamoto and H. Yamazaki, *Inorg. Chem.* **13**, 438 (1974).
383. P. M. Treichel and K. P. Wagner, *J. Organomet. Chem.* **61**, 415 (1973).
384. P. M. Treichel, K. P. Wagner, and R. W. Hess, *Inorg. Chem.* **12**, 1471 (1973).
385. I. S. Butler, F. Basolo, and R. G. Pearson, *Inorg. Chem.* **6**, 2074 (1967).
386. E. J. M. de Boer and J. H. Teuben, *J. Organomet. Chem.* **166**, 193 (1979).

387. R. J. H. Clark, J. A. Stockwell, and J. D. Wilkins, *J. Chem. Soc., Dalton Trans.* p. 120 (1976).
388. K. W. Chiu, R. A. Jones, G. Wilkinson, A. M. R. Galas, and M. B. Hursthouse, *J. Chem. Soc., Dalton Trans.* p. 2088 (1981).
389. A. Dormond and A. Dahchour, *J. Organomet. Chem.* **193**, 321 (1980).
390. J. D. Wilkins, *J. Organomet. Chem.* **67**, 269 (1974).
391. E. Klei, J. H. Telgen, and J. H. Teuben, *J. Organomet. Chem.* **209**, 297 (1981).
392. Y. Yamamoto and H. Yamazaki, *Inorg. Chem.* **16**, 3182 (1977).
393. W. R. Roper, G. E. Taylor, J. M. Waters, and L. J. Wright, *J. Organomet. Chem.* **157**, C27 (1978).
394. A. N. Nesmeyanov, L. I. Leont'eva, and L. I. Khomik, *Izv. Akad. Nauk SSSR, Ser. Khim.* p. 1656 (1976).
395. S. Otsuka and K. Ataka, *J. Chem. Soc., Dalton Trans.* p. 327 (1976).
396. K. P. Wagner, P. M. Treichel, and J. C. Calabrese, *J. Organomet. Chem.* **71**, 299 (1974).
397. Y. Yamamoto and H. Yamazaki, *Bull. Chem. Soc. Jpn.* **44**, 1873 (1971).
398. Y. Yamamoto and H. Yamazaki, *Inorg. Chim. Acta* **41**, 229 (1980).
399. G. Carturan, R. Zanella, M. Graziani, and U. Belluco, *J. Organomet. Chem.* **82**, 421 (1974).
400. R. Zanella, G. Carturan, M. Graziani, and U. Belluco, *J. Organomet. Chem.* **65**, 417 (1974).
401. R. D. Adams and D. F. Chodosh, *Inorg. Chem.* **17**, 41 (1978).
402. H. Werner, B. Heiser, and A. Kühn, *Angew. Chem., Int. Ed. Engl.* **20**, 300 (1981).
403. M. Tanaka and H. Alper, *J. Organomet. Chem.* **168**, 97 (1979).
404. H. Alper and M. Tanaka, *J. Organomet. Chem.* **169**, C5 (1979).
405. C. C. Yin and A. J. Deeming, *J. Organomet. Chem.* **133**, 123 (1977).
406. M. A. Andrews and H. D. Kaesz, *J. Am. Chem. Soc.* **101**, 7238 (1979).
407. M. A. Andrews, G. van Buskirk, C. B. Knobler, and H. D. Kaesz, *J. Am. Chem. Soc.* **101**, 7245 (1979).
408. S. J. Simpson and R. A. Andersen, *J. Am. Chem. Soc.* **103**, 4063 (1981).
409. H. C. Clark, C. R. Milne, and N. C. Payne, *J. Am. Chem. Soc.* **100**, 1164 (1978).
410. P. M. Treichel, J. J. Benedict, R. W. Hess, and J. P. Stenson, *J. Chem. Soc., Chem. Commun.* p. 1627 (1970).
411. Y. Yamamoto and H. Yamazaki, *Bull. Chem. Soc. Jpn.* **48**, 3691 (1975).
412. Y. Yamamoto and H. Yamazaki, *J. Organomet. Chem.* **90**, 329 (1975).
413. B. Crociani, M. Nicolini, and R. L. Richards, *J. Organomet. Chem.* **104**, 259 (1976).
414. B. Crociani, M. Nicolini, and T. Boschi, *J. Organomet. Chem.* **33**, C81 (1971).
415. B. Crociani, *Inorg. Chim. Acta* **23**, L1 (1977).
416. B. Crociani and R. L. Richards, *J. Organomet. Chem.* **154**, 65 (1978).
417. B. Crociani, M. Nicolini, and R. L. Richards, *J. Organomet. Chem.* **113**, C22 (1976).
418. B. Crociani, M. Nicolini, and R. L. Richards, *J. Chem. Soc., Dalton Trans.* p. 1478 (1978).
419. B. Crociani, M. Nicolini, and A. Mantovani, *J. Organomet. Chem.* **177**, 365 (1979).
420. B. Crociani, U. Belluco, and P. Sandrini, *J. Organomet. Chem.* **177**, 385 (1979).
421. S. Otsuka, A. Nakamura, T. Yoshida, M. Naruto, and K. Ataka, *J. Am. Chem. Soc.* **95**, 3180 (1973).
422. T. Boschi and B. Crociani, *Inorg. Chim. Acta* **5**, 477 (1971).
423. T. Kajimoto, H. Takahashi, and J. Tsuji, *J. Organomet. Chem.* **23**, 275 (1970).
424. G. Carturan, A. Scrivanti, and U. Belluco, *Inorg. Chim. Acta* **21**, 103 (1977).
425. J. L. Davidson, M. Green, J. Z. Nyathi, F. G. A. Stone, and A. J. Welch, *J. Chem. Soc., Dalton Trans.* p. 2246 (1977).

426. H. Yamazaki, K. Aoki, Y. Yamamoto, and Y. Wakatsuki, *J. Am. Chem. Soc.* **97**, 3546 (1975).
427. A. J. Carty, G. N. Mott, and N. J. Taylor, *J. Organomet. Chem.* **212**, C54 (1981).
428. C. J. Cardin, D. J. Cardin, and M. F. Lappert, *J. Chem. Soc., Dalton Trans.* p. 767 (1977).
429. R. Aumann and E. O. Fischer, *Angew. Chem., Int. Ed. Engl.* **6**, 879 (1967).
430. C. G. Kreiter and R. Aumann, *Chem. Ber.* **111**, 1223 (1978).
431. P. M. Treichel and K. P. Wagner, *J. Organomet. Chem.* **88**, 199 (1975).
432. K. Aoki and Y. Yamamoto, *Inorg. Chem.* **15**, 48 (1976).
433. Y. Yamamoto and H. Yamazaki, *Inorg. Chem.* **11**, 211 (1972).
434. Y. Yamamoto, K. Aoki, and H. Yamazaki, *J. Am. Chem. Soc.* **96**, 2647 (1974).
435. T-A. Mitsudo, H. Watanabe, Y. Komiya, Y. Watanabe, T. Takaegami, K. Nakatsu, K. Kinoshita, and Y. Miyagawa, *J. Organomet. Chem.* **190**, C39 (1980).
436. D. J. Yarrow, J. A. Ibers, Y. Tatsuno, and S. Otsuka, *J. Am. Chem. Soc.* **95**, 8590 (1973).
437. E. O. Fischer and W. Schambeck, *J. Organomet. Chem.* **201**, 311 (1980).
438. D. S. Gill, P. K. Baker, M. Green, K. E. Paddick, M. Murray, and A. J. Welch, *J. Chem. Soc., Chem. Commun.* p. 986 (1981).
439. D. F. Christian and H. C. Clark, *J. Organomet. Chem.* **85**, C9 (1975).
440. D. F. Christian, H. C. Clark, and R. F. Stepaniak, *J. Organomet. Chem.* **112**, 209 (1976).
441. D. F. Christian, G. R. Clark, W. R. Roper, J. M. Waters, and K. R. Whittle, *J. Chem. Soc., Chem. Commun.* p. 458 (1972).
442. G. R. Clark, J. M. Waters, and K. R. Whittle, *J. Chem. Soc., Dalton Trans.* p. 2556 (1975).
443. M. I. Bruce and R. C. Wallis, *J. Organomet. Chem.* **164**, C6 (1979).
444. M. I. Bruce and R. C. Wallis, *Aust. J. Chem.* **35**, 709 (1982).
445. J. A. S. Howell and P. Mathur, *J. Chem. Soc., Chem. Commun.* p. 263 (1981).
446. R. D. Adams and N. M. Golembeski, *J. Am. Chem. Soc.* **101**, 2579 (1979).
447. R. D. Adams and N. M. Golembeski, *J. Am. Chem. Soc.* **100**, 4622 (1978).
448. R. D. Adams and N. M. Golembeski, *J. Organomet. Chem.* **172**, 239 (1979).
449. E. Band, W. R. Pretzer, M. G. Thomas, and E. L. Muetterties, *J. Am. Chem. Soc.* **99**, 7380 (1977).
450. D. F. Christian and W. R. Roper, *J. Organomet. Chem.* **80**, C35 (1974).
451. D. F. Christian, H. C. Clark, and R. F. Stepaniak, *J. Organomet. Chem.* **112**, 227 (1976).
452. G. R. Clark, *J. Organomet. Chem.* **134**, 51 (1977).
453. W. Beck, K. Burger, and W. P. Fehlhammer, *Chem. Ber.* **104**, 1816 (1971).
454. W. P. Fehlhammer, T. Kemmerich, and W. Beck, *Chem. Ber.* **112**, 468 (1979).
455. W. P. Fehlhammer and L. F. Dahl, *J. Am. Chem. Soc.* **94**, 3370 (1972).
456. K. Bartel and W. P. Fehlhammer, *Angew. Chem., Int. Ed. Engl.* **13**, 599 (1974).
457. W. P. Fehlhammer, K. Bartel, and W. Petri, *J. Organomet. Chem.* **87**, C34 (1975).
458. K. R. Grundy and W. R. Roper, *J. Organomet. Chem.* **91**, C61 (1975).
459. Y. Fuchita, K. Hidaka, S. Morinaga, and K. Hiraki, *Bull. Chem. Soc. Jpn.* **54**, 800 (1981).
460. K. Hiraki, Y. Fuchita, and S. Morinaga, *Chem. Lett.* p. 1 (1978).
461. K. Hiraki and Y. Fuchita, *Chem. Lett.* p. 841 (1978).
462. L. S. Hegedus, O. P. Anderson, K. Zetterberg, G. Allen, K. Siirala-Hansen, D. J. Olsen, and A. B. Packard, *Inorg. Chem.* **16**, 1887 (1977), and references therein.
463. M. Fukui, K. Itoh, and Y. Ishii, *Synth. React. Inorg. Met.-Org. Chem.* **5**, 207 (1975).
464. Y. Ito, Y. Inubushi, T. Sugaya, and T. Saegusa, *J. Organomet. Chem.* **137**, 1 (1977).
465. F. Bonati and G. Minghetti, *J. Organomet. Chem.* **59**, 403 (1973).
466. A. Tiripicchio, M. T. Camellini, and G. Minghetti, *J. Organomet. Chem.* **171**, 399 (1979).
467. R. A. Michelin and R. Ros, *J. Organomet. Chem.* **169**, C42 (1979).
468. R. Richards and G. Rouschias, *J. Am. Chem. Soc.* **98**, 5729 (1976).

469. K. R. Grundy and W. R. Roper, *J. Organomet. Chem.* **113**, C45 (1976).
470. G. R. Clark, T. J. Collins, D. Hall, S. M. James, and W. R. Roper, *J. Organomet. Chem.* **141**, C5 (1977).
471. S. M. Boniface and W. R. Roper, *J. Organomet. Chem.* **208**, 253 (1981).
472. D. J. Miller and M. R. DuBois, *J. Am. Chem. Soc.* **102**, 4925 (1980).
473. G. Minghetti, F. Bonati, and G. Banditelli, *Synth. React. Inorg. Met.-Org. Chem.* **3**, 415 (1973).
474. G. Minghetti and F. Bonati, *Gazz. Chim. Ital.* **102**, 205 (1972), and references therein.
475. F. Bonati and G. Minghetti, *Gazz. Chim. Ital.* **103**, 373 (1973).
476. G. Minghetti, F. Bonati, and M. Massobrio, *Inorg. Chem.* **14**, 1974 (1975).
477. G. Banditelli, F. Bonati, and G. Minghetti, *Gazz. Chim. Ital.* **110**, 317 (1980).
478. G. Minghetti, F. Bonati, and G. Banditelli, *Inorg. Chem.* **15**, 2649 (1976).
479. F. Bonati and G. Minghetti, *J. Organomet. Chem.* **60**, C43 (1973).
480. H. Sawai, T. Takizawa, and Y. Iitaka, *J. Organomet. Chem.* **120**, 161 (1976).
481. B. Crociani, M. Nicolini, and R. L. Richards, *J. Organomet. Chem.* **101**, C1 (1975).
482. B. Crociani and R. L. Richards, *J. Chem. Soc., Chem. Commun.* p. 127 (1973).
483. M. Behnam-Dahkordy, B. Crociani, and R. L. Richards, *J. Chem. Soc., Dalton Trans.* p. 2015 (1977).
484. P. Brant, J. H. Enemark, and A. L. Balch, *J. Organomet. Chem.* **114**, 99 (1976).
485. G. M. Bancroft and P. L. Sears, *Inorg. Chem.* **14**, 2716 (1975).
486. B. Crociani and R. L. Richards, *J. Chem. Soc., Dalton Trans.* p. 693 (1974).
487. E. M. Badley, J. Chatt, R. L. Richards, and G. A. Sim, *J. Chem. Soc., Chem. Commun.* p. 1322 (1969).
488. J. Chatt, R. L. Richards, and G. H. D. Royston, *Inorg. Chim. Acta* **6**, 669 (1972).
489. B. Crociani, T. Boschi, G. G. Troilo, and U. Croatto, *Inorg. Chim. Acta* **6**, 655 (1972).
490. G. Banditelli, F. Bonati, and G. Minghetti, *Gazz. Chim. Ital.* **107**, 267 (1977).
491. G. Minghetti and F. Bonati, *J. Organomet. Chem.* **54**, C62 (1973).
492. R. Uson, A. Laguno, J. Vicente, J. Garcia, B. Bergareche, and P. Brun, *Inorg. Chim. Acta* **28**, 237 (1978).
493. G. Minghetti, L. Baratto, and F. Bonati, *J. Organomet. Chem.* **102**, 397 (1975).
494. W. M. Butler and J. H. Enemark, *Inorg. Chem.* **12**, 540 (1973).
495. D. J. Doonan and A. L. Balch, *J. Am. Chem. Soc.* **95**, 4769 (1973).
496. D. J. Doonan and A. L. Balch, *Inorg. Chem.* **13**, 921 (1974).
497. W. M. Butler, J. H. Enemark, J. Parks, and A. L. Balch, *Inorg. Chem.* **12**, 451 (1973).
498. L. Calligaro, P. Uguagliati, B. Crociani, and U. Belluco, *J. Organomet. Chem.* **92**, 399 (1975).
499. E. Rotondo, M. Cusamano, B. Crociani, P. Uguagliati, and U. Belluco, *J. Organomet. Chem.* **134**, 249 (1977).
500. L. Calligaro, P. Uguagliati, B. Crociani, and U. Belluco, *J. Organomet. Chem.* **142**, 105 (1977).
501. P. Uguagliati, B. Crociani, U. Belluco, and L. Calligaro, *J. Organomet. Chem.* **112**, 111 (1976).
502. G. A. Larkin, R. P. Scott, and M. G. H. Wallbridge, *J. Organomet. Chem.* **37**, C21 (1972).
503. B. Crociani, T. Boschi, M. Nicolini, and U. Belluco, *Inorg. Chem.* **11**, 1292 (1972).
504. R. Uson, A. Laguna, J. Vicente, J. Garcia, and B. Bergareche, *J. Organomet. Chem.* **173**, 349 (1979).
505. B. Crociani, P. Uguagliati, and U. Belluco, *J. Organomet. Chem.* **117**, 189 (1976).
506. D. H. Cuatecontzis and J. D. Miller, *J. Chem. Res., Synop.* p. 137 (1977).
507. T. Sawai and R. J. Angelici, *J. Organomet. Chem.* **80**, 91 (1974).
508. R. J. Angelici, P. A. Christian, B. D. Dombek, and G. A. Pfeffer, *J. Organomet. Chem.* **67**, 287 (1974).

509. P. R. Branson, R. A. Cable, M. Green, and M. K. Lloyd, *J. Chem. Soc., Dalton Trans.* p. 12 (1976).

510. J. Chatt, R. L. Richards, and G. H. D. Royston, *J. Chem. Soc., Dalton Trans.* p. 1433 (1973).

511. D. J. Doonan, J. E. Parks, and A. L. Balch, *J. Am. Chem. Soc.* **98,** 2129 (1976).

512. J. A. McCleverty and M. M. M. da Mota, *J. Chem. Soc., Dalton Trans.* p. 2571 (1973).

512a. J. E. Parks and A. L. Balch, *J. Organomet. Chem.* **71,** 453 (1974).

513. R. Greatrex, N. N. Greenwood, I. Rhee, M. Ryang, and S. Tsutsumi, *J. Chem. Soc., Chem. Commun.* p. 1193 (1970).

514. J. Altman and N. Welcman, *J. Organomet. Chem.* **165,** 353 (1979).

515. M. R. Churchill, B. G. DeBoer, and F. J. Rotella, *Inorg. Chem.* **15,** 1843 (1976).

516. R. D. Adams and N. M. Golembeski, *Inorg. Chem.* **18,** 2255 (1979).

517. S. Willis, A. R. Manning, and F. S. Stephens, *J. Chem. Soc., Dalton Trans.* p. 186 (1980).

518. S. Willis and A. R. Manning, *J. Chem. Soc., Dalton Trans.* p. 322 (1981).

519. S. Willis, A. R. Manning, and F. S. Stephens, *J. Chem. Soc., Dalton Trans.* p. 23 (1979).

520. S. Willis and A. R. Manning, *J. Chem. Res. Synopsis* p. 390 (1978).

521. S. Willis and A. R. Manning, *J. Organomet. Chem.* **97,** C49 (1975).

522. R. H. Quick and R. J. Angelici, *Inorg. Chem.* **20,** 1123 (1981).

523. A. J. L. Pombeiro and R. L. Richards, *Transition Met. Chem.* **5,** 281 (1980).

524. A. J. L. Pombeiro, R. L. Richards, and J. R. Dilworth, *J. Organomet. Chem.* **175,** C17 (1979).

525. A. J. L. Pombeiro, M. F. N. N. Carvalho, P. B. Hitchcock, and R. L. Richards, *J. Chem. Soc., Dalton Trans.* p. 1629 (1981).

526. J. Chatt, A. J. L. Pombeiro, and R. L. Richards, *J. Chem. Soc., Dalton Trans.* p. 1585 (1979).

527. U. Schöllkopf and W. Frieben, *Liebigs Ann. Chem.* p. 1722 (1980).

528. T. Tsuda, H. Habu, S. Horiguchi, and T. Saegusa, *J. Am. Chem. Soc.* **96,** 5930 (1974).

529. Y. Ito, T. Konoike, and T. Saegusa, *J. Organomet. Chem.* **85,** 395 (1975).

530. Y. Wakatsuki, O. Nomura, H. Tone, and H. Yamazaki, *J. Chem. Soc., Perkin Trans. 2* p. 1344 (1980).

531. R. Baker, R. C. Cookson, and J. R. Vinson, *J. Chem. Soc., Chem. Commun.* p. 515 (1974).

532. H. Breil and G. Wilke, *Angew. Chem., Int. Ed. Engl.* **9,** 367 (1970).

533. Y. Yamamoto, K. Aoki, and H. Yamazaki, *Proc. Int. Conf. Organomet. Chem., 9th, 1979* p. P12T (1979).

534. U. Schöllkopf, *Pure Appl. Chem.* **51,** 1347 (1979).

535. S. Otsuka, A. Nakamura, and K. Ito, *Chem. Lett.* p. 939 (1972).

536. D. Knol, C. P. A. van Os, and W. Drenth, *Rec. Trav. Chim. Pays-Bas* **93,** 314 (1974).

537. H. Sawai and T. Takizawa, *J. Organomet. Chem.* **94,** 333 (1975).

538. U. Schöllkopf, H. P. Porsch, and H. H. Lau, *Liebigs Ann. Chem.* p. 1444 (1979).

539. K. W. Henneke, U. Schöllkopf, and T. Neudecker, *Liebigs Ann. Chem.* p. 1370 (1979).

540. U. Schöllkopf, H. H. Lau, K. H. Scheunamann, E. Blume, and K. Madawinata, *Liebigs Ann. Chem.* p. 600 (1980).

541. Y. Yamamoto and H. Yamazaki, *J. Org. Chem.* **42,** 4136 (1977).

542. Y. Yamamoto and H. Yamazaki, *Synthesis* p. 750 (1976).

543. H. Sawai and T. Takizawa, *Bull. Chem. Soc. Jpn.* **49,** 1906 (1976).

544. J. D. McClure, U.S. Patent 3,739,005 (Cl. 260-453PC; C 017c) (1973); U.S. Patent Appl. 207,654 (1971).

545. N. B. Franco and M. A. Robinson, U.S. Patent 3,743,664 (Cl. 260-453PC; C07c) (1973); U.S. Patent Appl. 174,530 (1971).

546. T. Saegusa, Y. Ito, S. Tomita, and H. Kinoshita, *Bull. Chem. Soc. Jpn.* **45,** 496 (1972).

547. J. Langova and J. Hetflejs, *Collect. Czech. Chem. Commun.* **40,** 432 (1975).
548. Y. Suzuki and T. Takizawa, *J. Chem. Soc., Chem. Commun.* p. 837 (1972).
549. K. Kinugasa and T. Agawa, *J. Organomet. Chem.* **51,** 329 (1973).
550. P. B. J. Driessen and H. Hogeveen, *Tetrahedron Lett.* p. 271 (1979).
551. T. Saegusa, K. Yonzsawa, I. Murase, T. Konoike, S. Tomita, and Y. Ito, *J. Org. Chem.* **38,** 2319 (1973).
552. R. Baker and A. H. Copeland, *Tetrahedron Lett.* p. 4535 (1976).
553. R. Baker and A. H. Copeland, *J. Chem. Soc., Perkin Trans.* p. 2560 (1977).
554. Y. Yamamoto and H. Yamazaki, *Bull. Chem. Soc. Jpn.* **54,** 787 (1981).
555. H. Sawai and T. Takizawa, *Tetrahedron Lett.* p. 4263 (1972).
556. G. Minghetti and F. Bonati, *J. Organomet. Chem.* **73,** C43 (1974).

Index

Cumulative List of Contributors

Cumulative List of Titles

Acetylene and Allene Complexes: Their Implication in Homogeneous Catalysis, **14,** 245

Activation of Alkanes by Transition Metal Compounds, **15,** 147

Activation of Carbon Dioxide by Metal Complexes, **22,** 129

Alkali Metal Derivatives of Metal Carbonyls, **2,** 157

Alkali Metal–Transition Metal π-Complexes, **19,** 97

Alkyl and Aryl Derivatives of Transition Metals, **7,** 157

Alkylcobalt and Acylcobalt Tetracarbonyls, **4,** 243

Allyl Metal Complexes, **2,** 325

π-Allylnickel Intermediates in Organic Synthesis, **8,** 29

1,2-Anionic Rearrangement of Organosilicon and Germanium Compounds, **16,** 1

Application of ^{13}C-NMR Spectroscopy to Organo-Transition Metal Complexes, **19,** 257

Applications of ^{119m}Sn Mössbauer Spectroscopy to the Study of Organotin Compounds, **9,** 21

Arene Transition Metal Chemistry, **13,** 47

Arsonium Ylides, **20,** 115

Aryl Migrations in Organometallic Compounds of the Alkali Metals, **16,** 167

Basic Metal Cluster Reactions, **22,** 169

Biological Methylation of Metals and Metalloids, **20,** 313

Boranes in Organic Chemistry, **11,** 1

Boron Heterocycles as Ligands in Transition-Metal Chemistry, **18,** 301

Carbene and Carbyne Complexes, On the Way to, **14,** 1

Carboranes and Organoboranes, **3,** 263

Catalysis by Cobalt Carbonyls, **6,** 119

Catalytic Codimerization of Ethylene and Butadiene, **17,** 269

Catenated Organic Compounds of the Group IV Elements, **4,** 1.

Chemistry of Carbidocarbonyl Clusters, **22,** 1

Chemistry of Carbon-Functional Alkylidynetricobalt Nonacarbonyl Cluster Complexes, **14,** 97

Chemistry of Titanocene and Zirconocene, **19,** 1

Chiral Metal Atoms in Optically Active Organo-Transition-Metal Compounds, **18,** 151

^{13}C NMR Chemical Shifts and Coupling Constants of Organometallic Compounds, **12,** 135

Compounds Derived from Alkynes and Carbonyl Complexes of Cobalt, **12,** 323

Conjugate Addition of Grignard Reagents to Aromatic Systems, **1,** 221

Coordination of Unsaturated Molecules to Transition Metals, **14,** 33

Cyclobutadiene Metal Complexes, **4,** 95

Cyclopentadienyl Metal Compounds, **2,** 365

1,4-Diaza-1,3-butadiene (α-Diimine) Ligands: Their Coordination Modes and the Reactivity of Their Metal Complexes, **21,** 152

Diene-Iron Carbonyl Complexes, **1,** 1

Dyotropic Rearrangements and Related σ-σ Exchange Processes, **16,** 33

Electronic Effects in Metallocenes and Certain Related Systems, **10,** 79

Electronic Structure of Alkali Metal Adducts of Aromatic Hydrocarbons, **2,** 115

Fast Exchange Reactions of Group I, II, and III Organometallic Compounds, **8,** 167

Fischer–Tropsch Reaction, **17,** 61

Fluorocarbon Derivatives of Metals, **1,** 143

Fluxional and Nonrigid Behavior of Transition Metal Organometallic π-Complexes, **16,** 211

Free Radicals in Organometallic Chemistry, **14,** 345

Functionally Substituted Cyclopentadienyl Metal Compounds, **21,** 1

Olefin Metathesis, **17,** 449
Olefin Metathesis Reaction, **16,** 283
Olefin Oxidation with Palladium Catalyst, **5,** 321
Organic and Hydride Chemistry of Transition Metals, **12,** 1
Organic Chemistry of Copper, **12,** 215
Organic Chemistry of Gold, 20, 39
Organic Chemistry of Lead **7,** 241
Organic Complexes of Lower-Valent Titanium, **9,** 135
Organic Compounds of Divalent Tin and Lead, **19,** 123
Organic Substituted Cyclosilanes, **1,** 89
Organoantimony Chemistry, Recent Advances in, **14,** 187
Organoarsenic Chemistry, **4,** 145
Organoberyllium Compounds, **9,** 195
Organolanthanides and Organoactinides, **9,** 361
Organolithium Catalysis of Olefin and Diene Polymerization, **18,** 55
Organomagnesium Rearrangements, **16,** 131
Organometallic Aspects of Diboron Chemistry, **10,** 237
Organometallic Benzheterocycles, **13,** 139
Organometallic Chemistry: A Forty Years' Stroll, **6,** 1
Organometallic Chemistry: A Historical Perspective, **13,** 1
Organometallic Chemistry, My Way, **10,** 1
Organometallic Chemistry of Nickel, **2,** 1
Organometallic Chemistry of the Main Group Elements—A Guide to the Literature, **13,** 453
Organometallic Chemistry, Some Personal Notes, **7,** 1
Organometallic Complexes with Silicon–Transition Metal or Silicon–Carbon–Transition Metal Bonds, **11,** 253
Organometallic Nitrogen Compounds of Germanium, Tin, and Lead, **3,** 397
Organometallic Pseudohalides, **5,** 169
Organometallic Radical Anions, **15,** 273
Organometallic Reaction Mechanisms, **4,** 267
Organometallic Reactions Involving Hydro-Nickel, -Palladium, and -Platinum Complexes, **13,** 273
Organopolysilanes, **6,** 19
Organosilicon Biological Research, Trends in, **18,** 275
Organosulphur Compounds of Silicon, Germanium, Tin, and Lead, **5,** 1
Organothallium Chemistry, Recent Advances, **11,** 147
Organotin Hydrides, Reactions with Organic Compounds, **1,** 47
Organozinc Compounds in Synthesis, **12,** 83
Oxidative-Addition Reactions of d^8 Complexes, **7,** 53
Palladium-Catalyzed Organic Reactions, **13,** 363
Palladium-Catalyzed Reactions of Butadiene and Isoprene, **17,** 141
Pentaalkyls and Alkylidene Trialkyls of the Group V Elements, **14,** 205
Phase-Transfer Catalysis in Organometallic Chemistry, **19,** 183
Photochemistry of Organopolysilanes, **19,** 51
Preparation and Reactions of Organocobalt(III) Complexes, **11,** 331
Rearrangements of Organoaluminum Compounds and Their Group III Analogs, **16,** 111
α-π-Rearrangements of Organotransition Metal Compounds, **16,** 241
Rearrangements of Unsaturated Organoboron and Organoaluminum Compounds, **16,** 67
Recent Developments in Theoretical Organometallic Chemistry, **15,** 1
Redistribution Equilibria of Organometallic Compounds, **6,** 171